Lecture Notes in Computer Science 2598

Edited by G. Goos, J. Hartmanis, and J. van Leeuwen

Springer
Berlin
Heidelberg
New York
Barcelona
Hong Kong
London
Milan
Paris
Tokyo

Rolf Klein   Hans-Werner Six
Lutz Wegner (Eds.)

Computer Science in Perspective

Essays Dedicated to Thomas Ottmann

 Springer

Series Editors

Gerhard Goos, Karlsruhe University, Germany
Juris Hartmanis, Cornell University, NY, USA
Jan van Leeuwen, Utrecht University, The Netherlands

Volume Editors

Rolf Klein
Universität Bonn, Informatik I
Römerstraße 164, 53117 Bonn, Germany
E-mail: rolf.klein@uni-bonn.de

Hans-Werner Six
FernUniversität Hagen, Praktische Informatik III
Informatikzentrum, Universitätsstraße 1, 58084 Hagen, Germany
E-mail: Hw.Six@FernUni-Hagen.de

Lutz Wegner
Universität Kassel, Praktische Informatik - Datenbanken
34109 Kassel, Germany
E-mail: wegner@db.informatik.uni-kassel.de

Cataloging-in-Publication Data applied for

A catalog record for this book is available from the Library of Congress.

Bibliographic information published by Die Deutsche Bibliothek.
Die Deutsche Bibliothek lists this publication in the Deutsche Nationalbibliografie;
detailed bibliographic data is available in the Internet at <http://dnb.ddb.de>.

CR Subject Classification (1998): F, E.1-2, G.2, G.4, H.2, H.5, C.2, D.2, I.3

ISSN 0302-9743
ISBN 3-540-00579-X Springer-Verlag Berlin Heidelberg New York

Springer-Verlag Berlin Heidelberg New York
a member of BertelsmannSpringer Science+Business Media GmbH

http://www.springer.de

© Springer-Verlag Berlin Heidelberg 2003

Typesetting: Camera-ready by author, data conversion by Boller Mediendesign
Printed on acid-free paper   SPIN: 10872556   06/3142   5 4 3 2 1 0

Preface

This collection of contributions from leading German and international computer scientists celebrated Prof. Dr. Thomas Ottmann's 60th birthday in 2003. The title of the volume, *Computer Science in Perspective*, and the topics of these high-quality papers are indicative of the respect which Thomas Ottmann enjoys within the *Informatics* community as a leading researcher and devoted teacher.

German computer science ("Informatik") has a strong theoretical touch and has never denied its heritage from mathematics. In this vein is the love for rigorous proofs, the longing for brevity and efficiency of algorithms, and the admiration for a surprising solution. Unlike pure mathematics, *Informatics* can embody this beauty of thought in zeroes and ones and let them come to life on the computer screen.

Step-by-step theory influences practical tasks and forms the basis for application systems and commercial success. Quite often, the public underestimates the time-span in which the transfer of these technological processes happens. Although computer science is highly innovative, fundamental theories and basic results often take 20 or more years to become fully appreciated and reach the market. Edgar F. Codd (relational theory), Robert Metcalfe (Ethernet), Doug Engelbart (interactive graphical interface), Ole-Johan Dahl and Kristen Nygaard (Simula and object-orientation), and Claude Shannon (information theory) are but a few scientists to mention who shaped the industry decades ago.

Communicating the need for a sound theory to students and fellow scientists through well-written textbooks, computer-supported courses, or simply through teaching material placed onto the Web is another strong point of computer science. Hardly any other science can compete with such a large body of textbooks, most of it going well beyond simple handbooks or "cooking almanacs." Modern forms of teaching and advanced techniques for collaboration were first designed, developed, and applied in computer science.

Having outlined this view of *Informatics* as a science with deep theoretical roots and far-reaching consequences for very practical applications, we feel that Thomas Ottmann is a prototypical example of a scientist who can span this spectrum. It is thus not a pure coincidence that the well-known phrase "nothing is as practical as a good theory" – attributed to Kurt Lewin (1890–1947) – is often cited by him.

Educated as a mathematician specializing in logic and formal languages, he joined Hermann Maurer's Institute in Karlsruhe which in itself was (and is) an example of the bridge between theory and practice.

Soon after his appointment as a research assistant he decided to switch from formal languages to algorithms and data structures where he quickly accumulated a large number of outstanding publications. With the emerging field of computational geometry he widened his knowledge even more and added further highly respected results with applications to VLSI design to his curriculum vita.

The community recognized this and voted him into the DFG refereeing committee, asked him to chair the DFG-Schwerpunktprogramm (research focus) "data structures and efficient algorithms" and made him several offers of chairs in computer science which, after accepting in Karlsruhe first, eventually led him to

Freiburg where he continues to pursue – among other topics – his eLearning projects, in particular the authoring-on-the-fly development.

His experience in applying eLearning concepts to university education, visible through projects like VIROR and ULI, has made him an authority in the CAL field in Germany and has led, for example, to membership in the steering committee of the Swiss Virtual Campus. Needless to say he is a sought-after expert for advising on and evaluating new teaching projects in the academic world.

In all projects and proposals Thomas Ottmann strives for rigorous clarity and preciseness. At the same time he emphasizes the need to continuously promote the field and to change one's own interests at long intervals. Those who had the luck to have him as supervisor or referee – like ourselves – will fondly remember his advice to change one's own research topic after the Ph.D. or Habilitation, and so most of us did.

What he has to referee he examines with a scientific furor and what he manages to unmask as a weak or even false result is stamped "rubbish" and returned to the sender without mercy. Many a thesis has improved considerably under this scrutiny and those who have witnessed this strict "parental guidance" will try to pass it down to their own "Ph.D. and diploma" children.

On the other hand, Thomas Ottmann is quick to acknowledge a brilliant idea and to include it, with proper reference, into the next edition of his books or lectures. He will praise a young talent and whoever asks him for a judgement gets a straight answer, which is rare in the academic tar pits.

All in all it is probably fair to say that computer science in Germany today owes Thomas Ottmann a lot for keeping up the standards and seeing to it that the industry's tendency to fall for "vaporware" does not carry over to its scientific promoters.

If Thomas Ottmann perfectly represents computer science at its best, then so should the papers in this volume. They span the spectrum from formal languages to recent results in algorithms and data structures, from topics in practical computer science like Software Engineering or Database Systems to applications of Web Technologies and Groupware. Not surprisingly, eLearning assumes a prominent place. We hope they are a fitting tribute to an extraordinary scientist, teacher, and a personal friend to many.

Our sincere thanks go to all authors who spontaneously agreed to contribute to this collection. Maybe it is another proof of the rigorousness Thomas Ottmann instills in his followers that none of the authors who initially promised to submit bailed out in the end. At the same time we gladly acknowledge the support from Ms. Mariele Knepper who did the editorial polishing and handled the administrative tasks. A special thanks goes to Mr. Ziad Sakout for handling the TeXnicalities. Finally, we are grateful to Springer-Verlag for help and advice in preparing this special volume.

Bonn, Hagen, Kassel   Rolf Klein
November 2002   Hans-Werner Six
                                                                          Lutz Wegner

Table of Contents

Multimedia Lectures for Visually Oriented Medical Disciplines

Jürgen Albert and Holger Höhn

Universität Würzburg, Lehrstuhl für Informatik II,
D-97074 Würzburg, Germany

Abstract. Many scientific disciplines with an intrinsically high degree of multimedia content can be found among the life-sciences. Among the various faculties of medicine, especially dermatology is renowned for extensive use of clinical images in lectures and in textbooks. Training the "diagnostic eye" is an important part of the medical education. Furthermore, large collections of images, traditionally organized in numerous cabinets of slides, can be maintained and distributed a lot easier when their digitized counterparts are stored in an image database on some PC in the clinic's network. Thus, ideal conditions should hold for introduction of computer-based multimedia-lectures for dermatology and the distribution of teaching material on CD-ROMs or over the World Wide Web. This paper gives an account of an interdisciplinary multimedia project of three dermatology-departments, an institute for medical psychology and our chair. Since most of the implemented tools have been field-tested in recent semesters, we also can report about usability and acceptance among lecturers and students.

1 Introduction

If one tries to estimate the "natural amounts of multimedia" needed for teaching various scientific disciplines, it might be a good idea just to compare common textbooks by the percentage of pure text vs. graphical illustrations and photographs appearing in their pages. While abstraction and textual representation of knowledge is to be expected in every science, there are vast differences in how much image-material belongs to a core curriculum. Let us compare e.g. biology with economics or mathematics. Then probably everybody will agree that teaching botanics or other fields in biology without large collections of image-material seems unthinkable, whereas blackboard and chalk can be quite sufficient for operations research or algebra. And for the medical disciplines it is of vital interest (especially for patients) that education comprises analysing and interpreting X-rays, ultrasound- or NMR-images, just to name a few. In dermatology, for example, students have to learn to distinguish harmless naevi from developing malignant melanoma. Of course, bed-side teaching in the clinic is a regular part of the curriculum in dermatology, but there are obviously strict limitations how many patients can actually be visited by the several hundreds of students in a dermatology course per semester. Over 500 diagnosis-groups are to be considered relevant for clinical practice and there are normally great variations in the appearances of a single dermatological disease depending on age, sex, affected skin-region and race of a patient. Thus, large image-collections play a foremost role in the visual training of the students. It is not

R. Klein et al. (Eds.): Comp. Sci. in Perspective (Ottmann Festschrift), LNCS 2598, pp. 1–12, 2003.

uncommon that during one lecture 50 clinical pictures or more will illustrate a group of dermatological diseases.

Though we are stressing here that lecturing in dermatology is heavily relying on image-material, readers active in any other fields of multimedia-content generation will notice many similarities, when project-goals and priorities are to be specified, and when it comes to the phase of structuring and storing the learning objects. For example, "Authoring on the Fly" [1] takes a much more general approach to presentation and recording of lectures as we do. But in both cases intentionally lecturers can be spared from working out the detailed classroom notes, which is clearly an important point for acceptance. And equally provisions for a drastic reduction of the data-volumes are made, such that given lectures can also be offered online over the internet.

In the following chapters we will try to analyse first the current situation found in most dermatological departments, where traditional slide archives prevail for "multimedia-education" in the classroom. Then we will describe the migration-paths taken in the DEJAVU-project [3] to base teaching and learning on digital learning objects in effective ways for both lecturers and students. The scanned images from the slide archives still constitute the majority of learning objects and form the "backbone" of the system in many respects. In chapter 3 an outline of the implementations is presented, and the other types of learning objects like learning texts or 3D-animations are discussed. We conclude with a short report about the experiences gathered so far with applying the DEJAVU-packages by our project-partners.

2   Multimedia in Dermatological Education

The daily routine in dermatological clinics yields large amounts of images taken from patients for various reasons. Controlling the growth of some suspicous spot, supervising the healing-process after a surgery or during any kind of medical therapy, documenting the current general health-constitution of a patient, all these and many other reasons can lead to new images. Many clinics employ their own photographers and provide equipment for special zooms (oil-immersion) and microscopy for skin-tissue samples. From this abundance of slides lecturers can select those of highest technical quality and educational value. In general, this selection is done jointly and discussed by all lecturers in regular meetings, not only to supply the common lectures but also to prepare conference contributions or workshops.

These clinical slide archives often contain thousands of pictures and have been chosen over decades from some tens of thousands of stored photographs. Naturally, these collections are considered "treasures" of a clinic and are usually organized in special big transparent cabinets according to groups of diagnoses. The individual picture is labelled by the cabinet-name, its position number therein and additionally by the photographers task-number. There might be handwritten remarks on the frames as well. Thus, images are organized hierarchically and – at least in principle – can be related indirectly to the patients' records through the task-number. This kind of organization is practiced in the dermatology in W"urzburg and can certainly differ in other clinics. It is reasonably efficient for looking up the image-material available for diagnosis-groups; lecturers have to walk to the archive-room and decide with a magnifying glass, whether a special slide is useful for their lecture.

But image-retrieval already becomes difficult for secondary diagnoses, for example eczema and ulcerations after a HIV-infection. Or imagine, a lecturer wanted to look up all available pictures, where skin-diseases at fingers or toes appear. Other severe drawbacks, due to the large volume of those archives and the costs of producing slide copies, are the limitations for borrowing, the danger of damage or loss of a valuable original. Therefore, access to specific images may be a severe bottleneck for lecturers, and for the given reasons access is in general not possible for students. This situation for students is especially unfortunate, since during the lectures there is no chance for them to write down in sufficient detail, what they see in the images. On the other hand, diagnosing given images is a required part of their exams. Other types of educational multimedia like videos or animations on films also do exist for dermatology-lectures but compared to the slide collections their number and importance is negligable. For the students their own classroom-notes, learning texts of the lecturer, classical textbooks for dermatology and especially collections of exam-questions (typically multiple-choice) are still the most common forms of information sources [12].

It is more than obvious, that this situation can be greatly improved by providing archives of digital teaching material including image-databases in the clinics' intranets and (with restricted access) on the internet or on CD-ROMs for both lecturers and students [5], [6], [7].

2.1 Objectives in DEJAVU

In DEJAVU (Dermatological Education as a Joint Accomplishment of Virtual Universities) the dermatology departments of Jena, the Charité in Berlin and W"urzburg are providing the input for a hypertext-based, interactive learning system. Since spring 2000 DEJAVU has been funded by the national Ministerium for Education and Research (BMBF). Learning objects comprise image archives, complete multimedia lectures, learning texts, lecture notes, case reports and multiple choice questions. Our computer science chair is contributing several authoring and conversion tools as well as interfaces to the learning objects database for lecturers and students. These interfaces and the resulting teaching objects are evaluated with respect to their intuitive usability and effectiveness by our project-partner, the institute for Medical Psychology in Jena [3].

On the highest design level DEJAVU follows a multi-centre approach: Each dermatology-partner has his own intranet-server and manages generation, categorization and archiving of images and other learning objects independently. This was necessary for specific variations of the database-fields. Besides the core-fields like diagnoses or localizations, for which there exist established international standards, there was a need to express the different organizational forms, e.g. how and where the originals of digitized images or videos are physically stored. This is considered an important aspect of flexibility and extensibility and in practice indeed proved helpful. For cooperation and communication among the partners a project-server is located at our computer science chair. Partners can upload their completed learning objects in full or just the meta-data accompagnied by thumbnail-pictures for indexing. They specify access-rights for lecturers of the partner-clinics, for students at their own or a partner's faculty and for the general public. This may seem too complicated, but patients' privacy rights concerning records or images are to be respected and the local clinic-servers are strictly guarded by firewalls. Redundancies in the generation of learning objects are avoided whenever possible. Especially for the work-intensive video-clips and teaching texts, at an early project-stage agreements were made to share the work-load and to assign the relevant topics, also according to current research areas at the participating clinics.

2.2   Implementation of Standards

Experiences from previous projects with medical content had shown that it is worthwhile to build an open and extensible system of cooperating modules based on established international standards, wherever these are available [2], [4].

In a bottom-up view from the database-level the first applicable and established standard is IEEE's Learning Object Modelling Standard (LOM WD 6.1 [14]) which defines a convenient framework for hierarchically composing lectures and quizzes from images, texts, audio- and video-sequences together with their administrative data e.g. author's names, clinical categories, time stamps, etc. We had no difficulties to fit our object categories into the LOM concepts, often the predefined values were just the ones needed for our learning objects. Users most of the time do not have to worry about the LOM-datafields, since these can be derived from the specified users' roles or contain default values.

The next level of standards comes already from medical international organizations like the World Health Organization (WHO) or the National Library of Medicine (NLM) which also hosts the online version of the well-known MEDLINE database [15], containing the most important scientific journals for the life-sciences. The ICD-10-standard from WHO for coding medical diagnoses is actually implemented in a refined form, where each ICD-10-code can be extended by two additional digits to describe dermatological diagnoses with a higher precision. This originates from a national de facto standard, the "Dermatologische Diagnosenkatalog" [11], which is by construction upward compatible with ICD-10. Similar relations hold for the coded localizations, i.e. skin regions [13]. Whenever the authors felt the need to deviate from the given ICD-0-DA standard, at least a simple upward embedding into a standard category is guaranteed. Thus, retrieval operations on the categories can not miss potentially relevant learning objects.

The ICD-codes also open the doors for another network for navigation over images and texts. ICD and many other medical standards and thesauri are incorporated in the Unified Medical Language System (UMLS [17]), which essentially provides a unique concept identifier for each relevant medical term and lists the existing relations to other terms. Through this an enormous semantic network can be explored starting with the ICD-codes. Semantic networks are discussed as a next stage for information retrieval beyond linked HTML-documents (see e.g. [18], [21], [22]).

Another well-known part of UMLS are the so-called MeSH-terms, the Medical Subject Headings [16], which categorize the contents for all the contributions to journals in the MEDLINE database. After extracting the relevant subsets concerning dermatology from UMLS and MeSH, the navigation along these thesauri via hyperlinks and the support for efficient literature research in MEDLINE come more or less for free.

There is still another level of standards where decisions had to be taken. These could be called import and export formats, and the selection criteria were general availability, costs, efficiency, following this order in priorities. It will be no surprise, that MS-Word, MS-PowerPoint, baseline JPEG, mySQL, HTML were chosen. Appropriate data formats and players for videos and sound have also been tested and ranked. Currently, mp3 for sound from lectures and for other audio-clips from voices is preferred. For video-clips divX gave the best impression of the observed quality/compression ratios. For 3D-animations QuickTime VR proved superior to Freedom VR with respect to memory consumption and response times. But it has to be pointed out, that in general these are not the archiving formats of the originals. For example, the scanned slides can be transformed without further quality reductions to JPEG2000-format from their originals in bitmap-format, which have been stored in a lossless way on more than hundred CD-ROMs. Thus, improvements of technological platforms (network bandwidth, storage capacity of hard-disks or DVDs) can be exploited without going back to the time consuming, costly first steps of generating images, videos or audio-streams.

3   Support of Workflows

Having a software package installed with all parts functioning correctly is of course at most equally important for user-acceptance than to have convincing and measurable improvements in some steps of a workflow. Quite often users find it risky or too much initial effort to change established workflows. Also in our case it would not have been possible without external funding to build a digitized image archive from scratch and to convince lecturers to switch from their existing slide collections to multimedia-presentations. By and large for students the situation is comparable. There are certainly enough learning texts in the university libraries, so for students it will be not interesting to obtain yet another source of information. Instead they want to find the material they need for their preparation for exams quickly and in comfortable and compact form.

3.1   Previous Projects

Experiments with early prototypes of multimedia lectures in the three dermatology departments in Jena, Berlin and W"urzburg would not have been possible without software modules and experiences from previous projects. Most helpful were the results of the SENTIMED project (SEmantic Net Toolbox for Images in Medical EDucation, [8]), in which our chair had cooperated with the local department of dermatology in W"urzburg. This work had been supported by the Bavarian government initiative "Multimedia in Education" (MEILE). It had emphasized the preprocessing for multimedia-lectures: developing digitized image archives containing over 4000 high resolution images, generating, presenting and recording lectures and quizzes for training of the "diagnostic eye". An even earlier project proved useful for the basic stucturing and presenting dermatological learning objects in HTML-format. In 1997 our chair had converted a dermatological lexicon into a CD-ROM and an online-version for a well-known medical publisher [9]. And there were further JAVA-packages which could be adapted from a current CD-ROM-project for a 12-volume pharmaceutical manual [10], to support navigation by speedtyping or fast fulltext search, bookmarking and highlighting parts of the texts and to attach own notes at any chosen text position. An overview of the existing modules for authoring lectures, learning texts etc. is given in Fig. 1.

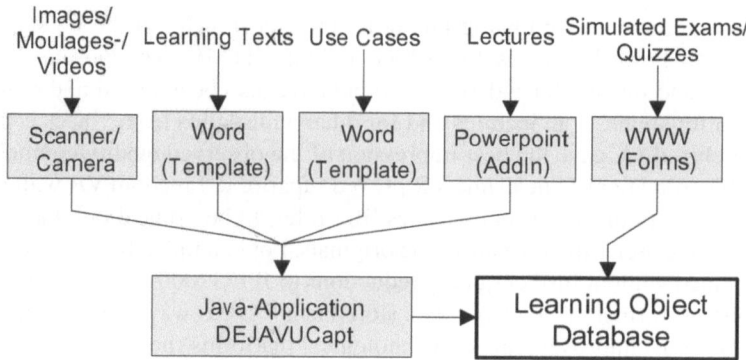

Figure 1. Authoring tools and import filters for the database of learning objects

3.2   Accessing Image-Archives

When analyzing the lifecycle of digitized dermatological images the following phases defined by the workflow and usage scenarios of lecturers can be identified:

- selection of slides that meet the demands of both high educational content and high technical quality
- generation of the "raw" digital form with best available resolution
- review and adjusting contrast and colour, cleaning, clipping, ... to produce a digital master form
- categorization and storage of the master-files together with "thumbnails" for referential purposes
- navigation in the database and selection of images or other teaching material for lectures, learning texts, quizzes, ...

Categorization of the image-data is supported by a JAVA application with extensible hierarchical and alphabetical pick-lists for diagnoses, localizations, etc. By this, data input is accelerated and typos are excluded. As explained above, the lists closely follow international standards like ICD-10 for diagnoses or ICD-0 for localizations. This interface can also be employed to retrieve any images from the database again, but the usual and more convenient way for lecturers is to activate a special PowerPoint Addin, to retrieve images or also video-clips etc. and to integrate them directly into presentations, as shown in Fig. 2.

The equivalent AddIn exists for MS-WORD to integrate images into the learning texts. Chosen images will then be displayed as thumbnails which are linked to the high-resolution versions of the corresponding images. Only lecturers and clinic personnel access the image-database directly. For students image-material is already embedded into didactical context, be it PowerPoint text-slides or multiple choice questions and answers in quizzes. The digital recording of a given presentation may be postprocessed with separation of the audio-stream, categorized and cross-linked with teaching texts and quizzes and thus become an independent learning object.

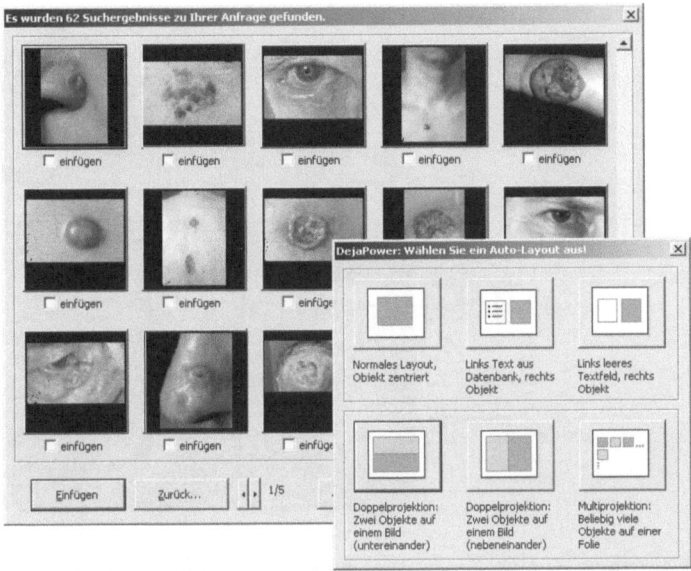

Figure 2. PowerPoint AddIn: For the database queries pick-lists for diagnoses, localizations etc. help to select images. One of the auto-layouts then has to be chosen to insert individual images into the PowerPoint presentation

3.3   Texts, Case Reports, and Quizzes

As images are but just one type of objects in dermatological teaching, the phases listed in the section above have to be generalized and adapted for learning texts, quizzes, audio- and video-sequences. In these cases the concepts and data-structures carry over nearly unchanged while in the terminology "thumbnails" have to be replaced by "summaries" for written or spoken text or by "keyframes" for video-clips and 3D-animations, respectively. As indicated in Fig. 1 learning texts, use cases and case reports etc. are composed according to master document formats. The authors can apply those master formats either as template documents in MS-Word (generally the preferred way) or as an HTML-form online. This avoids confronting lecturers with real XML-editing, which was considered as unacceptable for the authors. These input files are then converted to XML before integrating them into DEJAVU. From the XML-base different exports can be handled easily via XSLT. One of the export-formats can make use of the MS Learning iNterchange (LRN, see [20]), which is embedded into the MS Internet Explorer. Multiple choice questions are categorized according to given learning objectives, typically topics treated in one of the lectures [19].

Fig. 3 illustrates the DEJAVU web-interface for students to access previous lectures and to replay the slides. The same interface also provides other learning objects for the students. Through JAVA applets students can actually work on the teaching texts by highlighting important words or paragraphs, setting bookmarks or adding notes. The quizzes and multiple choice questions are mainly derived from written exams of the

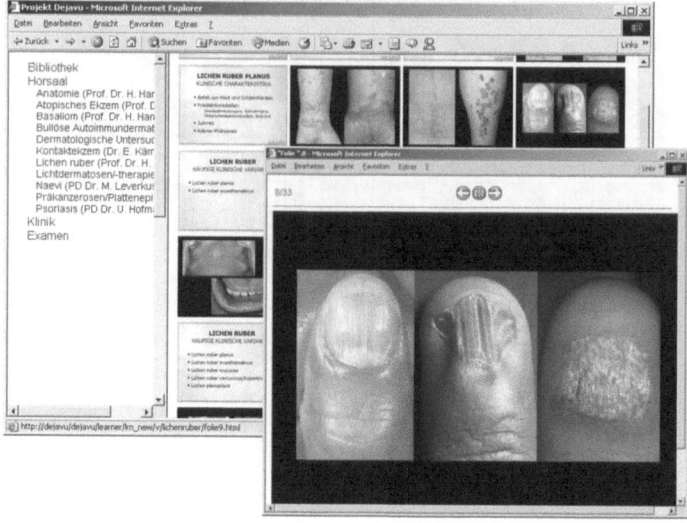

Figure 3. Lectures can be accessed by students after registration over the WWW

past semesters. Their content and style is chosen in such a way, that these tests can serve as simulated exams (see [12] and Fig. 4). The level of difficulty is configurable, e.g. how many hints for solutions are given. Thus, also graduate students and doctors can test and extend their dermatological knowledge independent of time and place. The extensive image database, currently containing more than 10.000 images, becomes a convenient tool for training of the "diagnostic eye".

3.4   Video-Clips and 3D-Animations

While large image databases are a must for dermatological education, the next steps to videos or animated sequences are often considered as too costly. Although, beyond the common usage of images, realism in lectures for dermatology should be greatly enhanced by adding a further dimension to images as learning objects, e.g. spatial depth in 3D-animations or the temporal dimension as in video-clips. Both forms can share useful features when embedded into hypertext, like synchronization with audio-streams or providing regions of interest (ROIs) linked to other learning objects. Nevertheless, they should normally be applied for different areas. In video-clips actions and reactions, i.e. causal relations, will play the foremost role. In 3D-animations a state of a patient at one moment in time can be displayed in great detail. In DEJAVU video generation is distributed along educational topics among the three clinics. More than a dozen videos have already been produced. It is planned to have about 20 to 30 video-clips available in spring 2003. These are to be integrated into lectures and texts as thumbnails just like single pictures. But for replay several options will control resolution, audio-integration, displaying of subtitles etc. to cover the different replay-contexts in lectures, in the clinic intranet or over the internet.

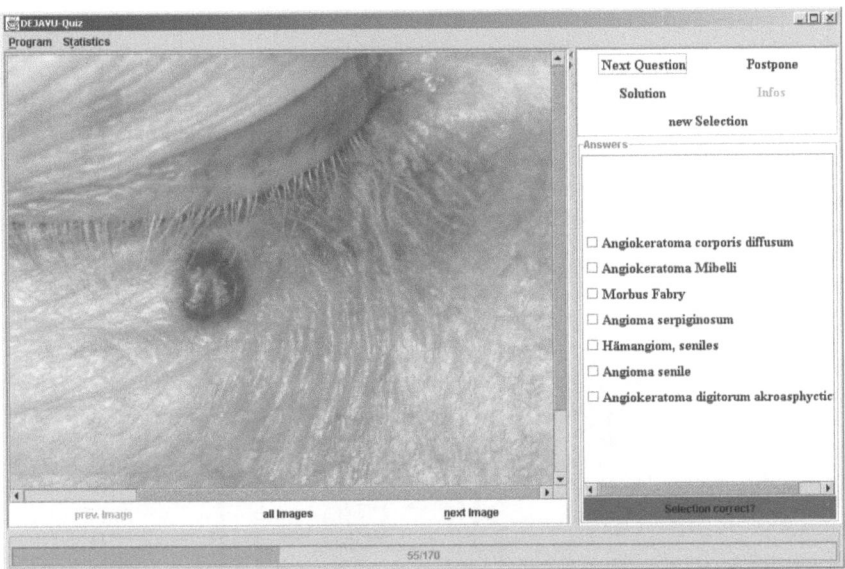

Figure 4. Quiz as part of simulated exams

As a recent, experimental part of the DEJAVU-project 3D-animations for dermatological moulages have been generated. Moulages are painted wax-models often taken as plaster-impressions from real patients. They were quite common in dermatological education around 1900, but their importance dwindled with the increasing quality and accuracy of colour-photographs. For us they serve now as a valuable and natural stepping stone for future "virtual 3D-patients". The usability of virtual 3D-moulages has been tested and compared with the interaction-types available for video-clips.

Our experimental set-up (see Fig. 5) includes the mechanical constructions to obtain well-defined geometrical positions for object and camera as well as complete software control for the camera-shots, for the step-motor moving the rotating plate to which a moulage is fixed and for the computer which stores the image-series in a raw format first. All camera-parameters are stored in log-files for reproducibility and local corrections of single images. The post-processing tools then take care of the separation between object and background, defining ROIs (see Fig. 6) and exporting the image-series to virtual reality data-formats, i.e. either Freedom VR or QuickTime VR.

It could be shown that it is not very complicated to produce attractive 3D-animations. The basic implementation was completed in a students' project within 3 months. As data-format QuickTime VR proved clearly superior to Freedom VR in file-size and handling. 3D-animations can complement video-clips and offer additional forms of interaction for students. The current software-prototype and the sample animations motivate further experiments with up-scaling the physical dimensions and replacing moulages by volunteering patients.

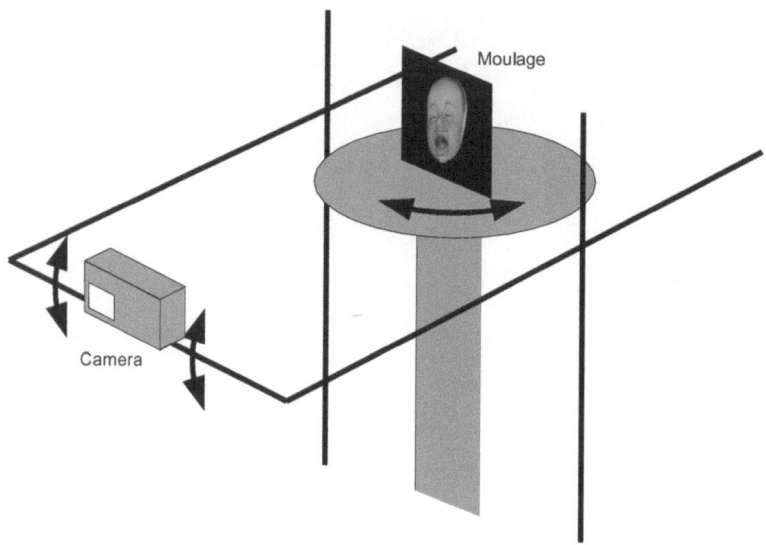

Figure 5. Moulages are scanned by a digital camera from well-defined distances and angles

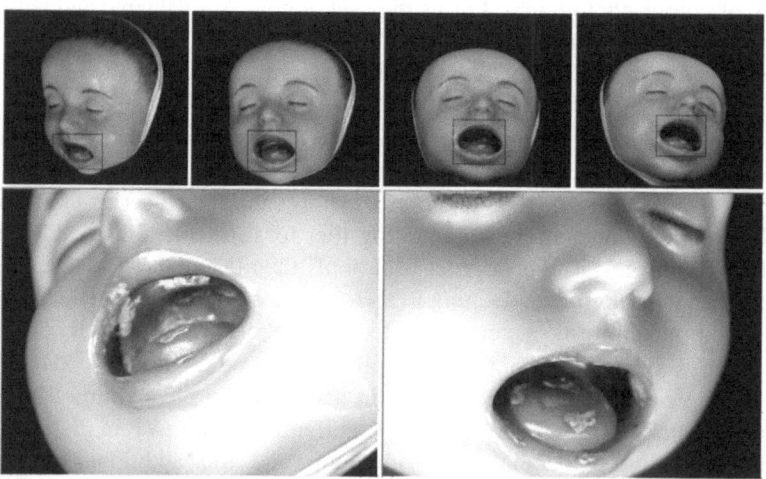

Figure 6. A few of the several hundred single images taken for one moulage, together with the ROIs and zooms

4   Lessons Given and Lessons Learned

On top of large digital image archives a powerful tool-set was created which efficiently supports lecturers in dermatology to prepare and present lectures or conference contributions. Several stages of the authoring tools have been field-tested in recent semesters in the participating clinics. The dermatological department in Jena has switched to the DEJAVU-tools completely for the lectures since summer 2000, in Berlin and W"urzburg it is still left to the lecturers whether they prefer the traditional way, though the clinic directors explicitly encourage them to try out the DEJAVU-tools. In discussions with lecturers it turned out, that all technical possibilities of traditional slide collections have to be covered too. Most notably the mixing of privately owned slides with those from the clinic for the composition of lectures. Seen from the PowerPoint-interface this is trivial to import arbitrary images in common formats, but some provisions concerning ownership had to be taken to cover this special case when placing those lectures in the web. Now we also offer to all lecturers that they can organize their own slide collections on their home-computers by the DEJAVU-tools.

Another obstacle had been the common use of two slide projectors in lectures, which is frequently important for comparing image contents directly e.g. for differential diagnoses. This is now handled by our PowerPoint AddIn too, which allows the arrangement of two or more pictures and the combination with text. Fig. 3 shows such examples.

Amazingly for us, one expected problem never occurred in practice: Neither lecturers nor students ever complained about reduced image-quality or brilliance of the digitized pictures compared to the original slides. To the contrary, in one of our first surveys a high percentage of students rated the digitized images as of equal or higher quality.

Some of the inherently given advantages of the multimedia-lectures will be effective only in the longer run. For example, existing lectures can be archived independently of the clinic's collection and thus be easily reused, modified or brought up-to-date. The technological shift to digital archives will enhance the accessibility of teaching material for students drastically. Except for replacing a slide projector by a multimedia projector, no principal reorganization of lecturing is required. Integration of further multimedial components like video-clips of diagnostic or therapeutic procedures or graphical animations is easy to accomplish. The technical quality of digital cameras and camcorders for some affordable price has improved recently at a great pace. Since already now some dermatological clinics in Germany are switching to digital camera equipment, one of the most time-consuming steps for the digital images archives will vanish in the near future.

Acknowledgements

We have to thank our DEJAVU-partners sincerely for their great patience in explaining to us, what really matters in teaching dermatology. And we are grateful for their endurance to try out and field-test our early software prototypes. Our thanks also go to the project-members and students in computer science for their numerous contributions in graduate-projects, diploma-theses and as programmers for DEJAVU. Many of them have learned more about dermatology than they had expected, and many also saw more high-quality pictures of dermatological diseases than they would have wished.

References

1. Authoring on the Fly, http://ad.informatik.uni-freiburg.de/aof/
2. Projektträger NMB: Lernplattformen: Standardisierungs-Initiativen, http://www.medien-bildung.net/lernplattformen/lernplattformen_5.php/ spezialthemen/lernplattformen
3. DEJAVU - Dermatological Education as Joint Accomplishment of Virtual Universities, http://www.projekt-dejavu.de/
4. Höhn, H.: Multimediale, datenbankgestützte Lehr- und Lernplattformen, Dissertation, Lehrstuhl für Informatik II, Universität Würzburg (2002)
5. Höhn H.: Combining Image Archives, Audio- and Video-Sequences for Teleteaching Lectures on Dermatology, 2nd European Symposium on Teledermatology, Zurich (2000)
6. Höhn H.: Digital aufbereitete Vorlesungen in der Dermatologie, Biomedical Journal, **56** (2000) 27–29
7. Höhn H., Esser W., Hamm H., Albert J.: Image archives, audio- and video-sequences for teleteaching. In Burg G. et al (Eds.): Current Problems Series: Telemedicine & Teledermatology, Karger-Publisher Basel, **32** (2002) 192–195
8. SENTIMED - Semantic Net Toolbox for Images in Medical Education, http://www2.informatik.uni-wuerzburg.de/sentimed/
9. Schmoeckel, Chr.: CD Klinische Dermatologie, Lexikon und Differentialdiagnose, Georg Thieme Verlag, Stuttgart (1997)
10. Blaschek W., Ebel S., Hackenthal E., Holzgrabe U., Keller K., Reichling J.: HagerROM 2002, Springer-Verlag, Heidelberg (2002)
11. Henseler T., Kreusch J., Tausch I. (Eds.): Dermatologischer Diagnosenkatalog, Georg Thieme Verlag, Stuttgart (1996)
12. Toppe, E.: GK 3 Dermatologie, Originalprüfungsfragen mit Kommentar, Schwarze Reihe, Georg Thieme Verlag, Stuttgart (2001)
13. Baumann R. P. et al.: Tumor-Histologie-Schlüssel ICD-O-DA, Springer-Verlag, Heidelberg (1978)
14. IEEE Learning Object Metadata (WD 6.1), http://ltsc.ieee.org/doc/wg12/LOM_WD6-1_1_without_tracking.pdf
15. National Library of Medicine: PubMed, http://www4.ncbi.nlm.nih.gov/PubMed/
16. National Library of Medicine: Medical Subject Headings, http://www.nlm.nih.gov/mesh/meshhome.html
17. National Library of Medicine: Unified Medical Language System, http://umlsinfo.nlm.nih.gov/
18. Sowa J.F.: Principles of Semantic Networks - Explorations in the Representation of Knowledge, Morgan Kaufmann Publishers, San Mateo (1991)
19. IMS Global Learning Consortium: IMS Question & Test Interoperability, http://www.imsproject.org/question/index.html
20. Microsoft Learning Resource iNterchange (LRN), http://www.microsoft.com/eLearn/
21. Berners-Lee T.: Metadata Architecture, http://www.w3.org/DesignIssues/Metadata
22. World Wide Web Consortium (W3C): Resource Description Framework (RDF), http://www.w3c.org/rdf

Combinatorial Auctions, an Example of Algorithm Theory in Real Life

Arne Andersson

Computing Science Department
Information Technology
Uppsala University
Box 337
SE - 751 05 Uppsala, Sweden
arnea@csd.uu.se

Abstract. In this article, we discuss combinatorial auctions, an interesting inter-disciplinary research field in Computer Science and Economics. In particular, we will (a) describe a set of real-world cases, (b) how to solve the associated computational problems, and (c) discuss the impact of the probability distributions chosen for benchmarking.

1 Introduction

Electronic trading is becoming one of the most successful applications of information technology. The application areas vary from electronic payments systems and bookkeeping to purchasing tools, e-procurement, e-sourcing, and electronic negotiation. Some areas of electronic trading combines fundamental and hard computational problems with a clear and direct connection to real-world applications. In order to have real impact, one needs a strong background in the foundations of Computer Science, Economics, and Game Theory as well as profound knowledge about real world-problems.

One interesting area is combinatorial auctions, which is a research field in its own right. A well-designed combinatorial auction enables bidders to express preferences across different offers (e.g. a seller can give a certain discount if a certain set of contracts is won). This possibility enables more efficient auctions - such as lower prices for the buyer and less risk for the suppliers. For example, conducting a procurement auction in this way can be seen as market-based lot formulation, where the lots are formed by cost structures, and not by the buyer's subjective beliefs.

To illustrate the concept, we give a mall example. Imagine yourself as a supplier:

- You can deliver product A at 40, product B at 50, but product A and B at 70. That is, you have synergies between the two products.
- A buyer is running an auction in which the current price for A is 33 and the current price for B is 43.
- Do you place a bid?

The risk with bidding in this case is obvious. In a lucky case, you bid A at 32 and B at 42, no further bids arrive and you can supply A and B at 74. But it could also be the

R. Klein et al. (Eds.): Comp. Sci. in Perspective (Ottmann Festschrift), LNCS 2598, pp. 13–21, 2003.

case that after your bids at 32 and 42, A drops down to 27, while your bid at B is still the lowest. Now, you are trapped into supplying B at a loss. Needless to say, auctions without the possibility to explicitly bid for A and B very poorly reflects underlying cost structures.

With a combinatorial auction the scene is changed. Here, bidders are allowed to place bids on combinations that they select themselves. The main drawback with combinatorial auctions lies in the complexity of the computational problems that need to be handled. Even in its simplest and most straightforward form, resolving a combinatorial auction is NP-hard.

2   Some Real-World Applications

2.1   Combinatorial Auctions in Public Procurement

For legal reasons, public procurement is not done via iterated bidding. Instead, a single shot sealed bid auction is performed. The procurer announces a request for quotation with specifications. He/she also states a deadline when offers should be received. Then, all offers are opened at the same time, quality parameters and price are evaluated, and the best offer is accepted. Even though an on-line electronic auction is not applicable, complex bid matching has a large potential.

Below, we illustrate with concrete cases:

Road painting, Swedish Road Administration 1999. 12 contracts were traded. One supplier, EKC, had the lowest price on 10 contracts. Initially, EKC was also allocated all these 10 contracts. However, EKC then declared that it was not capable of delivering ten contracts, it could deliver at most two. However, bidding on two contracts only, EKC would take the risk of bidding on "wrong" contracts; maybe someone else was beating them on these two contracts while they could have won some other contract.

As a result of this declaration from EKC, the road maintenance authorities decided to only give EKC two contracts (the alternative would have been to force the company into bankruptcy). Then, EKC was allocated the two contracts were they had the largest advantage in percentage. However, for the computational problem "minimize the cost such that EKC gets at most two contracts" the chosen solution was not the optimal one! In fact, the road administration could have done approximately 800 000 SEK better.

Of course, once modeled correctly, the computational burden of this optimisation problem is small for this problem size. Still, it is a very good illustration of the fact that proper computation is hard. Even more, good knowledge about computation needs to be combined with the design of bidding rules to avoid future problems like this one.

Paper material, Uppsala (regional procurement), 2002.   In this case, it was announced that combinatorial bids were allowed. Then, one supplier bid in the following way: "If I win these five specified contracts, I give 2% discounts on all contracts. And, if I also win these other four contracts, I give 4% discount on all contracts". Since, the number of commodities traded was approximately 300, these offers in combination with other combinatorial bids, could not be evaluated by hand. Furthermore, these bids were certainly not standard combinatorial bids.

With 300 commodities and with bids of the type above, we can clearly see the risks for very hard computational problems. Again, computational aspects need to be combined with sound market design.

Road covering, Swedish Road Administration 2002. In order to allow small suppliers to compete on as many contracts as possible, the following types of bids were allowed:

- Bids on combinations
- Capacity restrictions: A bidder is allowed to bid on more than its capacity. In order not to get too many contracts, the supplier is allowed to state a maximum total order volume that he can accept.

Suppliers utilized both possibilities, one of them sent in 27 different combinatorial bids. The possibility to express capacity restrictions helped small suppliers compete for many contracts, and it had a positive effect on the Road Administrations efforts to avoid cartels.

2.2   Combinatorial Bidding in Power Exchanges

We briefly describe the electricity consumption in Sweden Jan 30 - Feb 6 2001. On the Friday and Monday of that period, the consumption was about 10% higher than a "normal" cold winter day. That is, the increase in consumption was noticeable, but not dramatic. Still, the effect on electricity price was dramatic. At some occasion, the spot price on electricity at Nordpool (the Nordic energy exchange) increased by a factor of 7.

In media, this phenomenon was described as a problem of production capacity. It also led to a number of political decisions were the Swedish state decided to pay extra for spare production capacity, to be used in future similar cases. The cost for maintaining this spare capacity is now paid by the Swedish taxpayers. However, a closer look at the capacity problem reveals that it very well may be seen as a consumption problem or as a problem of an inefficient market. It turns out that a number of inefficiencies could have been removed, such as unnecessary use of electricity for heating, if a better market had been established. Also here, combinatorial trading, with the ability for both buyers and seller to express time-dependencies in their bids, would have made a dramatic difference. (On today's electricity market, it is only possible to trade hour-by-hour and only very limited dependencies between hours can be expressed.)

3   Computational Aspects

One of the main problems of combinatorial auctions in markets of any significant size is related to complexity; the determination of an optimal winner combination in combinatorial auctions is $\mathcal{NP}$-hard [13] (even if a certain approximation error can be tolerated [15, Proposition 2.3]). From a practical point of view this means that, in the general case, there is no known algorithm that can guarantee to find the optimal solution (or even a solution which is within some guaranteed tolerance from the optimal solution) in reasonable time as soon as the market is of any significant size. One way to overcome this problem is to restrict the combinations that bidders can request in such a way that efficient algorithms for determining the winner can be used [13]. Another way is to allow combinations on all bids and find efficient algorithms that as quickly as possible finds high quality solutions (and hopefully also the provably optimal solution) [11; 6; 15]. In this paper we look further into the topic of the latter approach. In particular, we point out:

- Many of the main features of recently presented special-purpose algorithms are rediscoveries of traditional methods in the operations research community.
- We observe that the winner determination problem can be expressed as a standard mixed integer programming problem, cf. [11; 16], and we show that this enables the management of very general problems by use of standard algorithms and commercially available software. This allows for efficient treatment of highly relevant combinatorial auctions that are not supported by special-purpose algorithms.
- The significance of the probability distributions of the test bid sets used for evaluating different algorithms is discussed. In particular, some of the distributions used for benchmarking in recent literature [6; 15] can be efficiently managed with rather trivial algorithms.

4   A Traditional Algorithm and Its Variants

Before discussing how very general versions of winner determination can be solved by general-purpose algorithms, we will investigate the basic case in which a bid states that a bundle of commodities, $\mathbf{q}_i = [q_{i0}, q_{i1}, \ldots, q_{ik}]$, $q_i \in \{0, 1\}$ (k is the number of commodities) is valued at $v_i \in \Re$. Given a collection of such bids, the surplus maximizing combination is the solution to the integer programming problem [16; 11]:

$$
\begin{aligned}
&\max \sum_{i=1}^{n} v_i B_i & \text{(The value of the allocation)} \\
&s.t. \sum_{i=1}^{n} q_{ij} B_i \leq 1, 1 \leq j \leq k & \text{(Feasibility constraint)},
\end{aligned}
\tag{1}
$$

where B_i is a binary variable representing whether bid i is selected or not.

This problem can be handled by a set partitioning algorithm introduced by Garfinkel and Nemhauser [7]. (For definitions of the set partitioning problem and related problems, cf. Balas and Padberg [5], and Salkin [14].) As pointed out by the originators, "the approach is so simple that it appears to be obvious. However, it seems worth reporting because it has performed so well". Indeed, some of the experimental results reported by Garfinkel and Nemhauser seem surprisingly good compared to recent experiments when taking the hardware performance at that time into account.

The principles of the Garfinkel-Nemhauser algorithm are as follows. The algorithm creates one list per row (i.e. commodity) and each set (column) is stored in exactly one list. Given an ordering of the rows, each set is stored in the list corresponding to its first occurring row. Within each list, the sets are sorted according to increasing cost (for example by keeping them in a binary search tree [1]). The search for the optimal solution is done in the following way:

1. Choose the first set from the first list containing a set as the current solution.
2. Add (to the current solution) the first disjoint set from the first list—corresponding to a row not included in the current solution—containing such a disjoint set, if any.
3. Repeat Step 2 until one of the following happens: (i) *The cost for the current solution exceeds the cost for the best solution:* this branch of the search can be pruned. (ii) *Nothing more can be added:* check if this is a valid solution/the best solution so far.
4. Backtrack: Replace the latest chosen set by the next set of its list and go to Step 2. When no more sets can be selected from the list, back up further recursively. If no more backtracking can be done, terminate.

Since the problem of Equation (1) is equivalent to the definition of set packing and the problems of set packing and set partitioning can be transformed into each other [5], the Garfinkel-Nemhauser algorithm can be used for winner determination in combinatorial auctions. It is also clear that it is trivial to modify the algorithm to be suited for set packing without any modification of the input; it is only a minor modification of the pruning/consistency test. Specifically, the sets need to be renamed as bids, cost has to be replaced by valuation, and item 3 needs to be replaced by: Repeat Step 2 until one of the following happens: (i) *The value for the current solution can not exceed the value of the best combination found so far:* this branch of the search can be pruned. (ii) *Nothing more can be added:* check if this is the best solution so far.

It turns out that some recently developed special-purpose algorithms are more or less rediscoveries of the algorithm above, with sme added features. For example, compared to the simple pruning described in the Garfinkel-Nemhauser algorithm, Sandholm's algorithm [15] and the CASS algorithm [6], use a more sophisticated technique which essentially is a rediscovery/specialization of the ceiling test [14].

5 Combinatorial Auction Winner Determination as a Mixed Integer Programming Problem

In this section we describe in some detail how very general optimal winner determination problems can be formulated as a mixed integer problem. For reading on mixed integer programming (MIP) in itself and its relation to set packing etc., we refer to the literature on combinatorial optimisation and operations research, e.g. [4; 5; 7; 8; 10; 14].

By properly formulating the problem, we get a large number of very attractive features. These include:

- The formulation can utilize standard algorithms and hence be run directly on standard commercially available, thoroughly debugged and optimised software, such as CPLEX[1].
- There may be multiple units traded of each commodity.
- Bidders can bid on real number amounts of a commodity, and place duplicate bids enabling them to express (approximate) linear segments of their preference space as a single bid.

---

[1] See www.cplex.com

- Bidders can construct advanced forms of mutually exclusive bids.
- Sellers may have non-zero reserve prices.
- There is no distinction between buyers and sellers; a bidder can place a bid for buying some commodities and simultaneously selling some other commodities.
- Arbitrarily complicated recursive bids with the above features can be expressed.
- Any constraints that can be expressed in linear terms can be used when defining the bids.
- Settings without free disposal (for some or all commodities) can be managed.

It should be pointed out that some of the generalizations achieved here could be expressed to fit algorithms for the basic case treated by Fujishima et al. and Sandholm [6; 15]. For example, mutually exclusive (XOR) bids are easily formulated by adding dummy commodities (e.g. "a XOR $b \Leftrightarrow (a \wedge c) \vee (b \wedge c)$"), but such transformations often give rise to a combinatorial explosion of bids [11].

The aspect of avoiding errors is important; for an illustrating example, see the recent discussion on caching in auction algorithms [3].

It is noteworthy that the formulation of Equation (1) can be run directly by commercially available software. (The formulations of this problem given by Rothkopf et al. [13], Fujishima et al. [6], and Sandholm [15], are however not suited as direct input for standard software.)

The formulation used here conforms with the formulations by Wurman [16] and Nisan [11]. Compared to these formulations, we observe that much more general combinatorial auctions than the ones treated so far can be expressed as mixed integer problems, and that they can be successfully managed by standard operations research algorithms and commercially available software.

With standard MIP methods, any constraint that can be expressed in linear terms can be used when defining a bid. Thus in the general case, the objective function will consist of terms representing the value of a (certain part of a) bid, times the degree to which it is accepted. That is, we need not restrict the auction only to binary choices. Correspondingly, the feasibility constraint need in the general case not be restricted to the cases where there is only one unit for sale of each commodity, free disposal can be assumed, etc.[2] It is also easy to see that the MIP approach can also be directly used for the minimal winning bid problem [13], i.e. the problem of replying to the question "If I request these and these amounts of these and these commodities, how much do I have to pay to get my bid accepted?".[3] Thus, there is a very general "bid assembler language" and powerful search engine for combinatorial auctions provided by standard MIP algorithms and software.

---

[2] The free disposal assumption [9] typically has a very drastic impact on any-time behavior; without the free disposal assumption, finding a feasible allocation is significantly harder.

[3] In fact, any winner determination algorithm can be used as a minimal winning bid algorithm: If there is already a bid in the winning set bidding for the same commodities as the request, then the problem is trivial. Else, construct a bid, b_i, with the requested commodities and a valuation, v_i, which is so high that the bid is guaranteed to win. Then compute the new winning combination. Compute the minimal valuation of the new bid as the surplus of the winning combination when b_i is not included minus the surplus of the winning combination when b_i is included plus v_i plus a minimal increase.

An example of a bid for which there is a straightforward linear formulation is the following: A buyer of electricity would like to buy power with the constraints that it wants

- power in four hours in a row out of six hours under consideration,
- 10.7–13.3kWh/h if the price per kWh is not higher than 0.12SEK,
- 10.7–14.2kWh/h if the price per kWh is not higher than 0.10SEK,
- a minimum average of 13.4kWh/h, and
- the total declared value not to exceed 5.90SEK.

Clearly, requiring that each bidder should give its bids as terms to be added to the objective function and the feasibility constraints together with a number of constraints may be a too heavy burden put on a bidder, cf. [11]. Therefore it makes sense to construct different high level bid grammars, which can be compiled into standard MIP formulations.

6   Does Mixed Integer Programming Work in Practice?

In [2] a set of experiments were presented, showing that a standard MIP solver like CPLEX, combined with some very simple heuristic algorithms, performs very well compared with special-purpose algorithms. This holds even if we run the experiments on benchmarking data designed by the inventors of the specfial-purpose algorithms. The added heuristics consist of

- A simple preprocessing, *column dominance checking* [14]
- A very simple breadth-first search over all bids, where each bid keeps a list of colliding bids.

It is generally recognized that it is most unfortunate that real-world test data are not available. As long as no such test data is available, it is perhaps reasonable to try different types of distributions and try to identify what types distributions are "hard" and which ones are "easy" for different types of algorithms. The empirical benchmarks are performed on the same test data as was given by Sandholm [15] and Fujishima et al. [6]. We use these tests not because we are convinced that they are the most relevant, but because they are the only ones for which we have data for competing algorithms. Our experience so far is that (seemingly small differences of) the bid distributions chosen have an *extreme* impact on the performance of the algorithms. Therefore one should be very careful when arguing about the practical usefulness of an algorithm without access to real data.

6.1   Discussion

In sum, CPLEX in combination with some simple preprocessing performs very well for the tested distributions. Under most reasonable assumptions of the collection of bids, the computation time is relatively small. Furthermore, if the bids are submitted in sequence under some amount of time, CPLEX can do subsequent searches starting from the best solution found up to the point of the arrival of a new bid.

From our experiments with different families of algorithms it is clear that if the probability distribution is known to the auctioneer, it is sometimes able to construct algorithms that capitalize significantly on this knowledge. Our main conclusion so far is that it is very important to obtain some realistic data and investigate whether it has some special structures that can be utilized by highly specialized algorithms (assuming that standard algorithms fall short on practically relevant instances).

Some probability distributions used for generating benchmark data may at a first glance may seem "hard" but turn out to be rather "easy". As a matter of fact, it is easy to construct very simple yet very efficient algorithms for these special cases. A good way to get hints for constructing specialized algorithms is to run a few experiments with a standard MIP software and then analyze the solutions.

More challenging is of course to construct more general and adaptive algorithms that perform excellently on all types of input. However, as pointed out by Sandholm [15, Proposition 2.3], we can not even hope for being able to construct approximate algorithms with reasonable optimality guarantees that are very efficient. Furthermore, constructing very general integer programming algorithms is of a much broader interest than the one of the e-commerce community, and such work should probably be published elsewhere; the e-commerce relevance of constructing general integer programming algorithms without having any idea about what real-world distributions may look like is not completely obvious. It may actually be that the bids of certain real-world auctions have very distinctive distributions (for example caused by bidding rules or by the nature of the preferences), for which very simple and efficient algorithms can be constructed. But it may also be completely different; real distributions may be very hard and the performance tests with artificially generated random input give no guarantee for their usefulness in real settings.

The construction of realistic probability distributions based on some of our main application areas—such as electronic power trade and train scheduling markets—together with some reasonable bidder strategies in certain attractive combinatorial auction models, such as iBundle [12] or the ascending bundle auction by Wurman [16] is important future work. However, one brief reflection on realistic probability distributions can be given already here; there are good reasons to believe that real distributions will be *much harder* than most of the artificial random data. For example, if the iBundle auction is used and we have bidders with the strategies that they only bid ε above the current prices (or taking the "ε-discount") we will have a very "tight" distribution; most bids are part of some combination which is close to optimal. This makes pruning drastically harder. But there are also other aspects of the hardness of real-world distributions [11]. Again this calls for gathering of real-world (or at least derivation of realistic) data, before focusing on heavily specialized algorithms.

7   Concluding Remarks

The usefullnes of combinatorial auctions is supported by reald-world data. The corresponding computational problem, usch as set partitioning and set packing, are hard. Nevertheless, there are efficient methods, such as Mixe Integer Programming, that show good behaviour for a very general class of problems related to combinatorial auctions.

As our final remark, we would like to state

It is more important to spend time on understanding what real world data look like than to design special-purpose algorithms for artificially generated data.

8 Acknowledgement

This article describes joint work with Mattias Willman (former Mattias Tenhunen) and Fredrik Ygge.

References

1. A. Andersson, Ch. Icking, R. Klein, and Th. Ottmann. Binary search trees of almost optimal height. *Acta Informatica*, 28:165–178, 1990.
2. A. Andersson, M. Tenhunen, and F. Ygge. Integer programming for combinatorial auction winner determination: Extended version. Technical report, Department of Information Technology, Uppsala University, July 2000. (Available from www.it.uu.se).
3. K. Asrat and A. Andersson. Caching in multi-unit combinatorial auctions. In *Proceedings of AAMAS*, 2002.
4. E. Balas. An additive algorithm for solving linear programs with zero-one variables. *The Journal of the Operations Research Society of America*, pages 517–546, 1965.
5. E. Balas and M. W. Padberg. Set partitioning: A survey. *SIAM Review*, 18:710–760, 1976.
6. Y. Fujishima, K. Leyton-Brown, and Y. Shoham. Taming the computational complexity of combinatorial auctions: Optimal and approximate approaches. In *Proceeding of the Sixteenth International Joint Conference on Artificial Intelligence, IJCAI'99*, pages 548–553, August 1999. (Available from robotics.stanford.edu/~kevinlb).
7. R. Garfinkel and G. L. Nemhauser. The set partitioning problem: Set covering with equality constraints. *Operations Research*, 17(5):848–856, 1969.
8. A. M. Geoffrion. An improved implicit enumeration approach for integer programming. *Operations Research*, 17:437–454, 1969.
9. A. Mas-Colell, M. Whinston, and J. R. Green. *Microeconomic Theory*. Oxford University Press, 1995.
10. T. Michaud. Exact implicit enumeration method for solving the set partitioning problem. *The IBM Journal of Research and Development*, 16:573–578, 1972.
11. N. Nisan. Bidding and allocation in combinatorial auctions. Working paper. Presented at the 1999 NWU Microeconomics Workshop. (Available from http://www.cs.huji.ac.il/ noam/), 1999.
12. D. Parkes. iBundle: An efficient ascending price bundle auction. In *Proceedings of the First International Conference on Electronic Commerce*, pages 148–157. ACM Press, Box 11405, New York, NY, November 1999. (Available from www.cis.upenn.edu/~dparkes).
13. M. H. Rothkopf, A. Pekeč, and R. M. Harstad. Computationally manageable combinatorial auctions. *Management Science*, 44(8):1131–1147, 1995.
14. H. M. Salkin. *Integer Programming*. Addison Wesley Publishing Company, Reading, Massachusetts, 1975.
15. T. W. Sandholm. An algorithm for optimal winner determination in combinatorial auctions. In *Proceeding of the Sixteenth International Joint Conference on Artificial Intelligence, IJCAI'99*, pages 542–547, August 1999. (Available from www.cs.wustl.edu/~sandholm).
16. P. Wurman. *Market Structure and Multidimensional Auction Design for Computational Economies*. PhD thesis, Department of Computer Science, University of Michigan, 1999. (Available from www.csc.ncsu.edu/faculty/wurman).

Systematic Development of
Complex Web-Based User Interfaces

Henrik Behrens, Thorsten Ritter, and Hans-Werner Six

Software Engineering, Dept. of Computer Science
FernUniversität, 58084 Hagen, Germany
{henrik.behrens, thorsten.ritter, hw.six}@fernuni-hagen.de

Abstract. Software components realising the graphical user interface (GUI) of a highly interactive system or the user interface of a Web application form an essential part of the entire implementation and significantly affect the effectiveness and maintainability of the software system. In this paper, we propose a software engineering based approach for the development of complex Web-based user interfaces (Web-UIs). The method comprises two basic tasks. The first task gathers and models the UI requirements. The second task maps the UI requirements model into a Web-UI software architecture. To this end, we propose DAWID (DIWA-based Web User Interface Development), an approach that adapts the DIWA framework for GUI development to Web-UIs. Since DIWA consequently applies fundamental software engineering (SE) principles and its adaption to Web-UIs requires only minimal changes, the resulting DAWID architecture also stays fully compliant with the SE principles.

1 Introduction

Up to now, many Web applications have failed in achieving basic software engineering (SE) standards because they have been developed by non-computing professionals and time to market has been more important than a systematic software engineering approach. At present, the increasing complexity of Web applications causes a growing concern about the lacking software quality of Web applications.

This paper addresses the systematic development of complex Web-based user interfaces. Usability and software quality of the user interface are of considerable importance for conventional applications with graphical user interfaces (GUIs) as well as for today's applications with Web-based user interfaces (Web-UIs). The Software components realising the graphical user interface of a highly interactive system or the user interface of a Web application form an essential part of the entire implementation and significantly affect the effectiveness and maintainability of the software system. In both kinds of applications, the UI software architecture as backbone of the UI implementation deserves particular attention and in-depth considerations.

We propose a software engineering based approach for the development of complex Web-UIs. The method comprises two basic tasks. The first task gathers and models the UI requirements. Among them are the functional requirements for the UI as well as the

R. Klein et al. (Eds.): Comp. Sci. in Perspective (Ottmann Festschrift), LNCS 2598, pp. 22–38, 2003.

"UI-specific requirements" like the static structure, the basic dynamic behaviour and usability aspects of the UI. The entire specification is compliant with the UML [8] and independent of the UI implementation platform, i.e. independent of GUI and Web-UI technology. The task modifies the SCORES+ approach [5] for a better integration of the functional requirements model and the related UI specific requirements model.

The second task maps the UI specific requirements model into a Web-UI software architecture. To this end, we propose DAWID (DIWA-based Web User Interface Development), an approach that adapts the DIWA framework for GUI development [9] to Web-UIs. DIWA applies fundamental software engineering principles and supports the construction of high quality GUI software architectures. The design rationale of DAWID is to take the DIWA architecture as heart of the Web-UI architecture paying special attention to keeping the necessary changes as minor and peripheral as possible. Since the adaption of DIWA to Web-UIs requires only minimal changes, the resulting DAWID software architecture also stays fully compliant with the SE principles.

The paper is organised as follows. Section 2 presents the modelling concepts for UI requirements. As useful preparation for the introduction of the DAWID architecture for Web-UIs, section 3 shortly recalls the design rationale of the DIWA architecture for GUIs. Section 4 proposes the DAWID software architecture and Section 5 concludes the paper.

2   User Interface Requirements

Modelling user interface requirements consists of two main steps. In the first step, the work units (use cases) the user performs with the help of the system are specified. (In the user interface community these work units are often called tasks.) In the second step, the description of this UI related functionality is complemented by a model of the fundamental characteristics of the user interface, among them the static structure, the basic dynamic behaviour, and usability aspects. Suitability and usefulness of the integrated specification depend on how well the functional and the complementing UI models fit together, in particular concerning abstraction level and granularity.

This section deals with the integrated modelling of UI related functional and complementing UI requirements. The approach refines our SCORES+ method [5] and all modelling elements are compliant with the UML [8]. Section 2.1 addresses the specification of UI related functional requirements, while sections 2.2 and 2.3 explain the modelling of the complementing UI requirements.

For simplicity reasons, we will often abbreviate UI related functional requirements to functional requirements and complementing UI requirements to UI requirements if no confusion can occur.

2.1 Functional Requirements

In this section, we briefly review the basic modelling elements of SCORES+ [5] which are devoted to the specification of functional requirements. They mainly comprise actors and use cases, activity graphs, and the domain class model. At the same time, we propose a modification of activity graphs for a better compliance with the UI modelling elements presented in Sect. 2.2.

Actors, Use Cases, and Activity Graphs. *Actors* characterise roles played by external objects which interact with the system as part of a coherent work unit (a use case). A *use case* describes a high level task which is related to a particular goal of its participating actors. Since the description of use cases is purely textual and vague, we take activity graphs for a more precise specification.

We modify SCORES+ by making activity graphs more compliant with the UI elements (see Sect. 2.2). We define five stereotypes for actions and a stereotype for associating actors with actions. Table 1 defines these stereotypes and their visualisations. Note that the *«decision interaction»* is a decision that an actor performs with the help of the system, while the decision PseudoState provided by the UML is executed by the system on its own.

Table 1. Activity graph elements as stereotypes

Stereotype	Description	Visualisation
«context action»	An ActionState stereotyped *«context action»* represents an action which is performed by an actor without the help of the system.	
«business interaction»	An ActionState stereotyped *«business interaction»* represents an action that is performed by an actor with the help of the system producing an observable result.	
«decision interaction»	An ActionState stereotyped *«decision interaction»* represents an action that is performed by an actor with the help of the system resulting in a decision on the subsequent action.	
«system action»	An ActionState stereotyped *«system action»* represents an action that is executed by the system on its own producing an observable result.	
«macro action»	A SubactivityState stereotyped *«macro action»* is an action that "calls" a subgraph (it reflects an «include» or an «extend» dependency).	
«actor in action»	An ObjectFlowState stereotyped *«actor in action»* depicts an actor which can be associated with some actions.	

Example: We take a simple Mail Tool as running example. The Mail Tool covers reading mail, composing new mail, editing a list of mail folders, a single mail folder with its inclosed messages, and a mail address book. Fig. 1 depicts the use case diagram for the Mail Tool. We identify the actors Mail Tool User and Mail Server and the use cases Read Mail, Compose Mail, Edit Folderlist, Edit Folder, New Folder, and Edit Addressbook.

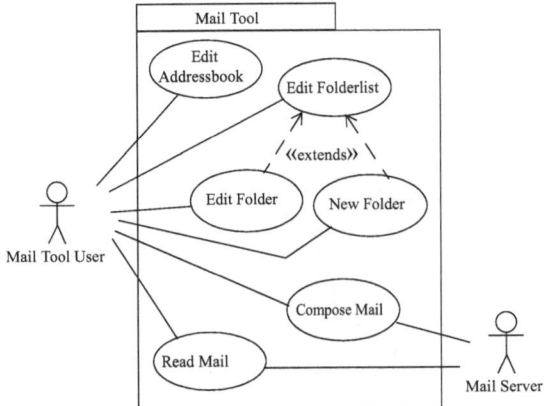

Fig. 1. Use case diagram for the Mail Tool functionality

Fig. 2 provides a textual description of the use case Edit Folderlist.

Use Case Edit Folderlist **in Model** Mail Tool
  Actors: Mail Tool User
  Main flow: The Mail Tool User selects a mail folder from a list of existing folders to open and
    view or edit inclosed mails.
    (**include**: Edit folder).
  Alternative flow The Mail Tool User creates a new folder.
    (**include**: New Folder)
  Alternative flow The Mail Tool User selects a mail folder from a list of existing folders and
    deletes the folder.
  End Edit Folderlist

Fig. 2. Textual description of use case Edit Folderlist

The activity diagram in Fig. 3 refines the use case Edit Folderlist. It starts with the *«decision interaction»* which decides on the subsequent action. In case of having chosen the *«business interaction»* Select Folder, the Mail Tool User can select a mail folder from the current list of mail folders and afterwards decide to edit the selected folder ("control flow" moves to the *«macro action»* Edit Folder starting the associated use case) or to delete it ("control flow" moves to the *«system action»* Delete Folder). In

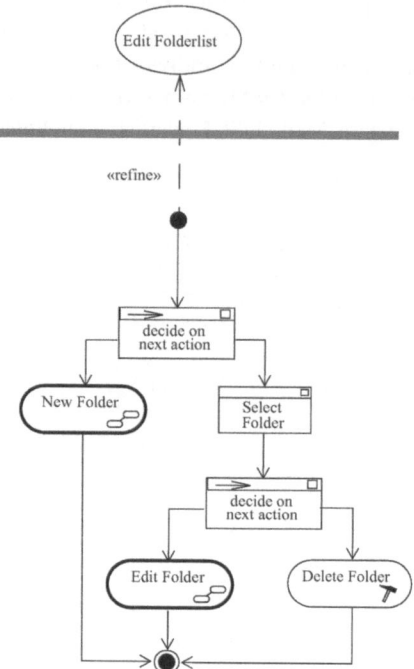

Fig. 3. Activity diagram of the use case Edit Folderlist

case of having chosen the *«macro action»* New Folder, the according use case is "called" and a new folder can be created and added to the current folder list.

Domain Class Model. The domain class model in the sense of Jacobson et al. [6] describes the most important types of objects which represent "things" that exist in the context in which the system works. For a use case and for each interaction and system action in the corresponding activity graph, we determine a so-called *class scope* denoting the set of (domain) classes involved in their execution, i.e. classes with instances being created, retrieved, activated, updated, or deleted. Furthermore, we balance activity graphs and the class model such that for each business interaction and system action in the activity graph a so-called *root class* exists that provides a *root operation* accomplishing ("implementing") the interaction, resp. system action, in the class model.

Fig. 4 depicts the central domain classes of the Mail Tool example. A Folderlist comprises a set of folders which contain folder entries. A Folder Entry can be either a Folder or a Mail. The *«system action»* DeleteFolder from the activity diagram depicted in Fig. 3 invokes the corresponding root operation deleteSelected() defined in the class Folder. This operation removes the selected folder from the folder list and deletes all contained folder entries.

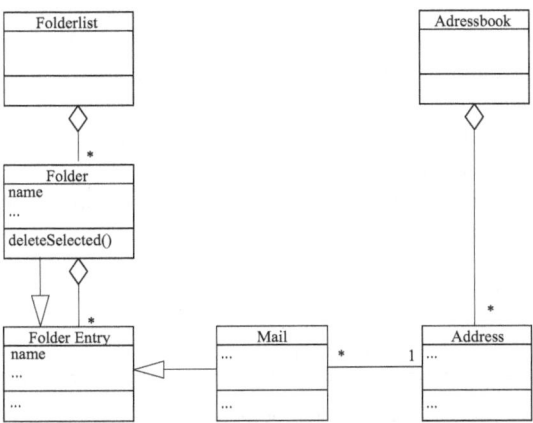

Fig. 4. Mail Tool domain class diagram

2.2 User Interface Elements

We now add UI elements to the activity graphs. The UI elements capture only the most important decisions concerning the user interface and specify neither the concrete interaction style nor the detailed UI design. The UI elements are modelled as classes and meet the abstraction level of activity diagrams. As with activity graphs for modelling functional requirements we modify the UI elements originally proposed in SCORES+.

Scenes and Class Views. The fundamental UI element is a *scene* which is modelled by a class stereotyped *«scene»*. There is a one-to-one relation between scenes and interactions of an activity graph. A scene is an abstraction of (a part of) the screen information which is presented to the user performing the according interaction.

A scene may comprise *class views* which are modelled by a class stereotyped *«class view»*. A class view is an abstract presentation of (properties of) instances and associations of a domain class (in the class scope) involved in the according interaction. A scene with no class view reflects no domain object information and is usually associated with a decision interaction. In cases where the abstract description of a class view does not sufficiently characterise the situation, e.g. if domain specific renderings are required, graphical drafts may additionally be attached.

Table 2 shows the visualisation of a scene and a class view together with a short description.

Scene Operations. *Scene operations* model the functionality provided by the user interface. We distinguish between two different kinds of scene operations.

An *«application operation»* is activated by the user during a *«business interaction»* and models an application function. It also stands for functions like editing or selecting instances of domain classes which are presented in a class view of the scene. After the execution of an «*application operation*» an implicit navigation to the next scene happens. Note that such an implicit navigation is triggered by the system and not activated by the user.

A «*navigation operation*» is activated by the user during a «*decision interaction*» and causes an explicit navigation to the selected scene.

The visualisations and short descriptions of the scene operations are provided by Table 2.

Table 2. UI elements as stereotypes

Stereotype	Description	Visualisation
«*scene*»	A class stereotyped «*scene*» represents an abstraction of (a part of) a screen.	Scene
«*class view*»	A class stereotyped «*class view*» is an abstract presentation of instances and relations of a domain class.	Class view
«*application operation*»	A scene operation stereotyped «*application operation*» is activated by the user during a «*business interaction*».	ApplicationOp
«*navigation operation*»	A scene operation stereotyped «*navigation operation*» is activated by the user during a «*decision interaction*».	NavigationOp

"Semanticless" user operations like window moving or re-sizing and scrolling are not considered. Like scenes and class views, scene operations abstract from concrete user interface details. For example, the particular interaction technique invoking a scene operation, e.g. menu item selection or drag and drop, is not specified. Likewise, low level interactions like mouse button presses or basic text editing operations are not considered.

Example: The right part of Fig. 5 depicts UI elements for the activity diagram of the use case Edit Folderlist (see Fig. 3). The Edit Folderlist scene is associated with a decision interaction that chooses either the interaction Select Folder or the macro action New Folder. It comprises the navigation operations New Folder and Select Folder but no class view. The scene Select Folder is associated with the interaction Select Folder and allows a user to select a folder from the current folder list which is depicted as a class view Folderlist. The scene Work With Folder is attached to a decision interaction that chooses either the system action Delete Folder or the macro action Edit Folder. The directed association subsequent between two scenes denotes that they are subsequent with respect to the "control flow" of the corresponding activity diagram.

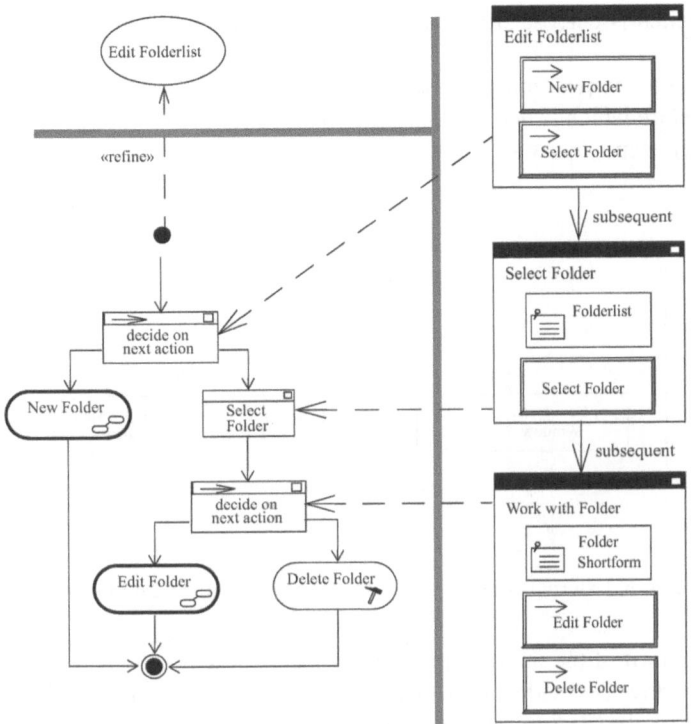

Fig. 5. Use case Edit Folderlist with activity diagram and scenes (actors supressed)

2.3 User Interface Structure and Navigation

User interfaces are decomposed into building blocks like e.g. windows or Web pages. They can again be divided into building blocks like e.g. panels or table cells. A UI building block is usually related to more than one interaction of a use case, sometimes even to the interactions of more than one use case, while a scene is associated with exactly one interaction. The next step of the UI requirements modelling therefore deals with the composition of building blocks from scenes.

We explain the process of composing building blocks from scenes for graphical user interfaces (GUIs). For Web-based user interfaces (Web-UIs), the composing task proceeds analogously.

The composition process for GUIs consists of three steps. The first step merges related "atomic scenes" into "superscenes" in an iterative process. In the next step, superscenes are composed to windows (internal window structure). The third step arranges the windows according to a hierarchical structure (external window structure).

When the composition process has completed, the final step of the UI requirements modelling determines the navigation between windows. We assume that a navigation between parent and child windows is provided by default. Hence, only addional navigation which is derived from domain requirements and usability aspects must be considered.

Applying the composition process to the Mail Tool example leads to an external window structure which is depicted in Fig. 6. The dotted arrows indicate the additional navigation between windows.

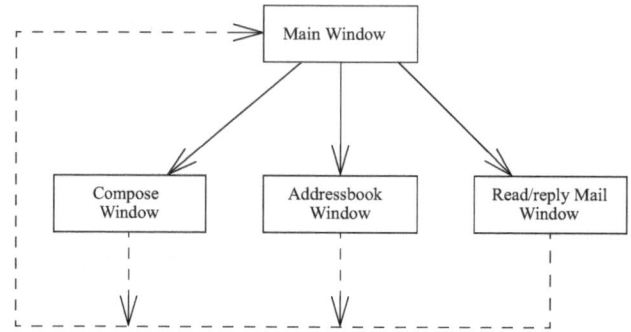

Fig. 6. External window structure and navigation for the Mail Tool

A possible draft of the internal structure of the main window is provided by Fig. 7. The main window is divided into three panels. The top panel contains the buttons "Compose" and "Edit Adressbook". The panel "Edit Folderlist" includes the two subpanels for the Folderlist and the three buttons below. Analogously, the panel "Edit Folder" comprises the two subpanels for the list of mails and the four buttons below.

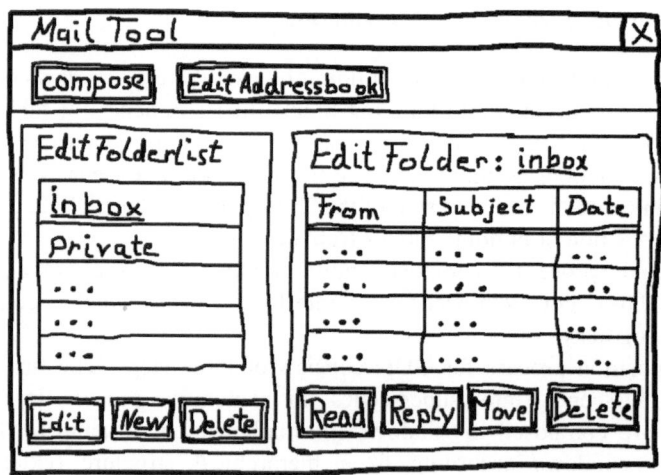

Fig. 7. Draft of the Mail Tool's main window

If the Mail Tool is to be implemented as Web application, the composition of the Web-UI structure and the determination of the additional navigation proceed analogously to the GUI case. Choosing Web pages instead of windows and e.g. cells of tables instead of panels, we end up with a Web-UI structure that matches the GUI structure. Because of the analogy of both UI structures we abstain from providing a depiction of the Web-UI structure.

3   The DIWA Architecture for Graphical User Interfaces

The DAWID software architecture for Web-UIs which will be proposed in Sect. 4 is based on a transplantation of the DIWA software architecture for GUIs into the Web environment. As useful preparation for the discussion of the DAWID approach, this section shortly recalls the DIWA design rationale [9]. DIWA is a framework for the development of high quality GUIs which is based on fundamental software engineering principles. This section ends up with an explanation of how the UI requirements model of the Mail Tool example is transformed into a DIWA software architecture.

3.1 Software Engineering Principles for User Interface Architectures

The logical separation of the user interface from the functional core of an application, usually realised by two separate layers or tiers, is a widely accepted software engineering (SE) principle. Additionally, the following mainstream SE principles have been proven particularly useful for UI architectures [9]:

Hierarchical structure. One of the oldest principles for controlling a large and complex system is its decomposition into a hierarchy of smaller components. Together with acyclic communication between the components, the hierarchical structure forms the central basis for controlling the overall complexity of the system.

Homogenous components. This principle relates to simplicity of concepts. Simple as well as complex components should be of the same structure and should be treated in the same way. In addition, the communication between components should follow the same protocol. Besides simplicity of concepts, homogeneity of components is an important prerequisite to composability and reusability.

Separation of concern. The responsibility of a component should be cohesive and clearly separated from the responsibilities of the other components. On the other hand, the responsibility of a component should be determined such that the number of its relationships with other components becomes minimal.

Acyclic communication. Acyclic communication, more precisely acyclic use dependencies, provide for better control of communication, easier testing, and improved maintainability of the software.

3.2 The DIWA Approach

We briefly recall the basic concepts of the DIWA architecture which we slightly modify for our purpose. For more details about the entire DIWA framework concerning e.g. the declarative language for the specification of dynamic dialog behaviour, the detailed responsibility of components, or the detailed communication between components see [9].

Global structure. DIWA provides a logical separation of the user interface from the functional core of the application and the decomposition of the user interface into homogeneous components, called user interface objects (UIOs). The hierarchical arrangement of UIOs often corresponds to geometric containment of the UIOs' screen rendering. Simple buttons as well as large composite components like complex windows are treated as UIOs thus being structured in the same way. Communication between UI and functional core of the application and also among UIOs is acyclic. The global DIWA organisation complies with the SE principles hierarchical structure, homogeneous components, and acyclic communication.

User interface object (UIO). A UIO encapsulates a certain part of the dialog behavior, the associated part of the screen layout, and the corresponding part for accessing the functional core of the application. As slight modification of the MVC design pattern [3], a UIO is composed of the three parts (classes) dialog control, presentation, and application interface. The parts of a UIO comply with the SE principles separation of concern and acyclic use dependencies.

Dialog control (DC). This class serves as interface of the UIO. It makes all decisions concerning the course of the dialog, i.e. it receives events and performs the appropriate reactions on them. An event is an object that can carry a user input (external event) or an internal message between UIOs (internal event). The dialog control's reaction on an event can be a call to the presentation (e.g. to change the screen rendering of the UIO), a call of application functions via the application interface, or the passing of an internal event to (the DC of) a direct sub-UIO.

Presentation (P). This class is responsible for drawing the UIO. The presentation retrieves the data to be presented on the screen by calling application functions via the application interface.

Application interface (AI). This class offers the presentation class access to application data and provides the dialog control with access to application functions.

The decomposition of a UIO into its three parts and the according events and use dependencies are illustrated in Fig. 8.

DIWA Software Architecture for the Mail Tool Example. In order to create a DIWA software architecture for the Mail Tool GUI, we take its window structure (see Sect. 2.3) as starting point. First, the hierarchy of windows and panels is mapped one-

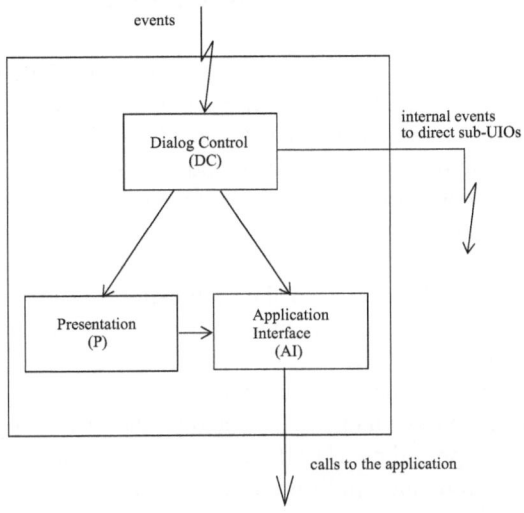

Fig. 8. DIWA UIO with events and use dependencies

to-one into a hierarchy of UIOs. Then UIOs are added for all widgets of each panel and inserted as sub-UIOs of their panel UIO.

Fig. 9 illustrates the DIWA software architecture for the main window of the Mail Tool. The root node of the UIO hierarchy corresponds to the main window. Most of the internal nodes are the panel UIOs, while the leaf nodes are the UIOs for the widgets contained in the panels.

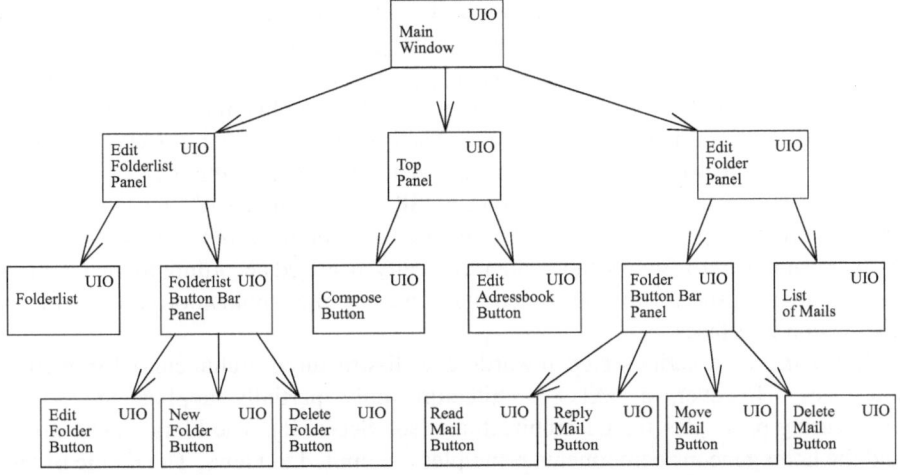

Fig. 9. Part of the DIWA architecture concerning the main window of the Mail Tool

4   The DAWID Architecture for Web-Based User Interfaces

A Web-UI is divided into a client tier and a server tier. The browser running on the client's computer is responsible for displaying the Web documents on the client's screen and for the communication between the client and the server tier. A Web document is built up from (static) HTML code, the communication protocol is HTTP.

The server tier comprises two main components: the Web server and the servlet engine. The Web server provides Web documents as response to HTTP requests sent by the client. The servlet engine is a framework that extends the functionality of a Web server by supporting the access to the application via servlets. Typically, a servlet transforms an HTTP-request to events which are sent to (server side) user interface components of the application.

At present, Web documents containing dynamic context are usually produced by Java Server Pages (JSPs) which are executed by the servlet engine. A JSP composes a Web document by providing HTML code for the static content and retrieving the dynamic content from the application. The composed Web document can then be delivered to the client.

4.1 Approaches to Web-Based User Interface Architectures

There exist a couple of approaches for a systematic structuring of Web-UI software architectures. One of the earliest attempts is the JSP Model 2 Architecture [4] which is a blueprint for a direct transformation of the MVC design pattern into the Web environment, where the C-part of the MVC is realised by a servlet, the M-part by a Java Bean and the V-part by a JSP. The Struts framework [1] is a sample implementation of the JSP Model 2 architecture. Both, the JSP Model 2 architecture and Struts propose a flat architecture that does not provide for a hierarchical decomposition of complex Web-UIs into (homogeneous) components.

Among the few approaches that appy the idea of hierarchically arranged components to Web-UI architectures, the Sun One Application Framework (JATO) [10] is the most prominent representative. The central component in JATO is a view. A view corresponds to a Web document or a part of it, e.g. a table cell or a text input field. For composing a Web document, a JSP is associated with the view. From an architectural viewpoint, a view can be regarded as a merge of the C- and V- part of the MVC design pattern. Views can be hierarchically arranged according to the Composite View pattern. Access to the functional core of the application is provided by application interfaces.

JATO makes a serious step towards a well-structured architecture for Web user interfaces. However, a JATO architecture only partially applies the software engineering principles for UI architectures (see Sect. 3.1). The hierarchical structure and the homogeneous components principles are limited to views. The merge of the C- and V-parts of the MVC design pattern into a single JATO view violates the separation of concern principle.

Summarising, a consequent application of sound GUI architecture principles to Web-UI architectures has not been fully achieved.

4.2 The DAWID Approach

In this section, we propose DAWID (DIWA-based Web User Interface Development), an approach that refines JATO for a better compliance with software principles for UI software architectures. The design rationale of DAWID is to taylor the DIWA architecture to the Web environment with as few and minor adaptions as possible.

The kernel of DAWID is a hierarchy of DAWID user interface objects (Web-UIOs). A Web-UIO is identical to a DIWA-UIO except for a slight modification of the presentation part due to the special conditions of the Web environment. In contrast to JATO, in DAWID a (part of a) Web document corresponds to a Web-UIO and not to a view.

A Web-UIO differs from a DIWA-UIO with respect to responsibility of the presentation part. Logically, the presentation of a Web-UIO comprises a JSP that is responsible for composing the corresponding (part of a) Web document, and a class Web-P that takes on the remaining responsibility of the original DIWA presentation part.

As in DIWA, the Web-UIOs communicate with each other via events. Furthermore, DAWID provides a servlet that transforms HTTP requests into events and sends the events to the root Web-UIO of the Web-UIO hierarchy.

The DAWID architecture depicted in Fig. 10 shows a clear separation of (the hierarchy of) Web-UIOs on the one side and the servlet and the JSPs on the other side. The servlet and the JSPs form a kind of a bridge between the servlet engine and Web-UIOs, allowing the latter to stay close to their DIWA archetypes.

Fig. 11 explains the collaboration of the architectural components using a scenario where a user sends an HTTP request by activating a button within a Web page. For simplicity reasons, we assume that the response is not a new Web document but only an updated version of the current one.

When the servlet has received the request from the Web server (1), the servlet transforms it into an event to (the DC of) the root Web-UIO (2). As with GUIs, the event is passed to (the DC of) the Web-UIO that is associated with the Web page containing the activated button (3). Now the DC passes the user input to the Web-P (4) and calls the AI for an invocation of the corresponding application function (5). When the execution of the application function has terminated, the DC notifies the Web-P to update itself (6) by calling the AI (7). Next, the DC calls the JSP (indirectly via the servlet engine) (8) which retrieves the modified data from the Web-P (9) and composes the updated version of the current Web document.

If the response to an HTTP request is not an updated version of the current Web document but a completely new Web document, then after step (5) the DC issues a forward event to the Web-UIO that is associated with the new Web document. This Web-UIO then proceeds with steps (6) to (9) accordingly.

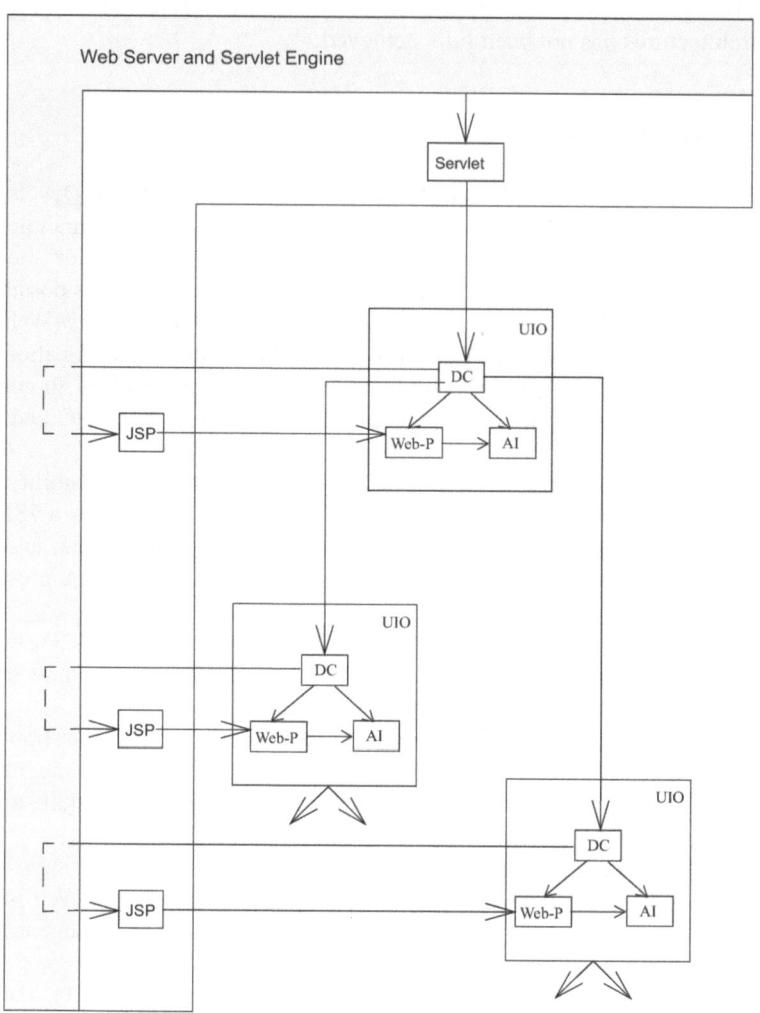

Fig. 10. DAWID architecture

DAWID Software Architecture for the Mail Tool Example. We map the hierarchy of Web pages and table cells of the Web-UI structure (see Sect. 2.3) one-to-one into a hierarchy of Web-UIOs. Then Web-UIOs are added for all widgets of each table cell and inserted as sub-UIOs of the table cell Web-UIO.

Since the GUI structure and the Web-UI structure of the Mail Tool and also the mapping of both UI structures to a software architecture are completely analog, the DAWID architecture is also analog to the DIWA architecture (see Fig. 9). We therefore abstain from providing a depiction of the DAWID software architecture.

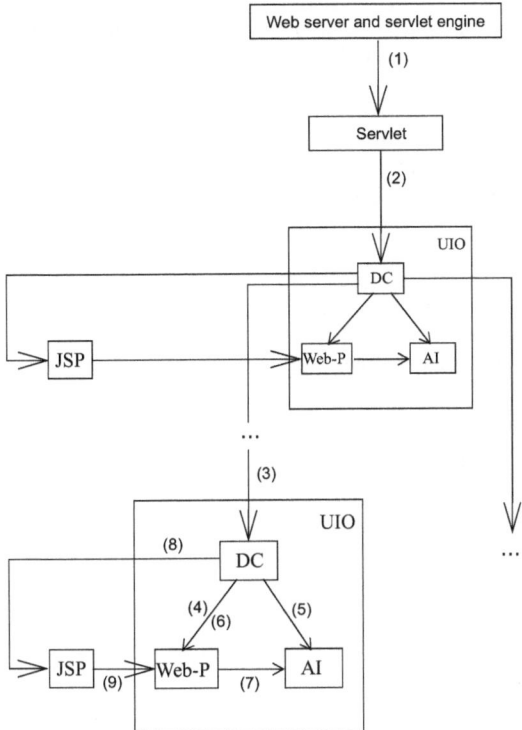

Fig. 11. Collaboration of the DAWID components

5   Future Work

Up to now, we have implemented a tool supporting SCORES+ and a framework for DIWA. Future work addresses the development of a DAWID framework, the refinement of the SCORES+ tool according to the modifications proposed in this paper, and the combination of both into one integrated tool. Besides tool development, we are going to validate our approach using a series of applications in various domains.

References

1.   Apache Software Foundation: Struts Jakarta Project, http://jakarta.apache.org/struts
2.   Booch, G., Rumbaugh, J., Jacobson, I.: The Unified Modeling Language Users Guide. Addison-Wesley/acm Press, Reading, MA (1999)
3.   Gamma, E., Helm, R., Johnson, R., Vlissides, J.: Design Patterns -- Elements of Reusable Object-Oriented Software, Addison-Wesley (1994)
4.   Seshadri, G.: Understanding JavaServer Pages Model 2 Architecture, JavaWorld, http://www.javaworld.com/javaworld/jw-12-1999/jw-12-ssj-jspmvc.html

5. Homrighausen, A., Six, H.-W., Winter, M.: Round-Trip Prototyping Based on Integrated Functional and User Interface Requirements Specifications. Requirements Engineering, Vol. 7, Nr. 1, Springer, London (2002)
6. Jacobson I., Booch G., Rumbaugh J.: The Unified Software Development Process. Addison-Wesley/acm Press, Reading, MA (1999)
7. Kösters, G., Six, H.-W., Winter M.: Coupling Use Cases and Class Models as a Means for Validation and Verification of Requirements Specifications. Requirements Engineering, Vol. 6, Nr. 1, Springer, London (2001) 3–17
8. Object Managment Group: Unified Modeling Language Specification. Version 1.3, OMG (1999)
9. Six, H.-W., Voss, J.: A Software Engineering Perspective to the Design of a User Interface Framework. Proc. IEEE CompSAC '92, Chicago (1992)
10. Sun Microsystems Inc.: Sun ONE Application Framework (JATO), http://developer.iplanet.com/tech/appserver/framework/

Teaching Algorithms and Data Structures:
10 Personal Observations

Hanspeter Bieri

Institute of Computer Science and Applied Mathematics, Neubrueckstrasse 10,
3012 Bern, Switzerland
Hanspeter.Bieri@iam.unibe.ch
http://www.iam.unibe.ch

Introduction

Since 1980, I have taught almost every year an elementary course on algorithms and data structures, first for the third and later for the second semester. The first performances - in PL/I times - were heavily based on [5]. Later the course became increasingly independent of any single textbook, and now makes use of a large number of sources, including, of course, the author's proper experiences. Nevertheless, my course is still very standard. It is known that many mathematics teachers at elementary schools get bored presenting the same proven stuff over the years, and more and more, therefore, teach "exotic" alternatives, in general not in favor of the students. This is a phenomenon I have always tried to avoid. Still I think that the elementary textbooks on algorithms and data structures have become too canonical too quickly, and that certain ways of presentation and also of selecting the contents are not as justified as they seem to be. In the following, I state some very personal observations which have arisen from the many performances of my standard course, in the hopes of challenging a little the textbook authors and teachers of algorithms and data structures. Each one of my 10 observations is concluded by a short statement put in a box and intended as kind of a summary and accentuation.

Thomas Ottmann is not only a renowned researcher in algorithms and data structures, but also a prominent innovator in teaching them to students of all levels. I am very pleased to dedicate this small contribution to him.

1 What Is a Data Structure?

Most textbooks on data structures do not define what a *data structure* is. They appeal to intuition and hope that discussing enough examples will suffice to make the notion familiar to the reader. This is not grave, but neither is it very satisfying. Can we do better? There are approaches that try to define data structures in an abstract and precise way [11], but the results are rather complicated and certainly not helpful to beginners. My favorite approach is that of [3] which reduces data structures to the more fundamental notion of *data types*:

> "To represent the mathematical model underlying an ADT (abstract data type) we use *data structures*, which are collections of variables, possibly of

R. Klein et al. (Eds.): Comp. Sci. in Perspective (Ottmann Festschrift), LNCS 2598, pp. 39–48, 2003.
© Springer-Verlag Berlin Heidelberg 2003

several different *data types*, connected in various ways. The *cell* is the basic building block of a data structure."

For instance, in order to represent an ADT stack, we can use a 1-dimensional array for the elements and an integer variable for the stack length. That is, we implement our ADT by means of two variables of a simpler data type. It is easier to define data types than data structures, but here the problem arises how to make a sufficiently clear distinction between *data model*, *data type* and *class* (in the sense of object-oriented programming).

In my course, I use to stress the fact that there are three aspects, i.e. the *mathematical*, the *data type*, and the *data structure aspect*, belonging to the same idea of how to structure data. A basic and very important example of how to structure data is the idea of the ordinary *list*: In mathematics, its concretization is the *finite ordered set*. With finite ordered sets, many properties of lists can well be explained, however a set is always *static* whereas lists are normally *dynamic*. It is interesting to note that Donald Knuth considers a list an ordered set [14], which is, to say the least, unprecise. In order to bypass this problem, [10] elegantly distinguishes between *mathematical sets* and *dynamic sets*. With data types, the concretization of the list idea is the *linear list*, i.e. a set of objects or elements together with a set of elementary operations. Finally with data structures, the most common concretizations of the list idea are the *simply-connected list* and the *doubly-connected list*, i.e. collections of connected variables which can be used to implement ADTs, such as e.g. an ADT binary tree. A linear list is not the same as a simply-connected list, and both are different from a finite ordered set. An analogous observation holds, of course, for any other idea of how to structure data.

> *Each basic idea of how to structure data has three aspects, that of mathematics, that of data types, and that of data structures. All three should be clearly distinguished although they are closely related.*

2   How Important Is Recursion?

When introducing algorithms and data structures, *recursion* is normally considered one of the most important and rather ambitious subjects. Recursion is certainly very significant in theoretical computer science and in functional programming . But in an introductory course based on an *imperative programming language*? And in practice? From an elementary point of view, recursion is the basis of *divide-and-conquer* and some well-known *functional programming languages*, such as Lisp; it is one of the chief motivations for the *stack*, and it is closely related to the *principle of mathematical induction*. Nevertheless, I have the feeling that its importance for practical computer science is normally exaggerated. Programmers in practice live normally well without using - or even knowing about- recursion, and profit from the fact that it is never absolutely necessary. Difficulty in understanding and reduced run-time performance are often heard arguments against recursion. They are not really valid, of course, but neither are they completely wrong.

I think that it is not by chance that in almost all textbooks on algorithms and data structures the same three examples occur again and again, namely the *factorial func-*

tion, the *Fibonacci numbers* and the *towers of Hanoi*. Should it not be very easy to find many more elementary examples if recursion were really so natural and fundamental? The three examples all have their well-known drawbacks: Due to the very limited range of the data type integer, computing n factorial by recursion makes practically little sense. Computing the Fibonacci numbers by ordinary binary recursion yields rather a good counterexample. And resolving the towers-of-Hanoi problem results in exponential time complexity which does not qualify this problem as being very appropriate for computer implementations. The special textbook [21] demonstrates well the wide range of elementary recursion. But it does not contain much material appropriate for a first course.

Recursion does belong to introductory computer science courses, but we should not spend too much time on it. It is one important subject beside several others. There are more relevant topics to discuss than e.g. its *elimination* or its *simulation*. And first-year students should not mainly be judged by their understanding of recursion.

> *In an introductory course, we should not overemphasize the importance of recursion.*

3   How to Teach an Algorithm?

There exist many good textbooks teaching algorithms, and there exist many bad ones too. Among the bad ones the most typical are those that almost exclusively list code in a specific programming language. These authors seem to believe that giving a complete implementation means automatically giving a complete explanation. And there seems even to exist a market for such an antididactic approach. Another approach that I do not consider very helpful consists in mainly providing a software package for *experimenting* with algorithms or for passive *algorithm animation*. This second approach reminds me of my former experiences with experimenting in physics. Either I did not understand what really happened, and in that case my understanding did not much improve by experimenting. Or I already understood what happened from the course, and in that case the experiment gave me perhaps a nice confirmation, but was certainly not a necessity. Of course, I agree that not all my colleagues in physics thought the same way...

According to my experience, using the *blackboard* - or a more modern medium with at least the same functionality - is still almost a prerequisite for teaching successfully an algorithm. Explaining an algorithm is quite similar to explaining a mathematical problem, and even modern mathematicians still like to present their ideas by using a blackboard. More concretely, I think that in order to teach an algorithm the following steps should be applied:

1. *Choose a small and simple, but typical example.*
2. *Formulate clearly the problem with the help of that example.*
3. *Apply the algorithm to that example, step by step, by performing directly the required modifications.*
4. *Repeating the same procedure for a second simple input is often a good next step.*
5. *Now algorithm animation may be very helpful, also in order to test special cases.*
6. *State the exact input and output in prose.*

7. *State the algorithm step by step in prose.*
8. *State the input, algorithm and output completely in pseudocode.*
9. *Give historical remarks and a reference to the original paper.*
10. *At the end additional experimenting may be very helpful.*

It is certainly not always necessary to cover all these steps. But presenting at least the small example and the main idea of the algorithm are certainly mandatory. Discussing the complete pseudocode is often not necessary and sometimes even disadvantageous. It is helpful to have software for animation and experimentation, but not a must.

Coming back to the blackboard, I would like to stress how important little ad hoc drawings are and how much more lively it is to make dynamically modifications to such drawings than to replace repeatedly a preproduced clean illustration by another such one. Blackboard and sponge, either real or virtual, form together an excellent support for performing directly small modifications.

> *Start simply and work with a small example. And do not forget the advantages of the blackboard.*

4   How to Design an Algorithm?

Most textbooks on algorithms and data structures pretend to teach how to design algorithms, and many bear even the word *design* in their title. Do they keep to their promise? Can algorithm design really be taught? Textbooks normally present general "design techniques", like *divide-and-conquer*, *backtracking*, or *greedy* approaches. These techniques are applied and adapted to a number of concrete problems for which they work well. A challenging example is [18] which follows this approach in an varied and innovative way. At the end of these books remains the hope that the reader has learned how to design algorithms... It is interesting to note that most books on the design of algorithms do not much differ from other introductory books on algorithms and data structures.

There has been much research and progress in the fields of *database* and *software design*, and several useful techniques have been developed that can well be taught and applied [25]. But *algorithm design* is different. It has more in common with *mathematical problem-solving* than with adapting a certain technique to new data and moderately new requirements. We thus arrive at the old question if mathematical problem-solving can be taught? There exist excellent texts on mathematical problem-solving [20], but they are more of a general than of a practical interest, and I doubt if they have much helped so far to solve non-trivial problems.

I do not mind teaching general "design techniques" and applying them to well-chosen problems. In this way students can learn a lot. I like especially books that also discuss wrong or suboptimal approaches and show how they can be improved [6]. And I just hesitate to talk about "algorithm design" in my introductory course. The main reason is that I think that the attribute "non-trivial" implies a lot: namely that it is impossible to simply apply a technique that one has learned before. "Non-trivial" means to me that some really good idea is probably required, and that is something that can not simply be taught or even automated.

> *Teaching general design techniques should not suggest that they work miracles. Non-trivial algorithms can seldom be designed in a straightforward way.*

5 How Much Analysis of Complexity Is Sensible?

Algorithm analysis is a difficult subject to teach, and in my course most yearly updates refer to it. But it is also an important subject and should not completely be postponed to more advanced courses. There has always been a lot of discussion on the pros and cons of the conventional *big-O approach* and its relevance in practice, and it is not the proper place here to continue this discussion.

I consider it important to teach that analysis of an algorithm does not only mean analysis of time and space complexity. That an algorithm is *correct* is more important and also, at least in practice, that it can be *readily implemented, understood,* and *modified*. I do introduce the big-O notation and am still undecided if $O(f)$ better stands for the right side of an equation or for a set [13]. Normally, some of my students are not quite sure, which order of magnitude is better, $O(nlogn)$ or $O(n^2)$. The pragmatic way to settle such questions is to obtain implementations for both cases and to compare the resulting running times. Although it is fun to spend some time on the big-O notation, I start believing that students should be allowed to absolve an introductory course on algorithms and data structures without a thorough understanding of it. I think that with algorithm analysis the old advise "keep it simple and stupid" is largely appropriate. It is possible to teach algorithms and data structures in a useful way with a minimum of analysis of complexity. So why risk the attractiveness of the course by investing too much effort in the big-O notation?

A special topic I like and consider instructive is to analyze the complexity of *divide-and-conquer algorithms*. In many cases it is easy to state the corresponding recurrence relation, but not so easy to solve it. Fortunately, there exists a "cookbook" method, sometimes called the *master method*, for easily solving a large class of recurrence relations [10]. Here students see that theory may be very practical, and they also recognize how small changes in the efficiency of the merge step can have big influences on the overall efficiency.

> *Teaching analysis of complexity could profit from less mathematics and more experimentation.*

6 What about Libraries?

It has always been an insight in practical computer science that optimal efficiency does not consist in reinventing the wheel again and again. This is in particular true with respect to solving a specific kind of problem algorithmically. Is it not a good idea to implement e.g. the heap sort once and for all, and then to use this same implementation whenever necessary?

There have been efforts from the beginning of using computers to build and provide *libraries*. Today, building good software libraries is an important matter of concern, and in the future it will become even more important. Prominent examples of very useful libraries are the *Standard Template Library* (STL) for C++, the *Java Collection Framework* (jcf), and the *LEDA library* for combinatorial and geometric computing [16].

The question I would like to address here is how to include the library concept when teaching algorithms and data structures. I think that libraries should be given more prominence in a first course than they are currently. Exemplary in this respect is e.g. the textbook [8]. In practice it is often very important to find out early in a project - or even before starting - what libraries are available and what contents they include. Both tasks are not trivial, and they should be taught even to beginners. Numerical analysis and image processing are just two very typical examples of the many fields where large libraries exist that are important in practice, but for non-experienced users are difficult to understand and apply.

According to my experience, it is helpful, interesting and not too expensive in a beginners' course to implement some of the algorithms in different ways, in particular with and without the help of a library. A pure blackbox approach, i.e. doing implementations only by leaning on libraries, should never be the goal. That is because many algorithms are based on very important and interesting ideas, and teaching these ideas is always essential, independently of the specific philosophy of the course.

> *Teaching how to use libraries is important, but should not be a core topic of the course.*

7   How Close Are Algorithms and Programming to Each Other?

Many authors publish several versions of their textbook on classic algorithms and data structures, i.e. first a version they pretend to be largely independent of any specific programming language, and then versions for Pascal, C, C++, and now Java. It is evident that this tactic can only work because these programming languages are not too different from each other - and close enough to the "nature" of the classic algorithms, i.e. they are all *imperative programming languages*. I for my part published once a "classic" data structure text for the APL2 language [7], which was quite a challenge because APL2 is only partly imperative. An example: In order to sort a list of numbers, the only thing we have essentially to do in APL2 is to apply to the given list one of the many primitive operators belonging to the language. Hence can it make sense to discuss heap sort or quick sort when the underlying programming language is APL2? It is not by chance, of course, that books on Lisp deal mostly with *applicative algorithms*, and books on Prolog with *deductive algorithms* [1],[22]. But I would not call these algorithms "classic". Textbooks on object-oriented programming largely omit the problem, since object-oriented programming occurs on a somewhat higher level. Certain application areas, especially artificial intelligence, use their own favorite programming language, mainly Lisp, or even a specialized language, such as a simulation language. But then the corresponding algorithms are typically specialized too.

It is a pity that most of the beautiful and clever classic algorithms we know are not more independent of the kind of programming languages most appropriate to implement them. I think that this gives the imperative programming languages somehow too much importance. But, of course, the first imperative programming languages, i.e. Fortran and Algol, were precisely designed to implement "classic" algorithms in as direct a way as possible. At any rate, we should not bypass the connection between algorithms and programming languages when teaching an introductory course on algorithms and data structures. An alternative would be to teach in the same course typical algorithms for different programming languages. The interesting new textbook [22] follows up this idea to some extent. It is an ambitious introduction, but succeeds well in making clear the connection between algorithms and programming languages.

> *Algorithms and programming languages are not as independent of each other as beginners normally believe. When teaching an introductory course this point should not be hidden.*

8   Why Are the Textbooks So Similar?

When I was studying mathematics, there was - at least in German - a very small market for textbooks. But the existing textbooks were all written very carefully, they had a long life and not too many changes between editions. Today the market is overfilled with computer books, many of which are just horrible. For today's readers it is very difficult to make a good choice, and the advice found on the web is often misleading rather than helpful. Many of these cheap books deal with algorithms and data structures. They typically contain little theory and a large percentage of code. Their success - and the number of stars they get on the web - is often inversely proportional to their quality. Unfortunately, the same holds to some degree for the excellent textbooks too. Quite annoying is that in most textbooks the same examples occur again and again. Computer science teaches that copying is often very efficient, but here it happens at the expense of originality.

There is not much controversy about the claim that Donald Knuth is the father of the field *algorithms and data structures* in computer science education. From 1968 - 1973 he published the first three volumes of his famous unfinished work *The Art of Computer Programming* which since then has defined the subject to a large extent [14]. Among the huge number of textbooks in English that have appeared so far, it seems to me that the short sequence [14],[2],[23],[10] of pioneering texts can be considered as kind of a "backbone" around which the other are positioned more or less closely. The sequence itself, being chronologically ordered, shows an evolutionary development: From one text to the next the selection and accentuation of the contents change in a significant way, but never very much. It could be worthwhile to analyse in detail this interesting phenomenon. In German, [19] is still the best reference on classic algorithms and data structures. [22] covers less classic material, but is broader in scope, and therefore a real alternative. But on the whole the approaches found in most textbooks continue to be quite similar.

Computer science is still a young discipline with a very fast and strong development. There has not been much time left for *didactic work* so far, which I consider the main reason for the similarity of most of today's textbooks. Compared to the textbooks on *calculus*, those on algorithms and data structures show little variety with respect to didactic sophistication. I believe that this dissimilarity is mainly a matter of time and has little to do with the subject.

> *Good introductory textbooks on algorithms and data structures do exist, but there is not yet a satisfactory variety. Real alternatives are still rare, but their appearance is just a matter of time.*

9   How Should a First Course on Algorithms and Data Structures Be Motivated?

I think that the purpose of an introductory course on data structures and algorithms is threefold: One purpose and maybe purpose one is to support *programming* and *application development*. Many of the students will become practical programmers, and their skills and maturity will much depend on their knowledge of the basic material of the field. The second purpose is to impart *culture*. In the same way as there are proofs from the BOOK [4], there are also many algorithms from the BOOK. Many truly great ideas can be discovered by studying algorithms and data structures, and the value of the acquired knowledge is largely independent of any specific usefulness in mind. And the third purpose, of course, is to get the students to a first contact with *research*: How does research look like in the field of algorithms and data structures, and what are some of the open problems?

A lot of research deals with non-classic algorithms (e.g. genetic algorithms, neural networks, and distributed algorithms [17],[15]) which should not be considered in a first course on algorithms and data structures . It may be a good idea however to introduce them already in a first course in computer science, in order to show the whole scope of the discipline. Other important research deals with the implementation of algorithms, in particular with robustness [16]. This too is not a subject for an introductory course, I think.

In the more restricted field of classic algorithms, finding *optimal* algorithms is still an important field of research: Can for a given problem an algorithm be found with a better time complexity. Or can it be proven that a certain time complexity can not be improved? Or can a more efficient *heuristic algorithm* be found that possibly only returns a good solution instead of the optimal one? It is especially rewarding to illustrate such topics by choosing examples from *computational geometry*. Much of computational geometry is elementary 2D-geometry from the mathematical point of view. It is not difficult to find a problem that can be explained very easily, that has an obvious solution equally easy to explain, but that becomes an interesting and challenging problem when the demands on the solution or on the time complexity are slightly changed [6].

> *When looking for motivations for learning data structures and algorithms, we should consider computational geometry.*

10   What Will Be the Future of the Classic Introductory Courses?

Although - or because - computer science is still fairly young, its current situation is more often and more intensively thought over than that of other disciplines. A prominent proof is the not yet completely finished *Computing Curricula 2001 report* (CC2001)

[9] which differs quite substantially from its predecessor (CC1991). It describes 6 possible implementation strategies for introductory computer science, i.e. imperative-first, objects-first, functional-first, breadth-first, algorithms-first, and hardware-first, thus showing that the diversity of possible points of view is still increasing. *Algorithmic thinking* and *programming fundamentals* are considered mandatory for all these approaches, including fundamental data structures and algorithms, recursion, and basic algorithm analysis. I agree and conclude that a classic introductory course on date structure and algorithms (of about 30 in-class hours) continues to be essential. The underlying programming language will continue to change every few years, but the essence of the course will probably remain fairly stable. However I think that a second or third general course on the subject will become less important than in the past. Many students - and practitioners - will just not require them, because their needs are not sufficiently demanding. Excellent libraries will be available and more easily to use than today's. Another reason is that more advanced or specialized fields, like e.g. image processing and artificial intelligence, introduce in their own special courses the techniques and algorithms important to them.

An introductory course teaches *classic algorithms*, but this notion is vague and changing with time. Even if the *Turing machine* continues to be the basis, not exactly the same algorithms will be considered "classic" as in the past. Many algorithms connected with *soft computing* (e.g. genetic algorithms and neural networks) [17] are increasingly changing from "exotic" to "classic". And who knows how "classic" algorithms for *quantum computers* will become some day [12]? I also believe that Peter Wegner is right with his very challenging paper [24] claiming that *interaction* is becoming more important than algorithms. We do not know what "algorithms" - and their importance - will be in the future, and consequently neither how they will have to be taught. But it may be expected that the changes will not be very fast. Therefore I intend to continue with my introductory course - and update it at about the same pace as hitherto.

> *Teaching classic algorithms and data structures will have a future. But the importance of the subject is somewhat decreasing, and therefore that of teaching it too.*

Final Remarks

After having stated my 10 personal observations, I hope that the reader still believes me that I am fond of teaching introductory courses on data structures and algorithms. My observations are mild, they do not ask for any substantially new approach. Since I am still a classic teacher in the sense that I share a real classroom with my students, questions concerning the use of *new media* have hardly been addressed. However, I am pleased that prominent researchers, like Thomas Ottmann, work on new media and try to improve computer science education by applying them [18]. Concerning course contents, I think that we should not consider them sacrosanct just because some famous people have once proposed them. Each curriculum has its proper needs, and each teacher has his own strenghts, experiences and preferences.

Finally I would like to repeat my opinion that in computer science *didactics* have been unduly neglected. Compared to mathematics, there has not been much research and much exchange of ideas in matters of didactics. I hope that this situation will start changing soon. There should now exist enough "free capacity" to do it, and the further development of computer science could profit a lot.

References

1. Abelson, H., Sussman, G.J., Sussman, J.: Structure and Interpretation of Computer Programs. 2nd Edition. MIT Press 1996.
2. Aho, A.V., Hopcroft, J.E., Ullman, J.D.: The Design and Analysis of Computer Algorithms. Addison-Wesley 1974.
3. Aho, A.V., Hopcroft, J.E., Ullman, J.D.: Data Structures and Algorithms. Addison-Wesley 1983.
4. Aigner, M., Ziegler, G.M.: Proofs from the BOOK. 2nd Edition. Springer 2001.
5. Augenstein, M.J., Tenenbaum, A.M.: Data Structures and PL/I Programming. Prentice-Hall 1979.
6. de Berg, M., van Kreveld, M., Overmars, M., Schwarzkopf, O.: Computational Geometry - Algorithms and Applications. 2nd Edition. Springer 2000.
7. Bieri, H., Grimm, F.: Datenstrukturen in APL2 - Mit Anwendungen aus der Künstlichen Intelligenz. Springer 1992.
8. Budd, T.A.: Data Structures in C++ Using the Standard Template Library. Addison-Wesley 1998.
9. Computing Curricula 2001 (CC2001). Steelman Report. IEEE Computer Society/ACM 2001.
10. Cormen, T.H., Leiserson, C.E., Rivest, R.L.: Introduction to Algorithms. MIT Press, 1990.
11. Guha, R.K., Yeh, R.T.: A Formalization and Analysis of Simple List Structures. In R.T. Yeh (Ed.): Applied Computation Theory, 150-182. Prentice-Hall 1976.
12. Hirvensalo, M.: Quantum Computing. Springer 2001.
13. Knuth, D.E.: Big omicron and big omega and big theta. SIGACT News, Vol. 8, 18-24 (1976).
14. Knuth, D.E.: The Art of Computer Programming. Volumes 1-3. Addison-Wesley 1998.
15. Lynch, N.A.: Distributed Algorithms. Morgan Kaufmann 1997.
16. Mehlhorn, K., Näher, S.: LEDA - A Platform for Combinatorial and Geometric Computing. Cambridge University Press 2000.
17. Munakata, T.: Fundamentals of the New Artificial Intelligence - Beyond Traditional Paradigms. Springer 1998.
18. Ottmann, T. (Hrsg.): Prinzipien des Algorithmenentwurfs. Spektrum Akademischer Verlag 1998.
19. Ottmann, T., Widmayer, P.: Algorithmen und Datenstrukturen. 4. Auflage. Spektrum Akademischer Verlag 2002.
20. Polya, G.: How to Solve It. 2nd Edition. Princeton University Press 1957.
21. Rohl, J.S.: Recursion via Pascal. Cambridge University Press 1984.
22. Saake, G., Sattler, K.-U.: Algorithmen und Datenstrukturen - Eine Einführung mit Java. dpunkt.verlag 2002.
23. Sedgewick, R.: Algorithms. Addison-Wesley 1983.
24. Wegner, P.: Why Interaction Is More Powerful Than Algorithms. Communications of the ACM 40 (5), 80-91 (1997).
25. Wirfs-Brock, R., Wilkerson, B., Wiener, L.: Designing Object-Oriented Software. Prentice-Hall 1990.

On the Pagination of Complex Documents

Anne Brüggemann-Klein[1], Rolf Klein[2], and Stefan Wohlfeil[3]

[1] Technische Universität München,Arcisstr. 21,D-80290 München, Germany
brueggem@informatik.tu-muenchen.de
[2] Universität Bonn, Römerstr. 164, D-53117 Bonn, Germany
rolf.klein@uni-bonn.de
[3] Fachhochschule Hannover, Ricklinger Stadtweg 118, D-30163 Hannover, Germany
stefan.wohlfeil@inform.fh-hannover.de

Abstract. The pagination problem of complex documents is in placing text and floating objects on pages in such a way that each object appears close to, but not before, its text reference. Current electronic formatting systems do not offer the pagination quality provided by human experts in traditional bookprinting.
One reason is that a good placement of text and floating objects cannot be achieved in a single pass over the input. We show that this approach works only in a very restricted document model; but in a realistic setting no online algorithm can approximate optimal pagination quality.
Globally computing an optimal pagination critically depends on the measure of quality used. Some measures are known to render the problem NP-hard, others cause unwanted side effects. We propose to use the total number of page turns necessary for reading the document and for looking up all referenced objects.
This objective function can be optimized by dynamic programming, in time proportional to the number of text blocks times the number of floating objects. Our implementation takes less than one minute for formatting a chapter of 30 pages. The results compare favorably with layouts obtained by Word, FrameMaker, or LATEX, that were fine-tuned by expert users.

1   Introduction

This work discusses the pagination problem of complex documents like scientific reports, journal or conference articles, books, or manuals. Such documents typically contain figures, tables, displayed equations, and footnotes that are *referenced* within the text. Lack of space or other reasons can prevent these objects from being placed at the very text positions where they are cited. Therefore, figures and other non-textual objects are usually allowed to *float* in the document, i.e. they may appear at a later position, even on a later page, than their first citations.

On encountering a reference to a figure, the reader usually needs to look up the figure before reading on, in order to understand the exposition. To provide high reading efficiency, the figure should be positioned close to the reference, ideally on the same page.

The pagination problem is in distributing the document's contents on pages in such a way that each referenced floating object appears close to, but not before, its text reference. If an object is cited more often than once, it should be placed close to its first citation, according to [4]. At the same time, other criteria observed in classical bookprinting

R. Klein et al. (Eds.): Comp. Sci. in Perspective (Ottmann Festschrift), LNCS 2598, pp. 49–68, 2003.

must be met. For example, pages should be close to one hundred percent full, no page should end with a single line of a new paragraph, etc.

One could be tempted to consider this problem somewhat antiquated. Are pages not just a passing state in the evolution from scrolls to scrollable windows? We do not think so. One reason is that reading from paper seems to be more efficient than reading from a computer screen. Hansen [10] found that users presented with a text on screen were facing greater difficulties to get a *sense of text* than those furnished with a paper version, in particular in reading a text critically, or with text that is rather long or conceptually difficult. More recent studies show that paging is faster than scrolling [5] and that readers tend to recall better when reading in a page environment as compared to a scroll environment [18].

In order to place the document's contents on pages, paragraphs must be broken into lines. This *line breaking* problem has been studied by Knuth and Plass [15]; it is handled superbly by systems like TEX. Therefore, we assume that the line breaking task is already finished when the pagination task starts. That is, the textual contents consists of a sequence of text lines with identical length[1]. Were it not for the floating objects, computing a pagination would now be easy: we would simply put the same number of lines on each page, maintaining identical base line distances.

With floating objects around, the pagination problem becomes complicated, as Plass [19] has shown during the development of TEX. He found that the algorithmic complexity of computing an optimal pagination critically depends on the *badness function* used for measuring pagination quality. Some badness functions cause the problem to become NP-complete while others allow an optimal pagination to be found in polynomial time. We feel that these badness functions also differ in their relevance to practice and in the side effects they may cause, like e.g. under-full pages.

In the early 80th, main memory was typically too small to hold several pages at the same time. Thus, none of the global optimization methods discussed by Plass [19] could be implemented in TEX. Instead, a simple first-fit pagination algorithm is employed until today in TEX as well as in LATEX [16]. In a single pass, every figure is put on the first page after its citation where it fits. However, it is by no means clear how good this strategy works; to our knowledge, there are no previous theoretical results on this problem. But many users share the experience that ever so often LATEX's placement of figures can be greatly improved on by manual intervention.

Since then the state of affairs in pagination has not changed very much. Asher [1] built a new system Type & Set based on TEX as the formatting engine. Type & Set operates in two passes. First, the text lines, figures and footnote lines are reconstructed from TEX's output file. Then, an optimizing pagination routine pastes the reconstructed material into pages. The measure to be optimized seems to mainly reflect page justification and balancing while figure placement is only a secondary concern. Kernighan [12] developed a page makeup program which postprocesses troff output. In order to keep the pagination algorithm as simple and efficient as possible, figure placement is also done in a single pass over the input.

---

[1] Note that this assumption does not hold for newspapers where figures are surrounded by text, so that the line length depends on the placement of floating objects.

The purpose of this paper is threefold. First, we investigate how good a pagination can be expected if the first-fit strategy for placing figures (that is used by current formatting systems) is used. The result is somewhat surprising. Whereas in the well-known *bin packing* problem the first-fit strategy is known to produce results at most seventy percent worse than optimal, no such guarantee can be given in the pagination problem. Only for a very restricted, unrealistic document model the first-fit strategy works well. Using competitive analysis we show that, under realistic assumptions, not only first-fit but *any* online pagination algorithm may produce results that are arbitrarily worse than necessary. This explains why so many people are not satisfied with paginations produced by LaTeX if no manual improvement is done.

Second, we introduce a new quality measure for paginations that is aimed at high reading efficiency. We propose to count the total number of pages the reader needs to turn while reading through the document and looking up the references. Minimizing this number keeps the document short and places floating objects close to their references in the text. Either of these goals may be emphasized by introducing weights. An optimal pagination with respect to this quality measure can be computed offline, in time proportional to the number of text blocks times the number of floating objects, by a dynamic programming algorithm.

Since the relevance to practice of this approach cannot be theoretically argued, our third contribution is in reporting on the practical results we have obtained so far. We decided to implement a prototype called XFORMATTER that employs the above pagination algorithm. It deals with two types of floating objects: figures and footnotes.

XFORMATTER not only minimizes the number of page turns; without adding to the algorithmic complexity, it also satisfies a number of additional pagination constraints commonly used in professional printing. For example, it avoids generating *widows* and *orphans*, and it justifies all pages to the same height within a specified range. XFORMATTER can also be used on double-sided documents. Here the two pages on the same spread are adjusted to exactly the same height. A floating object may be positioned on a left page even though its reference appears only on the subsequent right page [4]. Also, the user can specify that two figures should *not* appear on the same spread, a useful feature for exercises and solutions in textbooks.

We have run XFORMATTER on real-world documents that were previously formatted by Word, FrameMaker, or LaTeX, and carefully fine-tuned by expert users. The results are very encouraging. Not only did the quality of paginations by XFORMATTER favorably compare with the previous layouts; even the run time needed to achieve this quality seems quite acceptable. Typically, it took us less than one minute to optimally format a chapter of some 30 pages, on a standard Sun workstation.

The rest of this paper is organized as follows. In Section 2 we formally specify the pagination problem. Also, we introduce a new measure for the quality of a pagination. Section 3 contains the competitive analysis of online pagination algorithms. In Section 4 we show that our measure of quality can be optimized by dynamic programming because the *principle of optimality* is satisfied. Section 5 contains our practical results. Finally, we summarize, and suggest some further research.

2   The Pagination Task

In this section we specify the types of documents studied in this paper, and we define input and desired output of the pagination task. In general, we adopt the notations of Furuta et. al. [8].

2.1   The Document Model

We assume that the underlying document model provides for several *streams* of logical objects. Within each stream, all objects are of the same type, like text, figures, footnotes, etc. The relative order of objects of the same stream must be preserved in the output model.

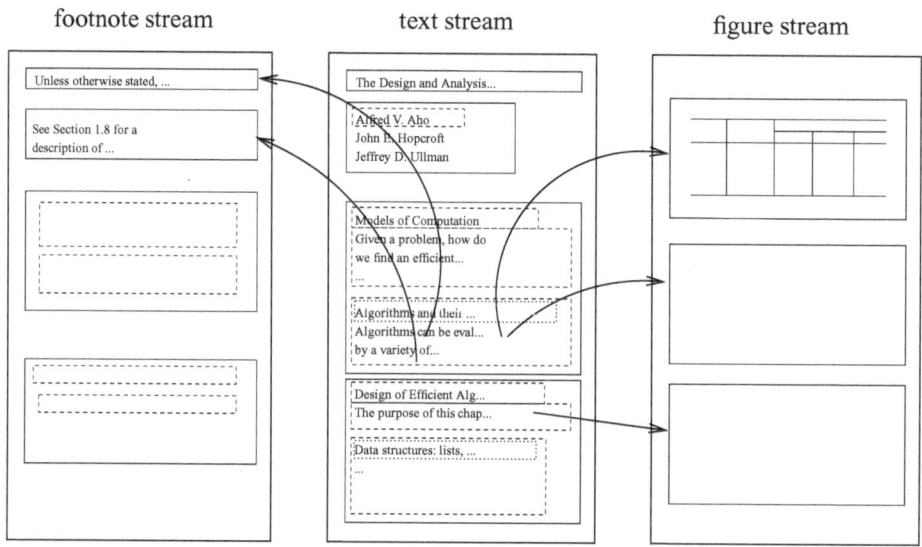

Figure 1. Logical structure of a document with three streams

Objects may refer to other objects of the same or of different streams; examples are cross references, bibliographical citations, footnotes, and references to figures or tables. For the pagination task the latter are of particular interest since they entail some freedom of choice. In fact, a footnote must start at the bottom of the same page that contains the reference to this footnote, but it may be continued at the bottom of subsequent pages. Figures and tables may float even more liberally; they may be printed on different pages than their referring text objects. An example document containing text, figure, and footnote streams is shown in Figure 1. As to the *text stream*, we assume that the line breaking task has already been completed. Afterwards, each paragraph consists of a sequence of line boxes and vertical glue boxes; compare [15; 20; 14].

All line boxes are of the same width, equal to the width of the text region. But their heights and depths may vary according to the line's textual contents. To make up for these differences, the vertical glue boxes between the lines (leading) usually have their heights so adjusted that all base line distances are the same; this regularity makes it easier to read the document. In general, each box is assigned a height and a depth as well as a capability to stretch or shrink that may depend on where, on a page, this box will appear, at the top, in the middle, or at the bottom.

Text boxes are denoted by t_i. We write $t_i < t_j$ if and only if t_i occurs earlier in the text stream than t_j does. The *figure stream* is a sequence of boxes, too. The size of a figure box depends on the size of the figure itself and the size of its caption. Distances between figure boxes are modelled by vertical glue boxes, as in the text stream. We will denote the elements of the figure stream by f_i. Figure boxes have the same properties as line boxes.

While a figure can be referred to many times, the first citation matters most for the pagination task, since this is where the figure should ideally be placed, according to [4]. We write $R(f_j) = t_i$ to express that the first reference to figure f_j occurs in text box t_i; function R is called the *reference function*. We assume in this paper that the reference function is monotonic. That is, the figures appear in their figure stream in the same order as their first citations occur in the text: $f_i < f_j$ holds if and only if $R(f_i) < R(f_j)$ is fulfilled. The *footnote stream* consists of text and glue boxes, n_i. It resembles the text stream in most aspects.

2.2   The Page Model

The basic properties of a page are its width and its height. We assume that each page contains but a single column. However, our approach can easily be generalized to pages with two or more columns of identical width. While discussing the potential of online pagination algorithms, in Section 3, we limit ourselves to single-sided documents, without loss of generality. In Section 4, we show how our offline pagination algorithm can be applied to double-sided documents, too.

Each page contains a number ≥ 0 of *regions* for each stream. A region represents an area where material of the corresponding stream will be typeset, in a top-down manner. We assume that all pages of the document belong to the same *page class*. In this paper, two different classes will be discussed.

The *simple* page class consists of pages that contain an arbitrary number of text and figure regions, in any order. Regions are placed directly on top of each other, and the full page height is available. This page class will be used only in Section 3, in exploring the reach of online pagination algorithms.

In practice, one does not usually place a figure region directly on top of a text region; to better separate the figure's caption from the subsequent text, some extra white space d must be inserted between them. Moreover, it is quite standard, and greatly simplifies the pagination task, to position all figures on the top part of a page. This leads to our second, so-called *extra glue* page class. It consists of pages that either contain only one text region, or only one figure region, or a figure region on top of a text region, with white space d between them. Examples are shown in Figure 2.

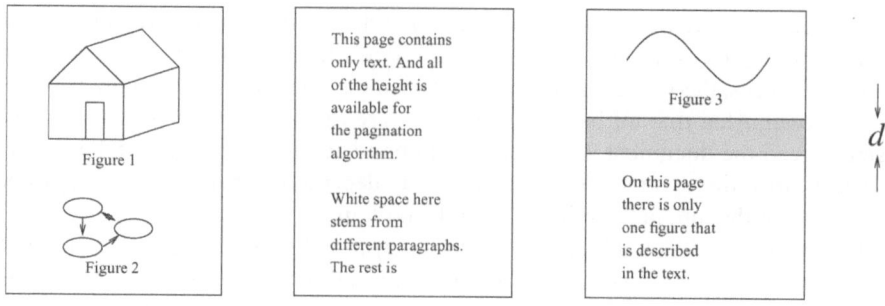

Figure 2. Some pages of the *extra glue* page class

In Section 3 we will see that the complexity of the pagination problem considerably increases by the need for extra glue between figures and text. Nonetheless, it is this more realistic page class our offline algorithm is designed to deal with.

2.3   The Pagination Output

Formally, a pagination P would be a mapping from the input boxes $b_i \in \{t_i, f_i, n_i\}$ to a set of triplets (p, x, y); here p denotes the page number, and (x, y) are the coordinates where the reference point of the box is placed on page p. $P : \{b_1, \ldots, b_n\} \rightarrow (p, x, y)$. In the sequel we are only interested in the number p of the page on which a box is placed; it will also be denoted by $P(b_i)$ as no confusion can arise.

2.4   The Pagination Goals

Pagination is difficult because a number of competing constraints and goals have to be satisfied simultaneously. In this section we will discuss these requirements. There are several style guides [4; 22; 21] that describe the goals of the pagination process in more detail.

Sequence: The pagination task has to maintain the order of the boxes within their respective streams. (This is the main difference between the pagination task and the *bin packing* problem, where a number of objects have to be placed in bins but their order does not matter.)

Widows and Orphans: The last line box of a paragraph which is set isolated at the top of a new page is called a *widow*. The first line box of a paragraph that is printed isolated at the bottom of a page is called an *orphan*. Widows and orphans may confuse the eye in reading [21] and may disturb a reader while parsing a page. Therefore, they should better be avoided.

Page Height: Each page should be perfectly full. This does not only look better, it is also more economical because fewer pages are needed. When widows and orphans must be avoided this may be too hard a constraint. Therefore the page depth may be adjusted a little bit [4], for example by allowing a page to run one line short. However, in documents printed on double-sided pages (like most books), the pages of one page spread must be equally full.

Floating Objects: Floating objects should be placed close to their *first* reference, but they must not be placed on a page before their citation. If a figure or table cannot be placed on the same page with its citation, it may be placed on a following page.

In documents that are printed on double-sided pages this constraint is relaxed. Here a floating object may be placed on the left-hand page while the citation is placed on the right-hand page. However, a floating object must not be placed on a page spread before its citation.

2.5   Measuring the Pagination Quality

Text that is intended to be read continuously should have a high *legibility*. Experimental work cited by Rubinstein [21] shows that legibility decreases whenever continuous reading is interrupted. Some interrupts occur at line ends, where the eye has to focus on the start of the next line.

If figures are not placed on the same page with their references, not only has the eye to focus on the figure, the reader even has to turn pages in order to locate the figure. Clearly, this interrupts continuous reading a lot more than re-focussing on the same page does. Therefore, Plass [19] suggested to count the page differences between each figure and all of its references, as a measure of the badness of a pagination. He defined two badness functions,

$$Lin(P) = \sum_{i=1}^{n} \sum_{j=1}^{r(i)} \left| P(f_i) - P(R_j(f_i)) \right| \text{ and } Quad(P) = \sum_{i=1}^{n} \sum_{j=1}^{r(i)} \left(P(f_i) - P(R_j(f_i)) \right)^2,$$

where $r(i)$ is the number of references to figure f_i from a line box, $P(b_i)$ denotes the number of the page where box b_i is placed, and $R_j(f_i)$ denotes the line box with the jth reference to f_i. Small values of $Lin(P)$ or $Quad(P)$ indicate a good pagination.

Plass proved that finding a pagination P, for an arbitrary given input, that minimizes $Quad$ is an NP-complete problem, by transforming the well-known problem *maximum 2-satisfiability* to the pagination problem.

We see two problems with these badness functions.

(1) They take into account all line boxes containing a reference to a given figure; but the Chicago Manual of Style [4] requires to place floating objects close to their *first* referencing line box.

(2) The quadratic badness function overvalues small improvements of very bad placements. Assume that pagination P_1 places 18 figures on the same page each with its reference, and one figure 10 pages behind its reference. Let us compare P_1 against a pagination P_2 of the same document, that places each of the 18 figures on the page after its reference and the last figure only 9 pages behind its reference. If measured with $Quad$, P_2 is better than P_1, because $Quad(P_1) = 18 \cdot 0^2 + 1 \cdot 10^2 = 100$ and $Quad(P_2) = 18 \cdot 1^2 + 1 \cdot 9^2 = 99$ But obviously, pagination P_1 is preferable.

Therefore, we suggest to directly measure what really interrupts the reader: the total number of page turns that are necessary while reading the document.

There are two reasons for turning a page: (1) The reader has finished a page and continues with the next one. (2) The reader encounters a reference to a floating object which is not placed on the current page; in that case one has to turn to the next page, or to a page even further away, view the floating object, and flip back, before reading can be continued.

In formatting a document, these reasons may carry different weights. For a technical manual, for example, it could be important that all figures appear on the same page with their references, so that the manual can be held in one hand. Then the page turns of type (2) should be minimized, perhaps at the expense of some extra pages. On the other hand, for a conference paper it could be mandatory not to exceed a given page limit, so type (1) page turns should be minimized. For these reasons, we propose the following quality measure.

$$Turn.S(\alpha, \beta, P) = \beta(p-1) + \sum_{i=1}^{n} \alpha\Big(P(f_i) - P(R(f_i)) \Big).$$

Here, p denotes the total number of pages, and α, β are non-negative weights that add up to 1. Weight α applies to the effort of thumbing from a reference to its figure and back, and β to the regular progression from page to page. As no figure f_i is placed on a page $P(f_i)$ before its referencing line box, all terms in the sum are positive.

Choosing $\alpha = 1$ and $\beta = 0$, we get Plass's original goal function if there is only one reference to each figure.

This measure can be extended to double-sided pages, where page turns of type (1) occur only after reading every second page. Let $S(x)$ denote the index of the spread containing page x; usually $S(x) = (x \text{ div } 2) + 1$ holds. The measure for the quality of such a pagination is

$$Turn.D(\alpha, \beta, P) = \beta(S(p) - 1) + \sum_{i=1}^{n} \alpha\Big(S(P(f_i)) - S(P(R(f_i))) \Big).$$

In principle, one could think of many ways to define a measure of pagination quality. But there is one requirement any sensible measure should fulfill.

Definition 1. *A measure of pagination quality is called* natural *if it has the following property. Increasing the page (or page spread) difference between a single figure and its referencing line box, while leaving all other such differences unchanged, increases the pagination badness.*

By this definition, the measures *Lin*, *Quad*, *Turn.S*, and *Turn.D* are all natural.

3   Online Pagination Algorithms

Most of today's document formatters like TEX, LATEX, troff, and pm use an *online* pagination algorithm that places all boxes on pages during a single pass over the input streams.

It is typical of an online algorithm not to know its whole input sequence in advance. Initially, it only knows the start of the input sequence; yet it has to act upon this incomplete knowledge immediately. Afterwards, the next input element is presented, and so on.

Clearly, an online algorithm A hampered by incomplete knowledge cannot perform as well as it could be if future inputs were known. By *competitive analysis* [7] its performance can be measured in the following way. Let I be an arbitrary input sequence. If I were fully known, the problem at hand could be solved optimally, incurring a minimal badness $C_{opt}(I)$. Now let $C_A(I)$ denote the badness caused by algorithm A on input sequence I. If there are constant numbers c and a such that, for each possible input sequence I, $C_A(I) \leq c * C_{opt}(I) + a$ holds then algorithm A is said to be *competitive* with *competitive factor* c. Roughly, this means that A behaves at most c times worse than necessary, whatever the input may be[2].

In this section we investigate how well the first-fit approach, or more general online algorithms, can work for the pagination problem. For the sake of simplicity, we restrict ourselves to a very simple model. The documents are single-sided. They consist of one text and one figure stream; all glue boxes are of height 0. Each box may appear as the last one on a page. Since all pages are requested to be 100 percent full, a valid pagination need not always exist; a necessary condition is that the sum of the heights of all boxes is a multiple of the page height.

3.1   Where Figure First Fit Works Well

The *FigureFirstFit* algorithm proceeds by placing line boxes on the page until a reference to a figure is encountered. In this case the line box is placed, and the figure referred to is appended to a figure queue. Each time a line box has been placed onto a page, the figure queue is served. Serving the figure queue means to repeatedly place the first figure of the queue on the current page as long as there is enough room; if the current page is completely full, a new page is started. Otherwise, placing line boxes is resumed. Using a *first in first out* queue guarantees that the figures are placed in the same order of their references.

Clearly, this algorithm works without lookahead. Once a page has been filled it is never changed, no matter what input will be read in the future. To show that the algorithm performs well in the restricted document model we assume the following. Text and figure boxes can neither stretch nor shrink; the line box height is fixed, and the page height, as well as all figure heights, are multiples thereof; pages are of the *simple page class* type. Under these assumptions we can prove the following.

Theorem 1. *If a pagination for a given input exists, then the* FigureFirstFit *algorithm is guaranteed to find a pagination. Under each natural measure of pagination quality, the pagination found is optimal.*

Proof. In this paper we only sketch the rather technical proof; details can be found in [23]. In order to show that the algorithm finds a pagination whenever there is one,

---

[2] Note that we are addressing the quality an online algorithm achieves, not its running time consumption.

we prove that any valid pagination can be transformed into the pagination constructed by *FigureFirstFit*. This can be accomplished by simply swapping a figure with a single preceding line box on the same page, or with a number of line boxes at the bottom of the preceding page whose total height equals the height of the figure, provided that the reference to the figure is not overrun.

To prove that a pagination compiled by *FigureFirstFit* is optimal with respect to any natural measure, we first show that each figure is placed, by *FigureFirstFit*, as close to the beginning of the document as possible. Now let M be an arbitrary natural measure of pagination badness, and let P be any valid pagination of the input document. Since in P every figure is at least as far away from its citation as in the pagination constructed by *FigureFirstFit*, the badness of P under M can only be larger, because M is natural.

In terms of competitive analysis, Theorem 1 states that *FigureFirstFit* is competitive with (optimum!) competitive factor 1, in the restricted document model introduced in this section.

3.2   Where No Online Pagination Algorithm Works Well

In this section we first investigate what happens if the simple page class is replaced with the more realistic *extra glue* page class introduced in Section 2.2. Otherwise, we adhere to the assumptions made in Section 3.1. Rather than arbitrary natural badness measures we are now using $C_A(I) = \sum_{i=1}^{n} P(R(f_i)) - P(f_i)$; This expression equals the badness function *Turn.S* introduced in Section 2.5 if we choose $\alpha = 1$ and $\beta = 0$.

Technically, it would be possible to adapt the *FigureFirstFit* algorithm to the extra gluc page class. But there is little to gain by this approach, as the following surprising result shows.

Theorem 2. *In the above model, no online algorithm for the pagination task is competitive.*

Proof. Let A be an online algorithm for pagination. Now consider the following pagination input. The text stream consists of $n + 16$ line boxes $\{t_1, \ldots, t_{n+16}\}$ that are of the same height, 1. The figure stream contains $n + 2$ figures $\{f_1, \ldots, f_{n+2}\}$ whose heights are as follows:

$$\text{figure } f_1 \ f_2 \ f_3 \ \cdots \ f_n \ f_{n+1} \ f_{n+2}$$
$$\text{height } 1 \ \ 1 \ \ 3 \ \ \quad 3 \ \ \ 1 \ \ \quad 1$$

The page height equals 5, and the extra glue between the figure region and the text region is of height 1.

When algorithm A starts, it sees a figure f_i just at the moment it processes its referencing text line, $R(f_i)$. Figures f_1 and f_2 are referred to in the first two line boxes, that is $R(f_1) = t_1$ and $R(f_2) = t_2$. Up to line box t_7 there are no more references to figures. Now algorithm A has essentially[3] two choices in filling the first two pages as shown in Figure 3.

The idea of the proof is as follows. No matter which of the two alternatives algorithm A chooses, we can always continue the input streams in such a way that the

---

[3] The algorithm could decide on not placing any figure on page 1 etc. But the resulting paginations would be even worse than the ones discussed.

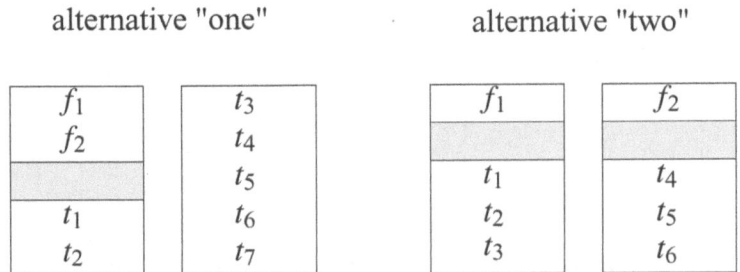

alternative "one"   alternative "two"

Figure 3. Two choices of an online pagination algorithm

resulting pagination becomes very bad. At the same time, there exists a different pagination of the same input that is tremendously better than what A has achieved.

1. If algorithm A chooses alternative 1, the referencing line boxes of $f_3, \ldots, f_{n+2}$ will be as follows

	figure	f_3	f_4	$\cdots$	f_i	$\cdots$	f_n	f_{n+1}	f_{n+2}
	referencing line box	t_{12}	t_{13}		t_{i+9}		t_{n+9}	t_{n+10}	t_{n+14}

1	2	3	4	$n+1$	$n+2$	$n+3$

Figure 4. Pagination choice 1 completed

Indeed, algorithm A cannot find a better pagination, because (1) figure f_3 cannot be placed on page 3, because in that case f_3 would be placed before its reference t_{12}, (2) each of the figures $f_3, \ldots, f_n$ needs a page of its own, (3) the remaining space on pages $4, \ldots, n+1$ therefore must be filled with one line box each, and (4) the last two pages are already optimally filled and f_{n+2} is on the same page as its reference.

The badness of this pagination is $n-1$ because f_1, f_2, and f_{n+2} are on the same page as their references and the remaining $n-1$ figures are one page after their references. On the other hand the pagination shown in Figure 5 is also valid—and much better!

In this pagination only f_2, f_{n+1}, and f_{n+2} are placed one page after their referencing line box; all other figures are optimally placed. Thus, a badness of 3 results.

1. If algorithm A chooses alternative 2, the referencing line boxes of $f_3, \ldots, f_{n+2}$ will be the following

	figure	f_3	f_4	$\cdots$	f_i	$\cdots$	f_n	f_{n+1}	f_{n+2}
	referencing line box	t_8	t_9		t_{i+5}		t_{n+5}	t_{n+8}	t_{n+9}

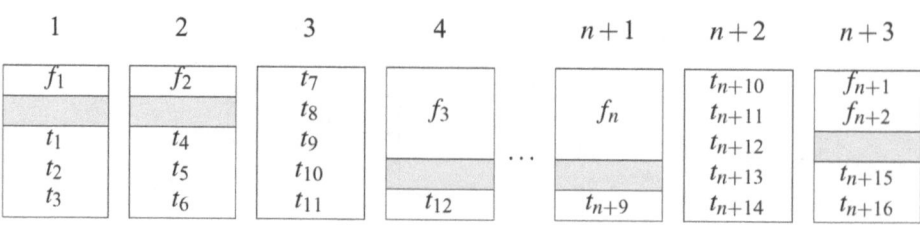

Figure 5. Optimal pagination for the given input

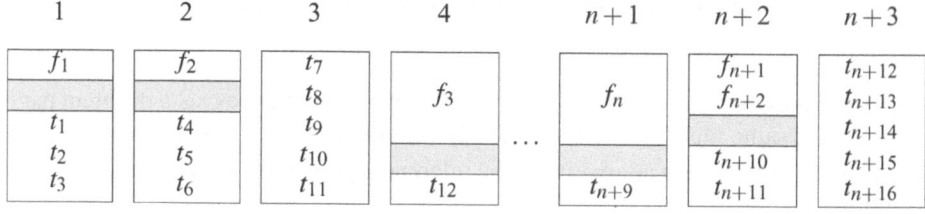

Figure 6. Pagination choice 2 completed

Again, there is only one way for the online algorithm to continue the pagination (see Figure 6). It is impossible for algorithm A to place f_3 on page three, because $R(f_3) = t_8$ would not also fit on page three. As no two of the figures $f_3, \ldots, f_n$ fit on the same page, there is only one way to fill the pages up to page $n+1$. Figures f_{n+1} and f_{n+2} are already referred to when page $n+1$ is completed, so it is optimal to place them immediately on page $n+2$. They cannot be placed on page $n+1$ because figure f_{n+2} would then be one page before its reference t_{n+9}. The page differences between a figure and its referencing line box are:

figure	f_1	f_2	f_3	f_4	f_5	f_6	$\cdots$	f_n	f_{n+1}	f_{n+2}
page difference	0	1	1	2	3	4		4	2	1

So the badness of this pagination is $4(n-5) + 10 = 4(n-3) + 2$. On the other hand the pagination in Figure 7 is also a valid pagination of the same input, but its badness is zero since all figures are placed on the same page with their referencing line boxes.

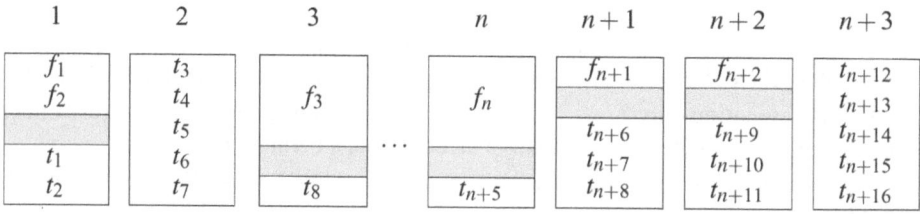

Figure 7. Optimal pagination for the given input

No matter which case applies, the pagination constructed by algorithm A is of badness at least $n-1$ whereas the optimal pagination incurs a badness of at most 3. Since the value of n can be arbitrarily large, algorithm A is not competitive.

The proof of Theorem 2 reveals another remarkable fact. Even if the pagination algorithm were allowed to inspect the input streams in advance and to check the sizes of all boxes, it could still not be competitive without knowing which figure is referred to by which line box, i.e. without knowing the reference function, R!

In Theorem 2 we have seen that online pagination algorithms do no longer work well if we replace the simple page class with the more realistic *extra glue* page class. The same happens if we adhere to the simple page class but relax the assumption that line boxes cannot stretch or shrink.

Theorem 3. *Given a document class as above, but with line boxes that may stretch or shrink, and pages of the simple page class. In this model no online algorithm is competitive.*

The proof is similar to the proof of Theorem 2; for details see [23].

4   Pagination by Dynamic Programming

In the preceding section we have seen that high pagination quality cannot be achieved in a single pass over the input. Therefore, we resort to offline methods, where all input streams are completely read before the pagination algorithm starts.

In Section 2.5 we have mentioned that the computational complexity of finding an optimal pagination depends on the underlying badness function. The main goal of this section is in proving that the measures *Turn.S* and *Turn.D* introduced in Section 2.5 do allow for efficient optimization, and to provide a suitable algorithm.

To this end, we are using an algorithmic technique known as *dynamic programming*; see [2]. The same technique has been successfully employed in the line breaking task; see [15]. For dynamic programming to work the *principle of optimality* must hold: The optimal solution of the main problem consists of optimal solutions of subproblems. This fact enables building up an optimal solution in a bottom-up way, by combining the optimal solutions of subproblems of ever increasing size.

We are working with the *extra glue* page class introduced in Section 2.2. First, we discuss single-sided documents; then we show how to extend our algorithm to double-sided pages.

4.1   Pagination for Single-Sided Pages

Suppose we are given a text stream of line boxes $t_1, \ldots, t_m$, and a figure stream $f_1, \ldots, f_n$. Let us assume that, for some indices k, l, we have already found a *partial pagination* $P_{k,l}$ that neatly places the line boxes $t_1, \ldots, t_k$ and figure boxes $f_1, \ldots, f_l$ onto p pages, in accordance with the pagination goals stated in Section 2.4. Such a partial pagination differs from a complete one in that there may be dangling references: Some textblock

t_h, where $h \leq k$, may contain a reference to a figure f_w, $l < w$, that has not been put on a page yet.

We adapt our measure *Turn.S* to also include dangling references in the following way. A figure f_w referred to in text box t_h, $h \leq k$, that does not appear on one of the p pages of partial pagination $P_{k,l}$ gives rise to the additive badness term $p + 1 - P_{k,l}(R(f_w))$, multiplied by the weight factor α. That is, instead of counting how many pages separate the citation from the figure—whose position is not yet known!— we charge the least number of page turns this figure is going to cause, namely if it were positioned on the very next page. With this extension, we can speak about the badness, *Turn.S*$(\alpha, \beta, P_{k,l})$, of a partial pagination $P_{k,l}$.

Note that for arbitrary values of k and l a partial pagination $P_{k,l}$ need not exist, because it may be impossible to completely fill a number of pages with the text and figure boxes given while complying with the pagination rules. But let us assume that it does, and that $P_{k,l}$ is optimal, with respect to the (extended) measure *Turn.S*, among all paginations of the first k line boxes and the first l figure boxes. If we remove the last page from $P_{k,l}$ we obtain a partial pagination $P_{i,j}$, for some indices $i \leq k$ and $j \leq l$. Now the following fact is crucial.

Theorem 4. *If partial pagination $P_{k,l}$ is optimal with respect to measure* Turn.S *then the partial pagination $P_{i,j}$ that results from removing the last page of $P_{k,l}$ is optimal, too.*

Proof. Let us consider how the values of *Turn.S*$(\alpha, \beta, P_{k,l})$ and *Turn.S*$(\alpha, \beta, P_{i,j})$ differ. Since there is one page more to turn in $P_{k,l}$, its badness has an extra additive term $\beta * 1$; compare the definition in Section 2.5. Suppose $P_{k,l}$ involves p pages. Each figure placed on the first $p - 1$ pages contributes identically to either badness value. The same holds for each figure possibly placed on page p, according to how *Turn.S*$(\alpha, \beta, P_{i,j})$ charges a dangling reference.

Next, let us assume that a figure f_w is cited, but not placed, in $P_{k,l}$. Then its contribution to the badness of $P_{k,l}$ is larger by $\alpha * 1$ because this pagination exceeds $P_{i,j}$ by one page. The same argument holds if the citation of f_w occurs on page p. Consequently, we have

$$Turn.S(\alpha, \beta, P_{k,l}) = Turn.S(\alpha, \beta, P_{i,j}) + \beta + \alpha * q(k,l),$$

where $q(k,l)$ denotes the number of those figures of index $> l$ whose citation occurs within a text box of index $\leq k$. One should observe that this quantity does not depend on the pagination $P_{k,l}$ itself but only on the indices k, l and on the input streams.

Now the claim of the theorem is immediate. If there were a partial pagination $P'_{i,j}$ of the same substreams $t_1, \ldots, t_i$ and $f_1, \ldots, f_j$ with a badness smaller than that of $P_{i,j}$ we could simply substitute it, and obtain a new pagination $P'_{k,l}$ whose badness would in turn be smaller than that of $P_{k,l}$, by the above formula. But this would contradict the assumed optimality of $P_{k,l}$.

In order to construct optimal paginations, we apply the dynamic programming technique in the following way. Suppose we want to determine if there exists a valid partial pagination $P_{k,l}$, and if so, compute an optimal one. At this time, all optimal partial paginations $P_{i,j}$, where $i \leq k, j \leq l$ and at least one index is strictly smaller, are already known. We check for which of them it is possible to place the remaining boxes

$t_{i+1},\ldots,t_k$ and $f_{j+1},\ldots,f_l$ on a single page without violating the pagination goals. Whenever this can be done we compute the badness of the resulting pagination $P_{k,l}$. Among all candidates $P_{k,l}$ hereby obtained, the best one is chosen, and stored for further use. Theorem 4 ensures that by this way we really obtain an optimal partial pagination $P_{k,l}$, if such a pagination exists. By proceeding with ever increasing values of k and l, we finally arrive at an optimal pagination $P = P_{m,n}$ of the whole document[4].

The asymptotic running time of this algorithm can be estimated as follows.

Theorem 5. *An optimal pagination of a document containing m text boxes and n line boxes can be compiled in time $O(nm)$.*

Proof. At first glance it seems as if the above algorithm had, for each pair of indices (k,l), to check all partial paginations with a smaller index pair (i,j). As these are quadratic in number, an overall running time in $O(n^2m^2)$ would result. But a lot of this work can be saved. Because there is a lower bound to the possible height of any text or figure box, at most a certain number of them—say w—can fit onto one page. Therefore, we need to check only those optimal partial paginations $P_{i,j}$ where $m - w \le i \le m$ and $n - w \le j \le n$ hold—a constant number! As such checks are carried out for each index pair (k,l), the claim follows.

The naive implementation of our pagination algorithm would use a $m \times n$ array for storing the relevant values of all optimal subpaginations $P_{i,j}$. How to save on the storage space needed while, at the same time, maintaining good performance has been explained in [23].

4.2 Pagination for Double-Sided Pages

With double-sided documents some careful modifications are necessary because we have two types of pages now, left ones and right ones, that must be treated differently.

When extending our badness measure *Turn.D* introduced in Section 2.5 to a partial pagination of substreams $t_1,\ldots,t_i$ and $f_1,\ldots,f_j$ we have to take into account the type of the last page. If it is of type left then all dangling references are ignored, for they may still be harmless if the corresponding figures are placed on the right page immediately following. But if the partial pagination ends with a right page, dangling references count as in the case of single-sided documents.

One should observe that we can no longer talk of *the* optimal partial pagination $P_{i,j}$ because there may be two of them, ending on a left and on a right page, correspondingly. Both of them must be considered by our dynamic programming algorithm.

Apart from these technical differences, the same results can be achieved as in Section 4.1.

---

[4] Or we learn that no pagination of the document is possible that does not violate the pagination goals; in this case human intervention is required, as will be discussed later.

Table 1. Badness of single- and double sided paginations

	Turn.S				Turn.D			
	doc1	doc2	doc3	doc4	doc1	doc2	doc3	doc4
Author's pagination	30	72	–	72	13	37	–	39
first-fit pagination	39	70	55	56	17	38	25	26
Turn.S-optimal pagination	31	65	47	52				
Turn.S-90% optimal pagination	–	59	38	47				
Turn.D-optimal pagination					13	30	25	25
Turn.D-90% optimal pagination					10	26	18	23

5   Practical Results

5.1   Pagination Quality

Measuring the time performance of an algorithm in terms of "Big-O" alone does not tell very much about its relevance to practice. Also, it is not clear how much better an "optimal" pagination really is, as compared to one compiled by e.g. LaTeX, without actually computing the two, and comparing them directly. To this end we have implemented our pagination algorithm and run some experiments on different real-world documents. In this section we present some of the results.

The documents we are reporting on have originally been produced with LaTeX, FrameMaker, or Microsoft Word. They were fine-tuned by expert users, with the intention of improving on the pagination quality. The first document, *doc1*, is a report for the government of Bavaria [6]; *doc2* is a chapter of a text book on Software Engineering [17]. The third document, *doc3*, is part of a Ph. D. thesis on school development research [11] in Germany. The last document, *doc4*, is a chapter of a book on Computational Geometry [13].

Table 1 compares the quality of a first-fit pagination, a *Turn.S*-optimal pagination, a *Turn.S*-optimal pagination where pages need not be exactly filled but may be only 90% full, and the pagination produced by the authors. This table shows that the quality of first-fit algorithms is rather unpredictable. In fact, it may be quite well or it may be very poor, depending on the document's contents. On the other hand, *Turn.S* optimal paginations are usually of a quality comparable to the author's pagination. But if we allow pages to remain a little under-full, the optimizing algorithm is able to find even better paginations.

If some pages are not 100% full, some lines of text or some figures have to be moved from the under-full page to a following page. This may lead to a pagination which needs more pages than a pagination which completely fills all pages. However, if the last page was very loosely filled before (say with only a few lines), the optimization may be possible without using another page; the last page just gets more material. Our examples have shown that a significant improvement may be achieved if only *one* more page is used.

Paginations computed for single-sided pages may be printed on double-sided pages and vice versa. But a *Turn.D*-optimal pagination need not be *Turn.S*-optimal if printed on single-sided pages; it may even be invalid because a figure might be placed on a page

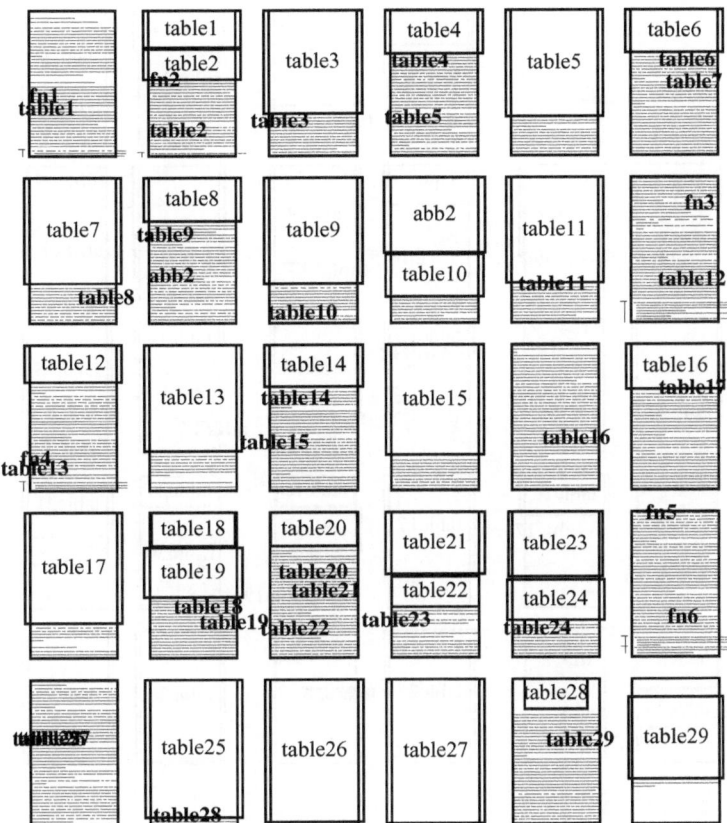

Figure 8. The first-fit pagination of *doc3*

before its reference! Conversely, *Turn.S*-optimal paginations are not *Turn.D*-optimal if printed on double-sided pages. Therefore, it is essential for the pagination algorithm to know on what kind of paper the document will be printed. The following table shows which improvement in quality was possible, with our example documents, by using double-sided pages, and by optimizing accordingly. Again we can see that the quality of the first-fit paginations varies a lot. Allowing pages to be only at least 90% full (but have pages of one page spread balanced!) has produced the best results. These paginations are substantially better than what the authors had been able to achieve manually. In our examples, it was not necessary to append more than one extra page to get the best results.

To illustrate the difference our algorithm can make we depict symbolically the status of *doc3* before and after optimization. Figure 8 shows what a first-fit algorithm (as built into Microsoft Word) had been able to achieve. On the other hand, Figure 9 shows the *Turn.D*-90% optimal pagination of *doc3*. It needs one page more but its quality is much higher. One can observe that pages 4 and 5 (at the top right corner) are not 100 percent full but balanced in page height.

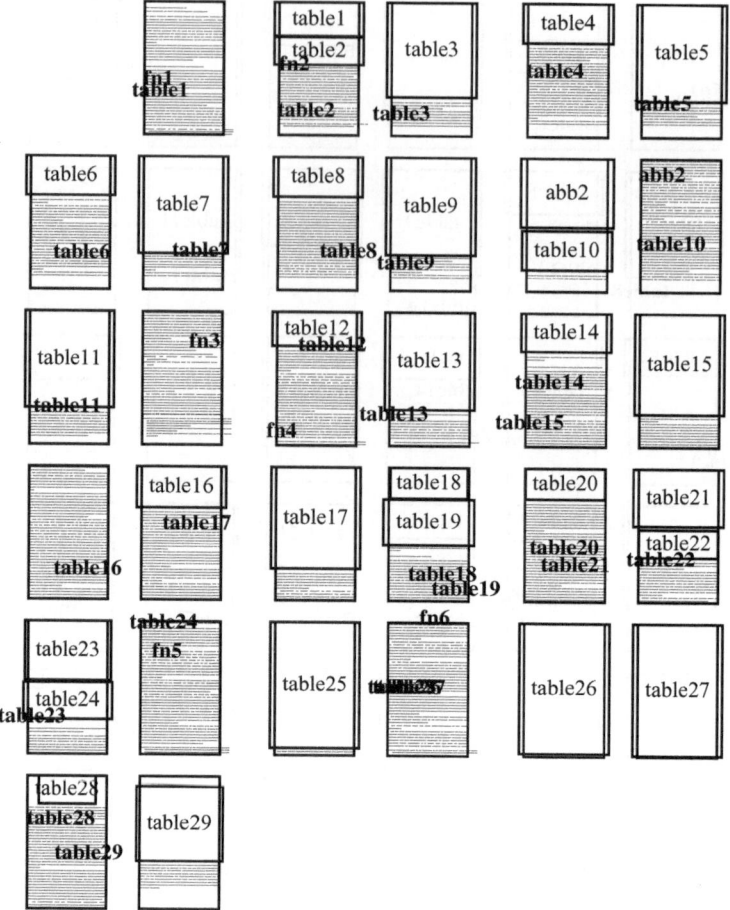

Figure 9. An optimal double-sided pagination of *doc3* with pages at least 90% full

This example also contains footnotes. Although not explicitly mentioned in this work, it is possible to extend the measure of quality and the pagination algorithm to cope with footnotes, too. Details about the extension may be found in [23].

5.2   Pagination Time

Needing an extra page is not the only price one has to pay to get better paginations. An optimizing algorithm needs more storage capacity and more running time to compute its results.

In [23] we have shown the techniques that were used to speed up the pagination program: (1) An efficient storage structure for the results of the subproblems. (2) A pruning heuristic to avoid computations that probably do not lead to the optimal solution. With these two techniques it was possible to compute a pagination in (at most) about one minute. In these experiments, a SUN Ultra I workstation with a 167 MHz processor and 256 MB of main memory has been used on the example documents. As compared to the many hours that authors today have to spent on manually improving the pagination of their documents, we think that our pagination algorithm is fast enough for daily use.

Although we did not paginate a whole book, our examples are still realistic. Namely, in practice there is no need to deal with a book as a whole because each chapter starts on a new page. Thus, the chapters may be paginated one by one, and then the resulting pages can be concatenated.

6   Conclusions and Further Work

In this work we have discussed the pagination problem of complex documents. From a theoretical point of view, we have proven why existing systems are bound to fail. Practically, we have presented a new measure of pagination quality, an optimizing pagination algorithm, and a prototype implementation that has shown very promising results.

An interesting question is if it will be technically possible to embed our algorithm into one of the major existing formatting systems; here we are widely open to suggestions and cooperation. Another approach could be to extend our prototype, XFORMATTER, into a fully-fledged formatting system.

Often the pagination of documents is not done by computer scientists but by people who have a background in arts. Therefore, the interface of a pagination program should be easy to use. This is all the more important as we do not expect that designers will always be satisfied with the very first pagination suggested by the system.

In such cases, the user should be provided with ways of interactively influence what the pagination algorithm does. This can be done by specifying constraints. It would be interesting to know if the two kinds of additional constraints currently implemented in XFORMATTER—under-full pages and ways to force two boxes on the same or on different spreads or pages— are sufficient. Here we hope to benefit from further experience.

References

1. ASHER, G. 1990. Type & Set: TEX as the Engine of a Friendly Publishing System. *Pages 91–100 of:* CLARK, M. (ed), *TEX: Applications, Uses, Methods.* Chichester, UK: Ellis Horwood Publishers. Proceedings of the TEX88 Conference.
2. BELLMAN, RICHARD. 1957. *Dynamic Programming.* Princeton, New Jersey: Princeton University Press.
3. BRÜGGEMANN-KLEIN, A., KLEIN, R., & WOHLFEIL, S. 1996. Pagination Reconsidered. *Electronic Publishing—Origination, Dissemination, and Design,* **8**(2&3), 139–152.
4. CHI82. 1982. *The Chicago Manual of Style.* The University of Chicago Press, Chicago.
5. DYSON, MARY C., & KIPPING, GARY J. 1997. The Legibility of Screen Formats: Are Three Columns Better Than One? *Computers & Graphics,* **21**(6), 703–712.
6. EFI95. 1995 (July). *Wissenschaftliche Information im elektronischen Zeitalter. Stand und Erfordernisse.* Bayerisches Staatsministerium für Unterricht, Kultus, Wissenschaft und Kunst, RB-05/95/14. Available on the WWW by the URL http://www11.informatik.tu-muenchen.de/EFI/. Bericht der Sachverständigenkommission zur elektronischen Fachinformation (EFI) an den Hochschulen in Bayern.

7. FIAT, AMOS, & WOEGINGER, GERHARD (eds). 1996. *On-line Algorithms: The State of the Art*. Springer-Verlag.

8. FURUTA, RICHARD, SCOFIELD, JEFFREY, & SHAW, ALAN. 1982. Document Formatting Systems: Survey, Concepts and Issues. *Pages 133–210 of:* NIEVERGELT, J., CORAY, G., NICOUD, J.-D., & SHAW, A.C. (eds), *Document Preparation Systems*. Amsterdam: North Holland.

9. GAREY, M. R., & JOHNSON, D. S. 1979. *Computers and Intractability: A Guide to the Theory of NP-Completeness*. New York, NY: W. H. Freeman.

10. HANSEN, WILFRED J., & HAAS, CHRISTINA. 1988. Reading and Writing with Computers: A Framework for Explaining Differences in Performance. *Communications of the ACM*, **31**(9), 1080–1089.

11. KANDERS, MICHAEL. 1996. *Schule und Bildung in der öffentlichen Meinung*. Beiträge zur Bildungsforschung und Schulentwicklung. Dortmund, Germany: IFS-Verlag.

12. KERNIGHAN, B. W., & WYK, CH. J. VAN. 1989. Page Makeup by Postprocessing Text Formatter Output. *Computing Systems*, **2**(2), 103–132.

13. KLEIN, R. 1997. *Algorithmische Geometrie*. Bonn: Addison-Wesley.

14. KNUTH, DONALD E. 1986. *The TeXbook*. Computer & Typesetting, vol. A. Reading, MA: Addison Wesley Publishing Company.

15. KNUTH, DONALD E., & PLASS, MICHAEL F. 1981. Breaking Paragraphs into Lines. *Software—Practice and Experience*, **11**, 1119–1184.

16. LAMPORT, L. 1986. *LaTeX: A Document Preparation System, User's Guide & Reference Manual*. Reading, MA: Addison-Wesley Publishing Company.

17. PAGEL, BERND-UWE, & SIX, HANS-WERNER. 1994. *Software Engineering; Band 1: Die Phasen der Softwareentwicklung*. Reading, Massachusetts: Addison-Wesley.

18. PIOLAT, ANNIE, ROUSSEY, JEAN-YVES, & THUNIN, OLIVIER. 1997. Effects of screen presentation on text reading and revising. *Int. J. Human-Computer Studies*, **47**, 565–589.

19. PLASS, M. F. 1981. *Optimal Pagination Techniques for Automatic Typesetting Systems*. Technical Report STAN-CS-81-870. Department of Computer Science, Stanford University.

20. PLASS, M. F., & KNUTH, D. E. 1982. Choosing Better Line Breaks. *Pages 221–242 of:* NIEVERGELT, J., CORAY, G., NICOUD, J.-D., & SHAW, A. C. (eds), *Document Preparation Systems*. Amsterdam: North Holland.

21. RUBINSTEIN, R. 1988. *Digital Typography: An Introduction to Type and Composition for Computer System Design*. Reading, MA: Addison-Wesley Publishing Company.

22. WILLIAMSON, H. 1983. *Methods of Book Design*. New Haven: Yale University Press.

23. WOHLFEIL, STEFAN. 1997 (Dec). *On the Pagination of Complex, Book-Like Documents*. Ph.D. thesis, Fernuniversität Hagen, Germany.

On Predictive Parsing and Extended Context-Free Grammars

Anne Brüggemann-Klein[1] and Derick Wood[2]

[1] Institut für Informatik, Technische Universität München,
Arcisstr. 21, 80290 München, Germany.
brueggem@informatik.tu-muenchen.de

[2] Department of Computer Science, Hong Kong University of Science & Technology,
Clear Water Bay, Kowloon, Hong Kong SAR.
dwood@cs.ust.hk

Abstract. Extended context-free grammars are context-free grammars in which the right-hand sides of productions are allowed to be any regular language rather than being restricted to only finite languages. We present a novel view on top-down predictive parser construction for extended context-free grammars that is based on the rewriting of partial syntax trees. This work is motivated by our development of ECFG, a Java toolkit for the manipulation of extended context-free grammars, and by our continuing investigation of XML.

1 Introduction

We have been investigating XML [3], the Web language for encoding structured documents, its properties and use for a number of years [5; 6; 15]. The language XML itself, the document grammars that XML defines, and various other Web languages are defined by extended context-free grammars; that is, context-free grammars in which the right-hand sides of productions are allowed to be any regular language rather than being restricted to only finite languages. Hence, we became interested in factoring out all grammar processing that Web applications are based on and need to perform, into a separate toolkit that we call ECFG.

A cornerstone of ECFG is to be able to generate parsers from grammars. We present in this paper the principles of generating strong LL(1) parsers from a subset of the extended context-free grammars that is pertinent to Web grammars. We call these parsers **eSLL(1) parsers**. In particular, we contribute a novel view to predictive parsing that is based on what we call partial syntax trees. The parser generator is intended to be one tool among many in our toolkit.

LaLonde [16] appears to have been the first person to seriously consider the construction of parsers for extended context-free grammars. The construction of LL(1)-like parsers for extended context-free grammars has been discussed by Heckmann [12], by Lewi and his co-workers [17], and by Sippu and Soisalon-Soininen [20]. Warmer and his co-workers [22; 23] and Clark [9] have developed SGML parsers based on LL(1) technology. Mössenböck [18] and Parr and Quong [19] have implemented LL(1) parser generators for extended context-free grammars.

R. Klein et al. (Eds.): Comp. Sci. in Perspective (Ottmann Festschrift), LNCS 2598, pp. 69–87, 2003.

We give, in Section 2, some general background information on extended context-free grammars and Web languages and discuss the level of support that is currently available for grammar manipulation. We then define the basic notation and terminology that we need, in Section 3, before introducing partial syntax trees, in Section 4. In Section 5, we describe a nondeterministic algorithm eNSLL to compute a sequence of leftmost partial syntax trees for a given input string. We define an extended context-free grammar to be an **eSLL(1) grammar** if and only if the algorithm eNSLL is actually deterministic and we characterize eSLL(1) grammars in terms of first and follow sets.

This paper is the full version of a conference paper [7]. We mention further results in the last section, the proofs of which are in an extended version of this paper [8].

2   The Grammar Toolkit

Starting with XML, we present in this section some general background information on extended context-free grammars and Web languages and discuss the level of support that is currently available for grammar manipulation. We focus on the facts that have led to our decision to develop the grammar toolkit ECFG and to equip it with a predictive-parser generator.

XML is defined by an extended context-free grammar. The XML grammar derives XML documents that consist of an optional Document Type Definition (DTD) and the document proper, called the document instance. The XML grammar describes the syntax for DTDs and instances in general terms. The DTD is specific to an application domain and not only defines the vocabulary of elements, attributes and references in the document but also specifies how these constructs may be combined. The DTD is again an extended context-free grammar.

There are a number of XML parsers that read a DTD and a document instance, and are able to determine whether both follow the general rules of the XML grammar and whether the instance conforms to the DTD. Furthermore, there are two well-established means for application programs to access XML data; namely DOM, a W3C standard, and SAX, an industry standard. XML parsers typically support both of these Application Programming Interfaces (APIs).

It is curious that none of the XML tools we are aware of provide API access to the DTD of an XML document. This limits and hinders the development of XML tools that are customized for the application domain and of tools that read and manipulate DTDs such as DTD-aware editors for document instances, DTD-browsers, DTD-analyzers and DTD-aware query optimizers for XML documents. State-of-the-art XML tools treat DTDs as black boxes; they may be able to handle DTDs but they do not share their knowledge!

For this reason we propose a more transparent approach to the development of XML tools that is based on ECFG, a Java toolkit for the manipulation of extended context-free grammars, that we are currently implementing.

Grammars are ubiquitous. In the Web context alone, we have not only the XML grammar and XML DTDs but also XML Schemas, the CSS grammar, the XPath grammar, and specific DTDs or quasi-DTDs such as MathML and XSLT. Web tools such as a CSS processor, an XPath query optimizer and an XML processor that validates a document instance against its DTD are very different applications. With ECFG we aspire to support any grammar manipulations that these diverse tools might need to perform.

The cornerstone of grammar manipulation is to generate parsers automatically from grammars. The theory of parser generators is well understood and has been explored, at least for the case of nonextended grammars, in a number of textbooks [1; 2; 21; 24]. Furthermore, this knowledge is embodied in generator tools such as lex, flex, bison, yacc, Coco/R and ANTLR. A parser generator constructs a parser from a given grammar. The parser in turn converts a symbol stream into a syntax tree or some other structure that reflects the phrase structure of the symbol stream; alternatively and more commonly, the parser does not expose the parse tree itself but triggers semantic actions that are coded into its grammar for each phrase that it recognizes in the symbol stream.

In the context of Web languages, the languages that a grammar derives are often grammars once more, as exemplified by the XML grammar that derives DTDs; that is, document grammars. These grammars not only need to be turned into parsers again, which would be in the scope of standard parser generators, but they also need to be analyzed and transformed in complex ways. The analysis requires, in some cases, computations that parser generators perform but do not expose; for example, computations of first and follow sets. For this reason, we have decided to design and implement our own parser generator in ECFG. In our domain of applications, grammars need to be first-class citizens.

The ECFG parser generator does not work with semantic actions; rather it builds the complete parse tree and exposes it to interested applications. There are a number of reasons for this design decision. Mainly, in general it seems advisable to separate the concerns of defining a grammar and of specifying how the grammar should be processed; semantic actions mix the two concerns. In addition, our application domain makes the separation of the two concerns all the more desirable, for two reasons: First, a number of applications that each require their own set of semantic actions will be based on a single grammar which, without separation, would have to include several sets of semantic actions. Second, grammars will themselves be generated or computed and classes of grammars will share the processing strategies that would have to be incorporated into semantic actions. It seems awkward to generate the semantic actions together with each grammar rather than separating them.

Semantic actions have worked well for traditional compiler compilers. Our design decision against the use of semantic actions is motivated by our application domain. In compiler construction, one is dealing with a single grammar, namely the grammar for the programming language, and a single task, namely to implement a compiler or a programming environment for that language. In the domain of Web applications we need to be able to handle very many grammars, of which many will not be handcrafted but will be generated, and a more diverse range of applications.

It is be a common assumption that XML documents are easy to parse; whereas this assumption has not been formally verified, is obviously true for document instances, which are fully bracketed. Hence, of all the alternative parsing strategies that are in use, our parser generator employs the simplest one, namely the strong LL approach to parsing with a one-symbol lookahead, generalizing it from "normal" to extended context-free grammars.

Interestingly enough, it turns out that only extended context-free grammars whose right-hand sides are one-unambiguous give rise to eSLL(1) parsers. In previous work [5], we characterized the rhs-languages of an XML document grammar to be exactly the one-unambiguous (or deterministic) regular languages. This fact confirms that our choice of parsing strategy is the right one for the domain of Web grammars.

3   Notation and Terminology

An **extended context-free grammar** G is specified by a tuple of the form (N, Σ, P, S), where N is a nonterminal alphabet, Σ is a terminal alphabet, P is a set of **production schemas** of the form $A \longrightarrow L_A$, such that A is a nonterminal and L_A is a regular language over the alphabet $\Sigma \cup N$, and S, the sentence symbol, is a nonterminal. Given a production schema $A \longrightarrow L_A$ such that α is a string in L_A, we say that $A \longrightarrow \alpha$ is a **production** of G. We call L_A the **rhs-language** of A.

The "normal" context-free grammars are a special case of the extended grammars in that their rhs-languages are finite. In contrast, extended grammars may have infinite rhs-languages which are, however, constrained to be regular.

A "normal" grammar G **derives** a string β from a string α, denoted by $\alpha \underset{G}{\longrightarrow} \beta$, in a single step using a production $A \longrightarrow \gamma$, if α can be decomposed into the form $\alpha = \alpha_1 A \alpha_2$ and β into the form $\beta = \beta_1 \gamma \beta_2$. To **derive** β from α, denoted by $\alpha \longrightarrow *_G \beta$, means to produce β from α in a sequence of single-step derivations. These notions of derivation translate naturally to the extended case.

For a grammar $G = (N, \Sigma, P, S)$, a nonterminal symbol A in N gives rise to two languages, namely:

1. The rhs-language L_A over $N \cup \Sigma$, that is assigned to A via P.
2. The language of all Σ-strings that can be derived from A using productions in G; as usual, we call this language the **language** of A and denote it by $L(A)$.

More generally, if α is a string over $N \cup \Sigma$, then the **language** $L(\alpha)$ of α is the set of all terminal strings that can be derived from α using productions in G. In particular, $L(S)$ is the **language** $L(G)$ of grammar G.

Since the rhs-languages of extended grammars are regular, extended and "normal" grammars derive the same class of languages, namely the context-free languages [13].

We represent the set of production schemas P of an extended context-free grammar $G = (N, \Sigma, P, S)$ as a transition diagram system [10; 11] which provides a nondeterministic finite automaton (NFA) for the rhs-language of each nonterminal. We require the automata to be of Glushkov type so that each state is labeled by a symbol in $N \cup \Sigma$ and each incoming transition bears that state's label.

In practice, the rhs-languages of an extended grammar are given as regular expressions. A regular expression can, however, be transformed into a Glushkov-type NFA with an output-sensitive algorithm in quadratic worst-case time; for the simple type of grammars that we consider in this paper, the transformation can even be achieved in linear worst-case time [4].

A **transition diagram system** $TS = (Q, \textbf{label}, F, \textbf{init}, \textbf{trans}, \textbf{belongs})$ over N and Σ has a set of states Q, a relation **label** $\subseteq Q \times (N \cup \Sigma)$ that maps each state to at most one symbol, a set of final states $F \subseteq Q$, a relation **init** $\subseteq N \times Q$ that assigns exactly one initial state to each nonterminal, a relation **trans** $\subseteq Q \times Q$ for the transitions between states and a relation **belongs** $\subseteq Q \times N$ that maps each state to the unique nonterminal to whose rhs-automaton the state belongs.

A relationship $p\,\textbf{trans}\,q$ implies that q has some label X (that is, $q\,\textbf{label}\,X$) and means that there is a transition from p to q on the symbol X. This notion of transition relation accounts for the Glushkov property.

A production $A \longrightarrow \alpha$ of the grammar G translates into a string α of the NFA $(N \cup \Sigma, Q, p_A, F, \textbf{trans})$ such that $A\,\textbf{init}\,p_A$. For simplicity's sake, we have assigned the full sets Q, F and **trans** to the automaton, although in practice only a subset of states in Q will be reachable by any initial state p_A. This is irrelevant for our complexity results since we'll always deal with full transition diagram systems.

Each state in the transition diagram system is uniquely assigned to some nonterminal via the relation **belongs**. A state that is reachable from a nonterminal's initial state must belong to that same nonterminal. When we construct the automata from the rhs-expressions of a grammar, the sets of states must hence be chosen to be disjoint.

Some of the relations in a transition diagram system are functions in reality. We have chosen to represent them as relations, so that we can employ the calculus of relational expressions [14] that has been propagated by Sippu and Soisalon-Soininen [20] when we prove our complexity results.

Relations are represented in the simplest-possible way, namely as repetition-free lists of the tuples that belong to them.

We note again that a transition diagram system can be computed efficiently from an extended grammar whose rhs-languages are represented by expressions by converting the expressions into Glushkov NFAs [4]. The time bound obviously includes the computation of the nonstandard relation **belongs**.

From now on we assume that every grammar $G = (N, \Sigma, P, S)$ is **reduced** in the sense that every nonterminal and every state is reachable and useful. We even require that each nonterminal A is **strongly useful** in the sense that its language $L(A)$ contains a string that is not empty.

We call a grammar **strongly reduced** if and only if it is reduced and if all its nonterminals are strongly useful, and we only consider strongly reduced grammars in this paper. With strongly reduced grammars, we can still represent all context-free languages, with the exception of two trivial languages, namely the empty language and the singleton language $\{\lambda\}$. The Web grammars that we know of are all strongly reduced.

For the remainder of the paper, let Σ denote a set of terminals and N denote a set of nonterminals; their union $\Sigma \cup N$ forms a **vocabulary**. An extended context-free grammar is given as $G = (N, \Sigma, P, S)$; its set of production schemes P is represented by the transition diagram system
$TS = (Q, \textbf{label}, F, \textbf{init}, \textbf{trans}, \textbf{belongs})$ over N and Σ.

Names A, B and C denote nonterminals, a and b denote terminals, X denotes a symbol in the vocabulary $\Sigma \cup N$, p denotes a state, α, β and γ denote strings over the vocabulary, whereas u, v and w denote strings over Σ. If we need additional names of any of these types, we use embellishments.

4   Partial Syntax Trees

We base our approach to parsing on the use of **partial syntax trees** whose nodes are **labeled** with symbols in the vocabulary $\Sigma \cup N$ and some of whose nodes are **annotated** with states in Q. Internal nodes are always labeled with nonterminal symbols from N; external nodes are labeled with symbols from either alphabet. Only nodes that have a nonterminal label may have an annotation. We call those nodes **active**.

A partial syntax tree represents the progress a parser has made in constructing a syntax tree for a given input string of terminals in a top-down manner. An active node represents a construction site where the parser may append further nodes to the list of child nodes, thus gradually expanding the active node. When the parser has completed the expansion work at a construction site, it will remove the annotation to make the node inactive. The goal is to construct a partial syntax tree without any active nodes such that its terminal-labeled leaves spell out the input string that is to be parsed. Leaves that are inactive and labeled with a nonterminal are not expanded and thus contribute the empty string to the input string.

A grammar, particularly its transition diagram system, constrains the work of a parser and determines if the partial syntax tree that is constructed **conforms** to the grammar: First of all, the tree's root must be labeled with the grammar's sentence symbol. Furthermore, the labels and annotations in the tree must conform to the grammar in the following way:

- For each inactive node v, the labels of the children of v must spell out a string in the rhs-language of v's label.
- For each active node v, its state annotation is reachable from the node label's initial state by the input string formed by the sequence of labels of v's children.

Particularly, each external active node must be annotated with the initial state of the node's label and the language of an inactive external node's label must contain the empty string. Furthermore, if a node has label A and annotation p, then p **belongs** A because p is reachable from A's initial state.

Since we wish to explore top-down left-to-right parsing, we are particularly interested in **leftmost** partial syntax trees in which the active nodes form a prefix of the rightmost branch of the tree. The frontier of a leftmost partial syntax tree is a sequence, from left to right, of inactive nodes, followed by at most one active node. The sequence of nodes in the frontier that have *terminal labels* yields a terminal string over Σ, which we call the **yield** of the leftmost partial syntax tree.

Nonterminal nodes in the frontier do not expand and, hence, contribute the empty string to the yield. This is consistent with their labels' rhs-languages containing the empty string when we are dealing with partial syntax trees for grammars.

We call a partial syntax tree on which all work has been completed (that is, which has no active nodes left) just a **syntax tree**. Whereas a partial syntax tree represents a parser's work-in-progress, a syntax tree represents the finished product that may then be exposed to application programs.

From a static point of view, the yields of a grammar's syntax trees form the language of the grammar, as demonstrated in the next lemma. We will consider the dynamic aspects of *constructing* syntax trees in the remainder of this section.

Lemma 1. *The yield of any syntax tree of an extended context-free grammar G is a string in* $L(G)$ *and, conversely, any string in* $L(G)$ *is the yield of some syntax tree of G.*

Proof. In a syntax tree of a grammar G (whose root must by definition be labeled with the grammar's sentence symbol S) there is a one-to-one correspondence between a node with a label A in N, whose children are labeled—character by character—with some α in $(\Sigma \cup N)*$, and an application of the production $A \longrightarrow \alpha$ in a derivation from S: In both cases, α must be in A's rhs-language. The correspondence is also valid in the special case that the syntax tree has an external node with label A in N and that $A \longrightarrow \lambda$ is a production in the grammar; that is, that the empty string λ belongs to the rhs-language of A.

In a syntax tree, each node in the frontier is labeled either with a terminal symbol or with a nonterminal symbol whose rhs-language contains the empty string. Hence, the yields of the syntax trees correspond exactly to the strings over the terminal alphabet that the grammar derives.   □

A partial syntax tree of grammar G can be constructed incrementally by beginning with an initial one-node partial syntax tree and then, step by step, adding nodes, changing the states that are associated with nodes and making active nodes inactive. Rather than applying a whole production $A \longrightarrow \alpha$, $\alpha = X_1 \cdots X_n$, in a single step to a node v with label A, we add n children to v one after the other, in n steps, labeling the children with $X_1, \ldots, X_n$ and keeping track in the state annotation of v of how far we have progressed. We view this process as a sequence of transformations on partial syntax trees.

A single **transformation** manipulates, nondeterministically, an *active node* v of a partial syntax tree in one of the following three ways, where we assume that v is labeled with a nonterminal A and is annotated with a state p that belongs to A:

1. If p is a final state, then v is made inactive by removing the state p from v. We call this transformation a **reduce step**.
2. If there is a transition from p to p' and if the label of state p' is a terminal a, then a new node v' is added as a new rightmost child of v. The new node is labeled with a and v is annotated with state p'. We call this transformation a **shift step**.
3. If there is a transition from p to p' and if the label of state p' is a nonterminal B, then an active node v' is added as a new rightmost child of v. The new node is labeled with B and is annotated with the initial state that is associated with B. Futhermore, v is annotated with state p'. We call this transformation an **expand step**.

Let us have a brief look at the kind of information we need if we wish to perform any of these tree transformations. For now we consider only the information that determines the *result* of a transformation, not the *conditions* under which a transformation is applicable. We assume that we know the relation **label** of the grammar's state transition diagram but nothing else of the grammar.

1. If we intend to perform a reduce step at any given node, we need no further information.
2. If we intend to perform a shift step at any given node, it suffices to know the new annotation of the node, since this determines the label of the new node.
3. If we intend to perform an expand step at any given node, it suffices to know the new annotation of the node, since this determines the label of the new node, and the annotation of the new node.

Later we will compress a parsing strategy into a table, whose indices determine the conditions under which transformations can be performed and whose entries determine the transformations that are admissable under those conditions. A transformation is then defined by its type and the at most two information items that we have already defined.

Note that the property of partial syntax trees to conform to a given grammar is invariant under the three transformations.

Lemma 2. *Each of the three transformations, when applied to a grammar-conformant partial syntax tree, results once more in a grammar-conformant tree.*

Proof. It is obvious how to prove the claim for each of the three transformation types. Note that the grammar conformance only has to be verified locally, at a construction site.   □

A **construction** of a partial syntax tree is a sequence of transformations that begins with the initial partial syntax tree whose one node is active, is labeled with the sentence symbol S and is annotated with the initial state of S and that ends with the partial syntax tree itself.

We illustrate these concepts with the grammar G that has the rules:

$$S \longrightarrow A * | ab,$$
$$A \longrightarrow a.$$

The grammar's transition diagram system is given in Figure 1.

A **leftmost construction** is a construction that begins with the initial partial syntax tree and consists only of leftmost partial syntax trees. (Note that the initial partial syntax tree and all syntax trees are leftmost.) A leftmost construction is the analog of a leftmost derivation for "normal" context-free grammars.

In any transformation step that transforms a leftmost partial syntax tree into another leftmost partial syntax tree, the active node that the transformation is applied to is uniquely determined: it is the deepest active node of the source tree, as stated in Lemma 3. The only choices lie in the type of transformation and in the transition of the grammar's transition diagram system that is to be applied. Hence, in a leftmost construction, we can at any time determine uniquely *the* active node to which the next transformation will be applied. Conversely, if each transformation step in a construction operates on the deepest active node of its source tree, then the construction is leftmost, as implied by Lemma 3 as well.

Lemma 3. *A transformation on a leftmost partial syntax tree results in another leftmost partial syntax tree if and only if it operates on the deepest active node of the source tree.*

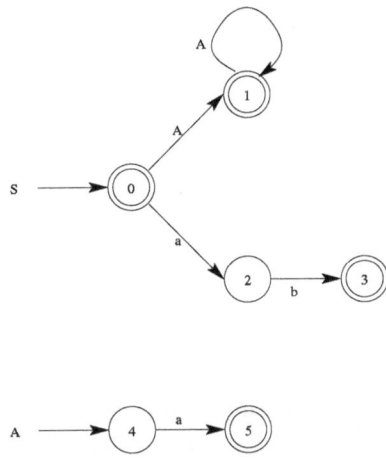

Figure 1. The transition diagram system for the grammar with production schemes $S \longrightarrow A * | ab$ and $A \longrightarrow a$: As usual, circles represent states and double circles represent final states; the arrows between the circles represent the relation **trans**, arrows between a nonterminal and a circle represent **init**; a state's label, if present, is defined unambiguously by the labels on the incoming arrows.

Proof. Any transformation either makes an active node inactive or appends a new node to the list of an active node's children. Therefore, if a transformation is applied to an active node that has an active child, the result tree is not leftmost. On the other hand, if a transformation is applied to the deepest active node of a leftmost partial syntax tree, the result is leftmost once more.   □

In the next few lemmas we build up a set of techniques that will also be useful in later proofs. We start by investigating pairs of leftmost partial syntax trees t_1 and t_2 such that t_1 can be transformed into t_2 by a sequence of transformation steps. Intuitively, such a transformation is possible if and only if t_1 can be "embedded" in t_2 in a way that we capture with our notion of prefix:

Let t_1 and t_2 be two leftmost partial syntax trees that conform to grammar G. Tree t_1 is a **prefix** of tree t_2, denoted by t_1 **prefix** t_2, if and only if there is a map μ that maps t_1's set of nodes *into* t_2's set of nodes and that satisfies the following conditions:

1. μ maps the root of t_1 to the root of t_2.
2. μ maps, for each node v of t_1, the ith child of v to the ith child of $\mu(v)$.
3. μ preserves labels; that is, for each node v in t_1, the labels of v and $\mu(v)$ are identical.
4. If node v of t_1 is not active, then node $\mu(v)$ is not active either; furthermore, v and $\mu(v)$ have the same number of children.
5. If some node v of t_1 and the node $\mu(v)$ are both active and have the state annotations p_1 respectively p_2, then the grammar's transition diagram system has a path from p_1 to p_2 on the labels of those children of $\mu(v)$ that are not in the range of μ.

6. If a node v of t_1 is active but the node $\mu(v)$ is not and if v has the state annotation p, then the grammar's transition diagram system has a path from p to some final state on the labels of those children of $\mu(v)$ that are not in the range of μ.

We call such a map the **prefix map** of t_1 and t_2. Obviously, for any two trees, there is at most one prefix map.

Lemma 4. *The* **prefix** *relation is a partial order on the set of leftmost partial syntax trees of a grammar.*

Proof. We must prove that **prefix** is reflexive, transitive and antisymmetric; we only demonstrate antisymmetry.

Let t_1 and t_2 be two leftmost partial syntax trees of some grammar such that t_1 **prefix** t_2 and t_2 **prefix** t_1. Furthermore, let μ be the prefix map of t_1 and t_2. We claim that $t_1 = t_2$.

Since t_2 is a prefix of t_1, the map μ is a bijection, and μ^{-1} is the prefix map of t_2 and t_1. Hence, t_1 and t_2 are structurally equivalent and have identical labels. Furthermore, for each node v of t_1, the two nodes v and $\mu(v)$ are either both active or both inactive. If v and $\mu(v)$ have state annotations p_1 and p_2 respectively, then the transition diagram system must have a path from p_1 to p_2 on the sequence of labels of those children of $\mu(v)$ that are not in the range of μ. Since μ is a bijection, this sequence is empty, which, together with the fact that state transition diagrams do not have null transitions, implies that $p_1 = p_2$. Hence, the annotations of t_1 and t_2 are identical, which completes the proof. □

Lemma 5. *Let t_1 and t_2 be leftmost partial syntax trees of some grammar. If t_1 is transformed into t_2 with one of the three transformation steps, then t_1 is a prefix of t_2.*

Proof. A simple case analysis of the three types of transformation. □

Lemma 6. *Let t_1 and t_2 be two partial syntax trees of some grammar such that t_1 is a proper prefix of t_2. Furthermore, let μ be the prefix map of t_1 and t_2. Finally, for each node v of t_1, let the two nodes v and $\mu(v)$ have the same number of children. Then, t_1 has at least one active node v whose image $\mu(v)$ is inactive.*

Proof. Since μ maps t_1's root to t_2's root, the assumptions imply immediately that t_1 and t_2 are structurally equivalent and have the same labels. Furthermore, if node v in t_1 and node $\mu(v)$ in t_2 are both active, with state annotations p_1 and p_2, respectively, then the grammar's transition diagram system transforms p_1 into p_2 on the empty input string; that is, p_1 and p_2 are the same. Hence, since $t_1 \neq t_2$, there must be an active node v in t_1 whose image $\mu(v)$ is inactive. □

Lemma 7. *A proper prefix of a partial syntax tree must have at least one active node.*

Proof. We prove the lemma by indirection.

Let t_1 and t_2 be two partial syntax trees such that t_1 **prefix** t_2 and let μ be the prefix map of t_1 and t_2. If t_1 has no active nodes, then neither has t_2. Therefore, for each node v of t_1, the nodes v and $\mu(v)$ have the same number of children. Hence, $t_1 = t_2$ by Lemma 6.   □

Lemma 8. *Let t_1 and t_2 be two distinct leftmost partial syntax trees of some grammar such that t_1 **prefix** t_2. Then, there is a leftmost partial syntax tree t that can be constructed from t_1 in a single transformation step and that is also a prefix of t_2.*

Proof. Let μ be the prefix map of t_1 and t_2. Lemma 7 implies that t_1 has at least one active node. We choose node v with state annotation p_1 as the deepest active node of t_1.

Case 1   If node $\mu(v)$ is inactive, the grammar's state transition diagram has a path from p_1 to some final state on the sequence of labels of those children of t_2 that are not in the range of μ. If, on the one hand, there are no such children, state p_1 itself is final and we may perform a reduce step at v. The resulting tree t is a prefix of t_2. If, on the other hand, some node v' with label X is the first child of $\mu(v)$ that is outside the range of μ, we may perform an expand or a shift step at v that labels the new node with X. The new state annotion p' of v may be chosen in such a way that the sequence of $\mu(v)$'s children that follow v' transform the grammar's transition diagram system from p' to some final state. Hence, the resulting tree t is a prefix of t_2.

Case 2   If node $\mu(v)$ is active, then each ancestor of $\mu(v)$ is active and, hence, cannot have any right siblings. Since μ maps the ith ancestor of v to the ith ancestor of $\mu(v)$, each proper ancester of v has the same number of children as its image under μ. Since t_1 and t_2 are distinct, Lemma 6 implies that for at least one active node of t_1, its image under μ must have strictly more children than the node itself. This node must be v, because t_1 is leftmost.

But $\mu(v)$ can have strictly more children than v only if $\mu(v)$ is active, with some state annotation p_2 such that the grammar's transition diagram system transforms p_1 to p_2 on the sequence of labels of those children of $\mu(v)$ that are not in the range of μ.

Again we choose node v' with label X as the first child of $\mu(v)$ that is outside the range of μ, and we perform an expand or a shift transformation at node v. The new node gets labeled X and the new state p' for v is chosen such that the sequence of labels of the siblings of v' transforms the grammar's transition diagram system from p' into p_2. Again, the new tree t is a prefix of t_2 □

We measure the amount of work that has been put into constructing a leftmost syntax tree t by its **weight**, denoted by **weight**(t), which we define to be the sum of the number of active nodes and twice the number of inactive nodes in t. We note the following properties of the weight:

Lemma 9. *Let t, t_1 and t_2 be leftmost partial syntax trees of some grammar. Then,* **weight** *has the following properies:*

1. **weight**$(t) \geq 1$.
*2. If t_1 **prefix** t_2, then* **weight**$(t_1) \leq$ **weight**(t_2).
*3. If t_1 **prefix** t_2 and* **weight**$(t_1) =$ **weight**(t_2), *then $t_1 = t_2$.*

4. *If t_2 is constructed from t_1 in a single transformation step, then* **weight**$(t_1) <$ **weight**(t_2).

Proof. We observe first that each prefix of a partial syntax tree t has fewer nodes and fewer inactive nodes than t. We now prove the four statements of the lemma:

1. Each tree has at least one node.
2. If t_1 is a prefix of t_2, then t_2 has more nodes than t_1 and each inactive node in t_1 corresponds to an inactive node in t_2. Hence, **weight**$(t_1) \leq$ **weight**(t_2).
3. If t_1 is a prefix of t_2 and **weight**$(t_1) =$ **weight**(t_2), the trees t_1 and t_2 must have the same number of nodes. Hence, if μ is the prefix map of t_1 and t_2, for each node v of t_1, the nodes v and $\mu(v)$ have the same number of children and the two nodes are either both active or both inactive. Then, Lemma 6 implies that $t_1 = t_2$.
4. A reduce step keeps the number of nodes constant but turns one active node inactive. Expand and shift steps add an additional node. Hence, all three types of transformations raise the weight.

□

We next establish that each leftmost partial syntax tree that conforms to a grammar is reached by a construction that only uses leftmost trees as intermediaries.

Lemma 10. *Let t_1 and t_2 be two leftmost partial syntax trees of some grammar. Then, t_1 is a prefix of t_2 if and only if there is a leftmost construction from t_1 to t_2.*

Proof. Since the prefix relation is transitive, the "if" claim is a corollary to Lemma 5.

We assume now that t_1 is a prefix of t_2. We prove the claim by induction on **weight**$(t_2) -$ **weight**(t_1), which is always at least 0, as stated in Lemma 9.

The base for the induction is the case that **weight**$(t_2) -$ **weight**$(t_1) = 0$; that is, **weight**$(t_1) =$ **weight**(t_2). We apply Lemma 9 once more and deduce that $t_1 = t_2$. Hence, t_2 can be constructed from t_1 with the empty sequence of transformation steps.

For the induction step, we assume that the claim holds for all pairs of trees whose weight difference is less than **weight**$(t_2) -$ **weight**(t_1). Furthermore, we assume that **weight**$(t_2) >$ **weight**(t_1); that is, t_1 is a proper prefix of t_2. We apply Lemma 8 and get a leftmost partial syntax tree t that can be constructed from t_1 in a single transformation step and that is also a prefix of t_2. By Lemma 9, **weight**$(t_1) <$ **weight**$(t) \leq$ **weight**(t_2). Hence, we complete the proof by applying the induction hypothesis to t and t_2.   □

Lemma 11. *For each grammar-conformant leftmost partial syntax tree, there is a leftmost construction.*

Proof. The claim follows from Lemma 10 since the initial partial syntax tree whose one node is labeled with the grammar's sentence symbol and is annotated with that symbol's initial state is a prefix of any partial syntax tree.   □

The previous lemmas culminate in the following theorem:

Theorem 1. *The language of a grammar consists of the yields of the grammar-conformant syntax trees. Furthermore, for each such syntax tree there is a leftmost construction.*

Proof. The theorem follows from Lemmas 1 and 11 since syntax trees are trivially leftmost.   □

5   Predictive Parsing

This section focuses on parsing; that is, for each terminal string u of a grammar, we construct a syntax tree whose yield is u. Our approach to parsing is top-down with a look-ahead of one symbol; that is, we do left-most constructions and read the input string from left to right, advancing at most one position at each transformation step. At each step, the choice of transformation is guided by the current input symbol.

A leftmost partial syntax tree is **compatible** with a given terminal string if its yield is a prefix of the string.

Lemma 12. *Any transformation step in a leftmost construction either appends one terminal symbol to the yield or leaves the yield unchanged.*

Proof. Reduce and expand steps leave the yield unchanged; a shift step appends one terminal symbol.   □

As our parsing strategy, we present a nondeterministic algorithm eNSLL to compute a leftmost construction in which each leftmost partial syntax tree is compatible with a given input string $a_1 \cdots a_n$ of terminals. The algorithm generalizes strong LL(1) parsing to extended grammars, which accounts for its name.

When given a leftmost partial syntax tree that is compatible with the input string, the algorithm expands the tree's deepest active node using one of the three transformation types given in Section 4.

When choosing a transformation type, the algorithm eNSLL is guided by the input symbol immediately to the right of the partial syntax tree's yield. We call this symbol the **current input symbol.** If the algorithms has moved beyond the end of the input string, we set the current input symbol to the empty string, thus signaling to the algorithm that it has read the input string completely.

We need to introduce a number of concepts to describe the behaviour of eNSLL:

First, for each pair of states p and p' such that $p\,\textbf{trans}\,p'$ let $\mathrm{L}(p,p')$ be the language of all strings over Σ that are the yields of syntax trees constructed from the one-node initial tree whose label is the symbol that belongs to p and whose state annotation is p such that the first transformation is to add a child to the root that is labeled with the label of p' and to annotate the root with state p'. Formally, $\mathrm{L}(p,p')$ is the union of all languages $\mathrm{L}(X\alpha)$ such that $p'\,\textbf{label}\,X$ and α in $(N\cup\Sigma)*$ moves the grammar's transition diagram system from p' to some final state.

Next, $follow(A)$ is the set of all terminals that can occur immediately after A in a string that G can derive from S. More formally, $follow(A)$ consists of all a in Σ such that G derives $\alpha A a\beta$ from S, for some α and β in $(N\cup\Sigma)*$. In addition, we add the empty string to $follow(A)$ for each A for which G derives some αA.

Finally, we consider every leftmost partial syntax tree whose deepest active node v is annotated with some state p and labeled with the nonterminal that p belongs to, and any construction whose first transformation step adds a new rightmost child to v and that annotates v with some state p' such that $p\,\textbf{trans}\,p'$ while labeling the new node with the label of p. The first terminals that are derived by such constructions form the set $first(p,p')$. To be precise, $first(p,p')$ consists of the first symbols of the strings in

$L(p, p')follow(A)$; note that we consider the empty string to be the first symbol of the empty string. Furthermore, $first(p)$ is the union of all $first(p, p')$ such that p **trans** p'.

The distinction between strong LL(1) parsing and LL(1) parsing is encapsulated in the definition of $first(p, p')$ in terms of $L(p, p')follow(A)$: We are dealing with an extension to *strong parsing* since $follow(A)$ lumps all the symbols together into one set that may follow A in any context; non-strong LL(1) parsing would have to be based on $L(p, p')follow(p)$ for some appropriately defined subset $follow(p)$ of $follow(A)$ that takes context p into account.

At this point, we state a fairly obvious lemma for later reference.

Lemma 13. *If p **trans** p' and p' **label** a, then $first(p, p') = \{a\}$.*

Proof. Let state p belong to nonterminal A. By definition of $L(p, p')$, the set $first(p, p')$ consists of all the first symbols of the strings in $L(a\alpha)follow(A)$ such that α moves the grammar's transition diagram system from p' to some final state. Since the grammar is reduced, we can find such an α and neither $L(a\alpha)$ nor $follow(A)$ are empty. Hence, since $L(a\alpha)follow(A)$ starts with a, we may conclude that $first(p, p') = \{a\}$.   □

Lemma 14. *If p **trans** p', then $L(p, p')$ contains an element that is different from the empty string.*

Proof. Let X be the label of p' and let α move the grammar's transition diagram system from p' to some final state. Since the grammar is reduced, such an α exists and neither $L(X)$ nor $L(\alpha)$ are empty. Furthermore, since the grammar is strongly reduced, $L(X)$ contains some string that is not empty. Thus, the fact that $L(X)L(\alpha)$ is a subset of $first(p, p')$ completes the proof.   □

Now let us have a closer look at the following scenario: We are given a leftmost partial syntax tree t that is compatible with some terminal string. Furthermore, we assume that t is a prefix of some syntax tree t' whose yield is the full terminal string. Thus, we can continue any leftmost construction of t until we reach a syntax tree for the terminal string. Let the tree's deepest active node v have annotation p. Finally, let a be the current input symbol. We are interested in only such a transformation at v that can be the first step of a continuation of a leftmost construction of t towards a syntax tree for the input string. If such a first transformation step is a reduce step, then a must be in $follow(A)$. In the case of a shift step that annotates v with some p', state p' must have a as its label. In this case, a is not the empty word and the input symbol is advanced. In the case of an expand step that annotates v with p', symbol a must be in $a \in first(p, p')$.

The algorithms eNSLL computes a leftmost construction for a syntax tree whose yield is a given input string of terminals. Hence, each partial syntax tree in this construction must be compatible with the input string. The algorithm starts with the initial one-node partial syntax tree that each construction starts with; this initial tree is compatible with any input string. At each step of the construction, it chooses nondeterministically a reduce, shift, or expand step to continue its computation, but the choice is constrained by the next input symbol and by the first and follow sets. More precisely, in a scenario as described above, the algorithm performs a reduce step only if a is in *follow*(*A*), it performs a shift step only if p' **label** a, and it performs an expand step only if a is in *first*(p, p'). Hence, the result tree is again compatible with the input string.

An eNSLL computation **terminates** if no further transformation steps are possible. The eNSLL algorithm **accepts** an input string if it terminates with a partial syntax tree that is, in fact, a syntax tree and whose yield is the complete input string.

We say that a grammar is **eSLL(1)** if and only if the eNSLL algorithm for the grammar is deterministic.

Theorem 2. *The eNSLL algorithm of a grammar accepts exactly the strings of the grammar.*

Proof. Given a word of the grammar as input, the algorithm eNSLL nondeterministically tries out all transformation steps that might result in a syntax tree for the input word; it leaves out only those alternatives that cannot possibly lead to a syntax tree for the given input. We complete the proof by applying Theorem 1.   □

Since we are primarily interested in deterministic parsers, we consider all possible conflicts that may cause eNSLL to be properly nondeterministic. There are exactly six types of conflict, which we describe in more detail furtheron: reduce-shift, reduce-expand, shift-shift, shift-expand, simple expand-expand, and expand-expand.

These are the six types of conflicts:

1. A **reduce-shift** conflict: The grammar's transition diagram system has a transition from some final state p which belongs to some nonterminal A to a state whose label is terminal and in *follow*(A); that is, for some p, p', a, A, the following conditions hold:
   (a) p is final.
   (b) p **belongs** A.
   (c) p **trans** p'.
   (d) p' **label** a.
   (e) $a \in$ *follow*(A).

2. A **reduce-expand** conflict: There is a transition from some final state p which belongs to some nonterminal A to a state p' with a nonterminal label such that *follow*(A) and *first*(p, p') have a non-empty intersection; that is, for some p, p', a, A, B, the following condition holds:
   (a) p is final.
   (b) p **belongs** A.
   (c) p **trans** p'.
   (d) p' **label** B.
   (e) $a \in$ *first*$(p, p') \cap$ *follow*(A).

3. A **shift-shift** conflict: There are transitions from some state p to two different state that have the same terminal label; that is, for some p, p', p'', a, the following conditions hold:
   (a) p **trans** p'.
   (b) p **trans** p''.
   (c) $p' \neq p''$.

    (d) p' **label** a.

    (e) p'' **label** a.

4. A **shift-expand** conflict: For some p, p', p'', a, B, the following conditions hold:

    (a) p **trans** p'.

    (b) p **trans** p''.

    (c) p' **label** a.

    (d) p'' **label** B.

    (e) $a \in first(p, p'')$.

5. A **simple expand-expand** conflict: There are two transitions from some state p to two different states that have the same nonterminal label; that is, for some p, p', p'', B, the following conditions hold:

    (a) p **trans** p'.

    (b) p **trans** p''.

    (c) $p' \neq p''$.

    (d) p' **label** B.

    (e) p'' **label** B.

6. A **expand-expand** conflict: For some p, p', p'', B, C, the following conditions hold:

    (a) p **trans** p'.

    (b) p **trans** p''.

    (c) p' **label** B.

    (d) p'' **label** C.

    (e) $first(p, p') \cap first(p, p'') \neq \emptyset$.

Theorem 3. *An extended context-free grammar is eSLL(1) if and only if it exhibits none of the six types of conflicts.*

Proof. Algorithm eNLL is nondeterministic if and only if it can encounter a scenario such as outlined previously, in which two different transformation steps are possible at the same time. With three types of transformation steps, this leaves, in principle, six combinations. We note first that no reduce-reduce conflict can occur: a reduce step at the deepest node of a leftmost syntax tree is alway unambiguous.

    In our classification of conflicts we split the combination of two expand steps into two cases, namely the simple expand-expand conflict, in which the expanding to a non-terminal leads to different states, and the other case, in which the current input symbol is not a sufficient base for a decision between an expansion to one or another nonterminal. Hence, our classification of conflicts is complete.

    Because we have assumed that the grammar is reduced, any state of the grammar's transition diagram system can occur as the annotation of the deepest active node of a leftmost syntax tree. Any of the conflicts that we have idenitified give rise to a scenario that eNLL can reach and in which it has a nondeterministic choice. Hence, our classification of conflicts is also correct.   □

Theorem 4. *An extended context-free grammar is eSLL(1) if and only if it satisfies the following two conditions:*

1. *For each final state p such that p **belongs** A, the sets first(p) and follow(A) are disjoint.*
2. *For each pair of different states p', p'' to which there are transitions from p, the sets first(p, p') and first(p, p'') are disjoint.*

Proof. The set *first*(p) is the union of all *first*(p, p') such that p **trans** p'. By Lemma 13, the first condition of the theorem is equivalent to the condition that the grammar has neither reduce-shift nor reduce-expand conflicts.

We establish now that the second condition of the theorem is equivalent to the grammar having none of the other four conflicts. We carry out a case distinction on p' and p''.

Three cases are obvious: The case that p' and p'' both have terminal labels corresponds to a shift-shift conflict. The case that one of p' and p'' has a terminal label and the other has a nonterminal label corresponds to a shift-expand conflict. The case that p' and p'' have different terminal labels corresponds to an expand-expand conflict.

Finally, we consider the case that p' and p'', while being different states, have the same nonterminal label B. Since the grammar is strongly reduced, $L(B)$ contains a nonempty string whose first symbol is in *first*(p, p') ∩ *first*(p, p''). Hence, the second condition of the theorem is violated and the grammar has a simple expand-expand conflict.

Taking the four cases together we have proved that a grammar satisfies the theorem's second condition if and only if it has none of the four types of conflicts that do not involve a reduce transformation.   □

We prove the next theorem in a companion paper [8] that focuses on parsing algorithms and complexity.

Theorem 5. *We can test if a grammar is eSLL(1) in worst-case time $O(|\Sigma| \cdot |G|)$.*

Theorem 6. *Let the rhs-languages of an extended context-free grammar be defined by regular expressions and let the grammar's transition diagram system be computed with the Glushkov construction [4]. If the grammar is eSLL(1), then the transition diagram system must be deterministic.*

Proof. An eSLL(1) grammar has no shift-shift and no simple expand-expand conflicts. Hence, the grammar's transition diagram system is deterministic.   □

Further Results

In addition to providing the proof for the complexity theorem in this paper, a companion paper [8] investigates two related topics:

First, it is straightforward to build a parsing table of parse actions from the first and follow sets that drive the eNSLL algorithm. There is at most one parse action in each table cell if and only if the grammar is eSLL(1). We can build the parse table in worst-case time $O(|\Sigma| \cdot |G|)$.

Second, "normal" context-free grammars are a special case of extended grammars. This carries over to strong LL(1)-grammars. In the companion paper, we characterize SLL(1) grammars in terms of eSLL(1) extended grammars.

We discuss implementation and application issues in a separate paper.

A Personal Note by the First Author

I came into Thomas Ottmann's research group at the University of Karlsruhe as a post-doc, having got a PhD in Mathematics from the same academic teacher as he did, the late Dieter Rödding. He introduced me to the emerging area of computer typesetting and document management which has long since broadened into content and knowledge management. I remember vividly the pioneer feeling we had when we finally got the Digiset photo typesetter at a local publishing house to produce TEX output in the days before PostScript.

Thomas Ottmann also gave me leave to work with Derick Wood for a number of longer and shorter periods. In his group, which ran under the then-apt nickname "Baum-schule" (tree nursery), and after having paid my dues in the form of a TEX macro package for drawing trees, I came into contact with formal languages, which has since then been the second leg on which my work stands and the basis of a long-standing collaboration.

Two things I appreciated most in Thomas Ottmann's research group: first, the spirit of academic freedom, second, the familiarity with up-to-date computer technology and its uses. I am proud and grateful to have been a member of the group.

Acknowledgements

The work of both authors was supported partially by a joint DAAD-HK grant. In addition, the work of the second author was supported under a grant from the Research Grants Council of Hong Kong.

References

1. A. V. Aho, R. Sethi, and J. D. Ullman. *Compilers: Principles, Techniques, and Tools.* Addison-Wesley Series in Computer Science. Addison-Wesley Publishing Company, Reading, MA, 1986.
2. J. Albert and Th. Ottmann. *Automaten, Sprachen und Maschinen.* Bibliographisches Institut, Mannheim, 1983.
3. T. Bray, J. P. Paoli, and C. M. Sperberg-McQueen. Extensible Markup Language (XML) 1.0. http://www.w3.org/TR/1998/REC-xml-19980210/, February 1998.
4. A. Brüggemann-Klein. Regular expressions into finite automata. *Theoretical Computer Science*, 120:197–213, 1993.
5. A. Brüggemann-Klein and D. Wood. One-unambiguous regular languages. *Information and Computation*, 140:229–253, 1998.
6. A. Brüggemann-Klein and D. Wood. Caterpillars: A context specification technique. *Markup Languages: Theory & Practice*, 2(1):81–106, 2000.
7. A. Brüggemann-Klein and D. Wood. On predictive parsing and extended context-free grammars, 2002. Proceedings of the International Conference CIAA 2002. To appear.
8. A. Brüggemann-Klein and D. Wood. On predictive parsing and extended context-free grammars: Algorithms and complexity results, 2002. Manuscript in preparation.
9. J. Clark, 1992. Source code for SGMLS. Available by anonymous ftp from ftp.uu.net and sgml1.ex.ac.uk.

10. D. J. Cohen and C. C. Gotlieb. A list structure form of grammars for syntactic analysis. *Computing Surveys*, 2:65–82, 1970.
11. D. Giammarresi and D. Wood. Transition diagram systems and normal form transformations. In *Proceedings of the Sixth Italian Conference on Theoretical Computer Science*, pages 359–370, Singapore, 1998. World Scientific Publishing Co. Pte. Ltd.
12. R. Heckmann. An efficient ELL(1)-parser generator. *Acta Informatica*, 23:127–148, 1986.
13. J. E. Hopcroft and J. D. Ullman. *Introduction to Automata Theory, Languages and Computation*. Addison-Wesley Series in Computer Science. Addison-Wesley Publishing Company, Reading, MA, 1979.
14. H. B. Hunt III, T. G. Szymanski, and J. D. Ullman. Operations on sparse relations. *Communications of the ACM*, 20:171–176, 1977.
15. P. Kilpeläinen and D. Wood. SGML and XML document grammars and exceptions. *Information and Computation*, 169:230–251, 2001.
16. W. R. LaLonde. Regular right part grammars and their parsers. *Communications of the ACM*, 20:731–741, 1977.
17. J. Lewi, K. de Vlaminck, E. Steegmans, and I. van Horebeek. *Software Development by LL(1) Syntax Description*. John Wiley & Sons, Chichester, UK, 1992.
18. H. Mössenböck. A generator for production quality compilers. In *Lecture Notes in Computer Science 471*, Berlin, 1990. Springer-Verlag. Proceedings of the Third International Workshop on Compiler-Compilers.
19. T. J. Parr and R. W. Quong. ANTRL: A predicated-LL(k) parser generator. *Software—Practice and Experience*, 25(7):789–810, 1995.
20. S. Sippu and E. Soisalon-Soininen. *Parsing Theory, Volume 1, Languages and Parsing, Volume 2, LL(k) and LR(k) Parsing,*. EATCS Monographs on Theoretical Computer Science. Springer-Verlag, Berlin, 1988.
21. P. D. Terry. *Compilers and Compiler Generators*. Out of print, available on the Web, 2000.
22. J. Warmer and S. Townsend. The implementation of the Amsterdam SGML parser. *Electronic Publishing, Origination, Dissemination, and Design*, 2:65–90, 1989.
23. J. Warmer and H. van Vliet. Processing SGML documents. *Electronic Publishing, Origination, Dissemination, and Design*, 4(1):3–26, March 1991.
24. R. Wilhelm and D. Maurer. *Compiler Design*. Addison-Wesley, Reading, MA, 1995.

Area and Perimeter Derivatives of a Union of Disks[*]

Ho-Lun Cheng and Herbert Edelsbrunner

Department of Computer Science, National University of Singapore, Singapore.
Department of Computer Science, Duke University, Durham,
and Raindrop Geomagic, Research Triangle Park, North Carolina.

Abstract. We give analytic inclusion-exclusion formulas for the area and perimeter derivatives of a union of finitely many disks in the plane.

Keywords. Disks, Voronoi diagram, alpha complex, perimeter, area, derivative.

1 Introduction

A finite collection of disks covers a portion of the plane, namely their union. Its size can be expressed by the area or the perimeter, which is the length of the boundary. We are interested in how the two measurements change as the disks vary. Specifically, we consider smooth variations of the centers and the radii and study the derivatives of the measurements.

We have two applications that motivate this study. One is topology optimization, which is an area in mechanical engineering [1; 2]. Recently, we began to work towards developing a computational representation of skin curves and surfaces [7] that could be used as changing shapes within a topology optimizing design cycle. Part of this work is the computation of derivatives. The results in this paper solve a subproblem of these computations in the two-dimensional case. The other motivating problem is the simulation of molecular motion in molecule dynamics [9]. The setting is in three-dimensional space, and the goal is to simulate the natural motion of biomolecules with the computer. The standard approach uses a force field and predicts changes in tiny steps based on Newton's second law of motion. The surface area and its derivative are important for incorporating hydrophobic effects into the calculation [4].

The main results of this paper are inclusion-exclusion formulas for the area and the perimeter derivatives of a finite set of disks. As it turns out, the area derivative is simpler to compute but the perimeter derivative is more interesting. The major difference between the two is that a rotational motion of one disk about another may have a non-zero contribution to the perimeter derivative while it has no contribution to the area derivative.

Outline. Section 2 introduces our approach to computing derivatives and states the results. Section 3 proves the result on the derivative of the perimeter. Section 4 proves the result on the derivative of the area. Section 5 concludes the paper.

[*] Research by both authors was partially supported by NSF under grant DMS-98-73945 and CCR-00-86013.

R. Klein et al. (Eds.): Comp. Sci. in Perspective (Ottmann Festschrift), LNCS 2598, pp. 88–97, 2003.

2   Approach and Results

In this section, we explain how we approach the problem of computing the derivatives of the area and the perimeter of a union of finitely many disks in the plane.

Derivatives. We need some notation and terminology from vector calculus to talk about derivatives. We refer to the booklet by Spivak [10] for an introduction to that topic. For a differentiable map $f : \mathbb{R}^m \to \mathbb{R}$, the derivative at a point $\mathbf{z} \in \mathbb{R}^m$ is a linear map $Df_{\mathbf{z}} : \mathbb{R}^m \to \mathbb{R}$. The geometric interpretation is as follows. The graph of $Df_{\mathbf{z}}$ is an m-dimensional linear subspace of $\mathbb{R}^{m+1}$. The translation that moves the origin to the point $(\mathbf{z}, f(\mathbf{z}))$ on the graph of f moves the subspace to the tangent hyperplane at that point. Being linear, $Df_{\mathbf{z}}$ can be written as the scalar product of the variable vector $\mathbf{t} \in \mathbb{R}^m$ with a fixed vector $\mathbf{u}_{\mathbf{z}} \in \mathbb{R}^m$ known as the *gradient* of f at $\mathbf{z}$: $Df_{\mathbf{z}}(\mathbf{t}) = \langle \mathbf{u}_{\mathbf{z}}, \mathbf{t} \rangle$. The *derivative* Df maps each $\mathbf{z} \in \mathbb{R}^m$ to $Df_{\mathbf{z}}$, or equivalently to the gradient $\mathbf{u}_{\mathbf{z}}$ of f at $\mathbf{z}$.

In this paper, we call points in $\mathbb{R}^m$ *states* and use them to represent finite sets of disks in $\mathbb{R}^2$. For $m = 3n$, the state $\mathbf{z}$ represents the set of disks $B_i = (z_i, r_i)$, for $0 \leq i \leq n - 1$, where $[\mathbf{z}_{3i+1}, \mathbf{z}_{3i+2}]^T = z_i$ is the center and $\mathbf{z}_{3i+3} = r_i$ is the radius of B_i. The perimeter and area of the union of disks are maps $P, A : \mathbb{R}^{3n} \to \mathbb{R}$. Their derivatives at a state $\mathbf{z} \in \mathbb{R}^{3n}$ are linear maps $DP_{\mathbf{z}}, DA_{\mathbf{z}} : \mathbb{R}^{3n} \to \mathbb{R}$, and the goal of this paper is to give a complete description of these derivatives.

Voronoi decomposition. A basic tool in our study of derivatives is the Voronoi diagram, which decomposes the union of disks into convex cells. To describe it, we define the *power distance* of a point $x \in \mathbb{R}^2$ from B_i as $\pi_i(x) = \|x - z_i\|^2 - r_i^2$. The disk thus contains all points with non-positive power distance, and the bounding circle consists of all points with zero power distance from B_i. The *bisector* of two disks is the line of points with equal power distance to both. Given a finite collection of disks, the *(weighted) Voronoi cell* of B_i in this collection is the set of points x for which B_i minimizes the power distance,

$$V_i = \{x \in \mathbb{R}^3 \mid \pi_i(x) \leq \pi_j(x), \ \forall j\}.$$

Each Voronoi cell is the intersection of finitely many closed half-spaces and thus a convex polygon. The cells cover the entire plane and have pairwise disjoint interiors. The *(weighted) Voronoi diagram* consists of all Voronoi cells, their edges and their vertices. If we restrict the diagram to within the union of disks, we get a decomposition into convex cells. Figure 1 shows such a decomposition overlayed with the same after a small motion of the four disks. For the purpose of this paper, we may assume the disks are in general position, which implies that each Voronoi vertex belongs to exactly three Voronoi cells. The *Delaunay triangulation* is dual to the Voronoi diagram. It is obtained by taking the disk centers as vertices and drawing an edge and triangle between two and three vertices whose corresponding Voronoi cells have a non-empty common intersection. The *dual complex K* of the disks is defined by the same rule, except that the non-empty common intersections are demanded of the Voronoi cells clipped to within their corresponding disks. For an example see Figure 1, which shows the dual complex

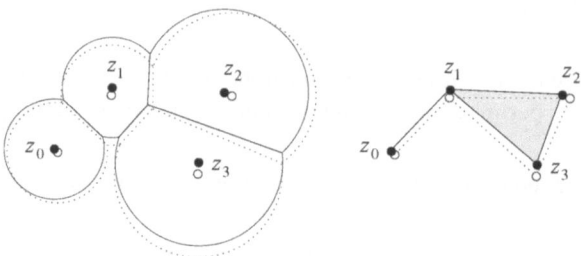

Figure 1: Two snapshots of a decomposed union of four disks to the left and the dual complex to the right.

of four disks. The notion of neighborhood is formalized by defining the *link* of a vertex $z_i \in K$ as the set of vertices and edge connected to z_i by edges and triangles,

$$\mathrm{Lk}\, z_i = \{z_j, z_j z_k \mid z_i z_j, z_i z_j z_k \in K\}.$$

Similarly, the link of an edge $z_i z_j$ is the set of vertices connected to the edge by triangles, $\mathrm{Lk}\, z_i z_j = \{z_k \mid z_i z_j z_k \in K\}$. Since K is embedded in the plane, an edge belongs to at most two triangles which implies that the link contains at most two vertices.

Measuring. We use fractions to express the size of various geometric entities in the Voronoi decomposition. For example, β_i is the fraction of B_i contained in its Voronoi cell and σ_i is the fraction of the circle bounding B_i that belongs to the boundary of the union. The area and perimeter of the union are therefore

$$A = \pi \sum r_i^2 \beta_i \quad \text{and} \quad P = 2\pi \sum r_i \sigma_i.$$

We refer to the intersection points between circles as *corners*. The corner to the left of the directed line from z_i to z_j is x_{ij} and the one to the right is $\bar{x}_{ji}$. Note that $x_{ji} = \bar{x}_{ij}$. We use $\sigma_{ij} \in \{0,1\}$ to indicate whether or not x_{ij} exists and lies on the boundary of the union. The total number of corners on the boundary is therefore $\sum \sigma_{ij}$. Finally, we define β_{ij} as the fraction of the chord $x_{ij} x_{ji}$ that belongs to the corresponding Voronoi edge.

Given the dual complex K of the disks, it is fairly straightforward to compute the β_i, σ_i, β_{ij}, and σ_{ij}. For example, $\sigma_{ij} = 1$ iff $z_i z_j$ is an edge in K and if $z_i z_j z_k$ is a triangle in K then z_k lies to the right of the directed line from z_i to z_j. We sketch inclusion-exclusion formulas for the remaining quantities. Proofs can be found in [6]. Define $B_i^j = \{x \in \mathbb{R}^2 \mid \pi_j(x) \le \pi_i(x) \le 0\}$, which is the portion of B_i of B_j's side of the bisector. Define $A_i^j = \mathrm{area}(B_i^j)$ and $A_i^{jk} = \mathrm{area}(B_i^j \cap B_i^k)$. Similarly, let P_i^j and P_i^{jk} be the lengths of the circle arcs in the boundaries of B_i^j and $B_i^j \cap B_i^k$. Then

$$\beta_i = 1 - \left(\sum_j A_i^j - \sum_{j,k} A_i^{jk} \right) / \pi r_i^2,$$

$$\sigma_i = 1 - \left(\sum_j P_i^j - \sum_{j,k} P_i^{jk} \right) / 2\pi r_i,$$

where in both equations the first sum ranges over all vertices $z_j \in \mathrm{Lk}z_i$ and the second ranges over all $z_j z_k \in \mathrm{Lk}z_i$ in K. Finally, let $c_{ij} = x_{ij}x_{ji}$ be the chord defined by B_i and B_j and define $c_{ij}^k = \{x \in \mathbb{R}^2 \mid \pi_k(x) \le \pi_i(x) = \pi_j(x) \le 0\}$, which is the portion of c_{ij} on B_k's side of the bisectors. Define $r_{ij} = \mathrm{length}(c_{ij})/2$ and $L_{ij}^k = \mathrm{length}(c_{ij}^k)$. Then

$$\beta_{ij} = 1 - \left(\sum_k L_{ij}^k \right) / 2r_{ij},$$

where the sum ranges are over all $z_k \in \mathrm{Lk}z_i z_j$ in K. The analytic formulas still required to compute the various areas and lengths can be found in [8], which also explains how the inclusion-exclusion formulas are implemented in the Alpha Shape software.

Motion. When we talk about a motion, we allow all $3n$ describing parameters to vary: each center can move in $\mathbb{R}^2$ and each radius can grow or shrink. When this happens, the union changes and so does the Voronoi diagram, as shown in Figure 1. In our approach to studying derivatives, we consider individual disks and look at how their Voronoi cells change. In other words, we keep a disk B_i fixed and study how the motion affects the portion of B_i that forms the cell in the clipped Voronoi diagram. This idea is illustrated in Figure 2. This approach suggests we understand the entire change as an accumulation

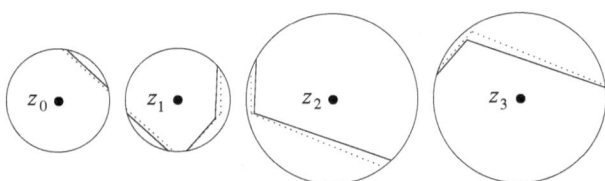

Figure 2: Two snapshots of each disk clipped to within its Voronoi cell. The clipped disks are the same as in Figure 1, except that they are superimposed with fixed center.

of changes that happen to individual clipped disks, and we understand the change of an individual clipped disk as the accumulation of changes caused by neighbors in the Voronoi diagram. A central step in proving our results will therefore be the detailed analysis of the derivative in the interaction of two disks.

Theorems. The first result of this paper is a complete description of the derivative of the perimeter of a union of disks. Let $\zeta_{ij} = \|z_i - z_j\|$ be the distance between two centers. We write $u_{ij} = \frac{z_i - z_j}{\|z_i - z_j\|}$ for the unit vector between the same centers and v_{ij} for u_{ij} rotated through an angle of 90 degrees. Note that $u_{ji} = -u_{ij}$ and $v_{ji} = -v_{ij}$.

PERIMETER DERIVATIVE THEOREM. The derivative of the perimeter of a union of n disks with state $\mathbf{z} \in \mathbb{R}^{3n}$ is $DP_{\mathbf{z}}(\mathbf{t}) = \langle \mathbf{p}, \mathbf{t} \rangle$, where

$$\begin{bmatrix} \mathbf{p}_{3i+1} \\ \mathbf{p}_{3i+2} \end{bmatrix} = \sum_{j \neq i} \left(p_{ij} \frac{\sigma_{ij} + \sigma_{ji}}{2} + q_{ij}(\sigma_{ij} - \sigma_{ji}) \right),$$

$$\mathbf{p}_{3i+3} = 2\pi\sigma_i + \sum_{j \neq i} r_{ij} \frac{\sigma_{ij} + \sigma_{ji}}{2},$$

$$p_{ij} = \frac{r_i + r_j}{r_{ij}} \left(1 - \frac{(r_i - r_j)^2}{\zeta_{ij}^2} \right) \cdot u_{ij},$$

$$q_{ij} = \frac{r_j - r_i}{\zeta_{ij}} \cdot v_{ij},$$

$$r_{ij} = \frac{1}{r_{ij}} \left(\frac{(r_i - r_j)^2}{\zeta_{ij}} - \zeta_{ij} \right).$$

If $z_i z_j$ is not an edge in K then $\sigma_{ij} = \sigma_{ji} = 0$. We can therefore limit the sums in the Perimeter Derivative Theorem to all z_j in the link of z_i. If the link of z_i in K is a full circle then the perimeter and its derivative vanish. This is clear also from the formula because $\sigma_i = 0$ and $\frac{\sigma_{ij} + \sigma_{ji}}{2} = \sigma_{ij} - \sigma_{ji} = 0$ for all j.

The second result is a complete description of the derivative of the area of a disk union.

AREA DERIVATIVE THEOREM. The derivative of the area of a union of n disks with state $\mathbf{z}$ is $DA_{\mathbf{z}}(\mathbf{t}) = \langle \mathbf{a}, \mathbf{t} \rangle$, where

$$\begin{bmatrix} \mathbf{a}_{3i+1} \\ \mathbf{a}_{3i+2} \end{bmatrix} = \sum_{j \neq i} a_{ij} \beta_{ij},$$

$$\mathbf{a}_{3i+3} = 2\pi r_i \sigma_i,$$

$$a_{ij} = 2 r_{ij} \cdot u_{ij}.$$

We can again limit the sum to all z_j in the link of z_i. If the link of z_i in K is a full circle then the area derivative vanishes. Indeed, $\mathbf{a}_{3i+3} = 0$ because $\sigma_i = 0$ and $\sum a_{ij} \beta_{ij} = 0$ because of the Minkowski theorem for convex polygons.

3   Perimeter Derivative

In this section, we prove the Perimeter Derivative Theorem stated in Section 2. We begin by introducing some notation, continue by analyzing the cases of two and of n disks, and conclude by investigating when the derivative is not continuous.

Notation. For the case of two disks, we use the notation shown in Figure 3. The two disks are $B_0 = (z_0, r_0)$ and $B_1 = (z_1, r_1)$. We assume that the two bounding circles intersect in two corners, x and $\bar{x}$. Let r_{01} be half the distance between the two corners. Then $\zeta_0 = \sqrt{r_0^2 - r_{01}^2}$ is the distance between z_0 and the bisector, and similarly, ζ_1 is the distance between z_1 and the bisector. If z_0 and z_1 lie on different sides of the bisector then

$\zeta = \zeta_0 + \zeta_1$ is the distance between the centers. We have $r_0^2 - r_1^2 = \zeta_0^2 - \zeta_1^2 = \zeta(\zeta_0 - \zeta_1)$ and therefore

$$\zeta_i = \frac{1}{2}\left(\zeta + \frac{r_i^2 - r_{1-i}^2}{\zeta}\right), \tag{1}$$

for $i = 0, 1$. If the two centers lie on the same side of the bisector, then $\zeta = \zeta_0 - \zeta_1$ is the distance between the centers. We have $r_0^2 - r_1^2 = \zeta_0^2 - \zeta_1^2 = \zeta(\zeta_0 + \zeta_1)$ and again Equation (1) for ζ_0 and ζ_1. Let θ_0 be the angle $\angle z_1 z_0 x$ at z_0, and similarly define $\theta_1 =$

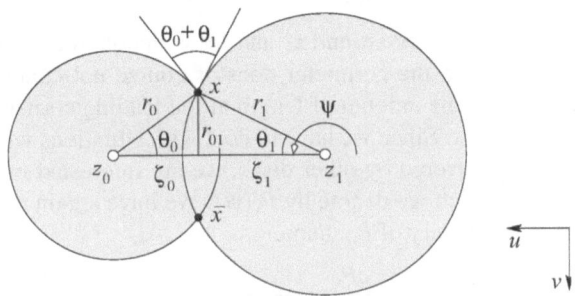

Figure 3: Two disks bounded by intersecting circles and the various lengths and angles they define.

$\angle z_0 z_1 \bar{x} = \angle x z_1 z_0$. Then

$$\theta_i = \arccos \frac{\zeta_i}{r_i}, \tag{2}$$

for $i = 0, 1$, and we note that $\theta_0 + \theta_1$ is the angle formed at the two corners. The contributions of each disk to the perimeter and the area of $B_0 \cup B_1$ are

$$P_i = 2(\pi - \theta_i)r_i, \tag{3}$$
$$A_i = (\pi - \theta_i)r_i^2 + r_{ij}\zeta_i, \tag{4}$$

for $i = 0, 1$. The perimeter of the union is $P = P_0 + P_1$, and the area is $A = A_0 + A_1$.

Motion. We study the derivative of P under motion by fixing z_1 and moving the other center along a smooth curve $\gamma(s)$, with $\gamma(0) = z_0$. At z_0 the velocity vector of the motion is $t = \frac{\partial \gamma}{\partial s}(0)$. Let $u = \frac{z_0 - z_1}{\|z_0 - z_1\|}$ and v be the unit vector obtained by rotating u through an angle of 90 degrees. We decompose the motion into a slope preserving and a distance preserving component, $t = \langle t, u \rangle u + \langle t, v \rangle v$. We compute the two partial derivatives with respect to the distance ζ and an angular motion. We use Equations (1), (2), and (3) for $i = 0$ to compute the derivative of P_0 with respect to the center distance,

$$\frac{\partial P_0}{\partial \zeta} = \frac{\partial P_0}{\partial \theta_0} \cdot \frac{\partial \theta_0}{\partial \zeta_0} \cdot \frac{\partial \zeta_0}{\partial \zeta}$$

$$= (-2r_0) \cdot \left(-\frac{1}{r_0 \sin \theta_0} \right) \cdot \left(\frac{1}{2} - \frac{r_0^2 - r_1^2}{2\zeta^2} \right)$$

$$= \frac{r_0}{r_{01}} \left(1 - \frac{r_0^2 - r_1^2}{\zeta^2} \right).$$

By symmetry, $\frac{\partial P_1}{\partial \zeta} = \frac{r_1}{r_{01}} (1 - \frac{r_1^2 - r_0^2}{\zeta^2})$. The derivative of P is the sum of the two derivatives, and therefore

$$\frac{\partial P}{\partial \zeta} = \frac{r_0 + r_1}{r_{01}} \left(1 - \frac{(r_0 - r_1)^2}{\zeta^2} \right). \tag{5}$$

To preserve distance we rotate z_0 around z_1 and let ψ denote the angle defined by the vector u. During the rotation the perimeter does of course not change. The reason is that we loose or gain the same amount of length at the leading corner, $\bar{x}$, as we gain or loose at the trailing corner, x. Since we have to deal with situations where one corner is exposed and the other is covered by other disks, we are interested in the derivative of the contribution near x, which we denote by $P_x(\psi)$. We have a gain on the boundary of B_1 minus a loss on the boundary of B_0, namely

$$\frac{\partial P_x}{\partial \psi} = \frac{r_1 - r_0}{\zeta}. \tag{6}$$

As mentioned above, the changes at the two corners cancel each other, or equivalently, $\frac{\partial P}{\partial \psi} = \frac{\partial P_x}{\partial \psi} + \frac{\partial P_{\bar{x}}}{\partial \psi} = \frac{r_1 - r_0}{\zeta} + \frac{r_0 - r_1}{\zeta} = 0$.

Growth. We grow or shrink the disk B_0 by changing its radius, r_0. Using Equations (1), (2), and (3) for $i = 0$ as before, we get

$$\frac{\partial P_0}{\partial r_0} = \frac{\partial P_0}{\partial \theta_0} \cdot \frac{\partial \theta_0}{\partial \zeta_0} \cdot \frac{\partial \zeta_0}{\partial r_0}$$

$$= \left(2(\pi - \theta_0) \frac{\partial \zeta_0}{\partial \theta_0} \frac{\zeta}{r_0} - 2r_0 \right) \cdot \frac{\partial \theta_0}{\partial \zeta_0} \cdot \frac{r_0}{\zeta}$$

$$= 2(\pi - \theta_0) + \frac{2(r_0^2 - \zeta \zeta_0)}{r_{01} \zeta}$$

because $\frac{\partial \theta_0}{\partial \zeta_0} = -\frac{1}{\sin \theta_0} \frac{r_0^2 - \zeta \zeta_0}{r_0^3}$ and $\frac{1}{\sin \theta_0} = \frac{r_0}{r_{01}}$. The computation of the derivative of P_1 is more straightforward because r_1 and ζ both remain constant as r_0 changes. Using Equations (1), (2), and (3) for $i = 1$, we get

$$\frac{\partial P_1}{\partial r_0} = \frac{\partial P_1}{\partial \theta_1} \cdot \frac{\partial \theta_1}{\partial \zeta_1} \cdot \frac{\partial \zeta_1}{\partial r_0} = (-2r_1) \cdot \left(-\frac{1}{r_1 \sin \theta_1} \right) \cdot \left(-\frac{r_0}{\zeta} \right) = -\frac{2r_0 r_1}{r_{01} \zeta}.$$

Note that $\frac{2r_0^2 - 2r_0 r_1}{\zeta} - \zeta_0$ is equal to $\frac{(r_0 - r_1)^2}{\zeta} - \zeta$. The derivative of the perimeter, which is the sum of the two derivatives, is therefore

$$\frac{\partial P}{\partial r_0} = 2(\pi - \theta_0) + \frac{1}{r_{01}} \left(\frac{(r_0 - r_1)^2}{\zeta} - \zeta \right). \tag{7}$$

The first term on the right in Equation (7) is the rate of growth if we scale the entire disk union. The second term accounts for the angle at which the two circles intersect. It is not difficult to show that this term is equal to $-2\cos(\theta_0 + \theta_1) - \frac{2}{\sin(\theta_0 + \theta_1)}$, which is geometrically the obvious dependence of the derivative on the angle between the two circles, as can be seen in Figure 3.

Assembly of relations. Let P be the perimeter of the union of disks B_i, for $0 \leq i \leq n-1$. By linearity, we can decompose the derivative along a curve with velocity vector $\mathbf{t} \in \mathbb{R}^{3n}$ into components. The i-th triplet of coordinates describes the change for B_i. The first two of the three coordinates give the velocity vector t_i of the center z_i. For each other disk B_j, we decompose that vector into a slope and a distance preserving component, $t_i = \langle t_i, u_{ij} \rangle u_{ij} + \langle t_i, v_{ij} \rangle v_{ij}$.

The derivative of the perimeter along the slope preserving direction is given by Equation (5). The length of the corresponding vector p_{ij} in the theorem is this derivative times the fractional number of boundary corners defined by B_i and B_j, which is $\frac{\sigma_{ij} + \sigma_{ji}}{2}$. The derivative along the distance preserving direction is given by Equation (6). The length of the corresponding vector q_{ij} in the theorem is that derivative times $\sigma_{ij} - \sigma_{ji}$, since we gain perimeter at the corner x_{ij} and loose at x_{ji} (or vice versa, if $\langle t_i, v_{ij} \rangle < 0$). The derivative with respect to the radius is given in Equation (7). The first term of that equation is the angle of B_i's contribution to the perimeter, which in the case of n disks is $2\pi\sigma_i$. The second term accounts for the angles at the two corners. It contributes to the derivative only for corners that belong to the boundary of the union. We thus multiply the corresponding term r_{ij} in the theorem by the fractional number of boundary corners. This completes the proof of the Perimeter Derivative Theorem.

4   Area Derivative

In this section, we prove the Area Derivative Theorem stated in Section 2. We use the same notation as in Section 3, which is illustrated in Figure 3.

Motion. As before we consider two disks $B_0 = (z_0, r_0)$ and $B_1 = (z_1, r_1)$, we keep z_1 fixed, and we move z_0 along a curve with velocity vector t at z_0. The unit vectors u and v are defined as before, and the motion is again decomposed into a slope and a distance preserving component, $t = \langle t, u \rangle u + \langle t, v \rangle v$. The distance preserving component does not change the area and has zero contribution to the area derivative. To compute the derivative with respect to the slope preserving motion, we use Equations (2) and (4) for $i = 0$ to get the derivative of A_0 with respect to ζ_0,

$$\frac{\partial A_0}{\partial \zeta_0} = -r_0^2 \frac{\partial \theta_0}{\partial \zeta_0} + \frac{\partial r_{01}}{\partial \zeta_0} \zeta_0 + r_{01} = \frac{r_0^2}{r_{01}} - \frac{\zeta_0^2}{r_{01}} + r_{01} = 2r_{01}.$$

Symmetrically, we get $\frac{\partial A_1}{\partial \zeta_1} = 2r_{01}$. The derivative of the area with respect to the distance between the centers is $\frac{\partial A_0}{\partial \zeta_0} \cdot \frac{\partial \zeta_0}{\partial \zeta} + \frac{\partial A_1}{\partial \zeta_1} \cdot \frac{\partial \zeta_1}{\partial \zeta}$, which is

$$\frac{\partial A}{\partial \zeta} = 2r_{01}, \tag{8}$$

because $\partial\zeta_0 + \partial\zeta_1 = \partial\zeta$. This result is obvious geometrically, because to the first order the area gained is the rectangle with width $\partial\zeta$ and height $2r_{01}$ obtained by thickening the portion of the separating Voronoi edge.

Growth. Using Equation (4) for $i = 0$ and 1, we get

$$
\frac{\partial A}{\partial r_0} = \frac{\partial A_0}{\partial r_0} + \frac{\partial A_1}{\partial r_0}
$$

$$
= 2(\pi - \theta_0)r_0 - \left(r_0^2 \frac{\partial\theta_0}{\partial r_0} + r_1^2 \frac{\partial\theta_1}{\partial r_0} \right)
$$

$$
+ \frac{\partial r_{01}}{\partial r_0}(\zeta_0 + \zeta_1) + r_{01}\left(\frac{\partial\zeta_0}{\partial r_0} + \frac{\partial\zeta_1}{\partial r_0} \right).
$$

The right hand side consists of four terms of which the fourth vanishes because $\partial\zeta_0 + \partial\zeta_1 = 0$. The third term equals $\frac{r_0}{r_{01}\zeta}\zeta_1\zeta$. The second term is $r_0^2(\frac{\zeta_0\zeta - r_0^2}{r_{01}r_0^2})\frac{r_0}{\zeta} + r_1^2\frac{r_0}{r_{01}\zeta}$. The second and third terms cancel each other because $r_0^2 - r_1^2 + \zeta_1\zeta - \zeta_0\zeta = 0$. Hence,

$$
\frac{\partial A}{\partial r_0} = 2(\pi - \theta_0)r_0. \tag{9}
$$

This equation is again obvious geometrically because to the first order the gained area is the fraction of the annulus of width ∂r_0 and length $P_0 = 2(\pi - \theta_0)r_0$ obtained by thickening the boundary arc contributed by B_0.

Assembly of relations. Let A be the area of the union of disks B_i, for $0 \le i \le n - 1$. We decompose the derivative into terms, as before. The derivative along the slope preserving direction is given by Equation (8). The length of the corresponding vector a_{ij} in the theorem is this derivative times the fractional chord length, which is β_{ij}. The derivative with respect to the radius is given by Equation (9). It is equal to the contribution of B_i to the perimeter, which in the case of n disks is $2\pi r_i \sigma_i$. This completes the proof of the Area Derivative Theorem.

5   Discussion

Consider a finite collection of disks in the plane. We call a motion that does not change radii and that at no time decreases the distance between any two centers a *continuous expansion*. The Area Derivative Theorem implies that the derivative along a continuous expansion is always non-negative. The area is therefore monotonously non-decreasing. This is not new and has been proved for general dimensions in 1998 by Csikós [5]. The more restricted version of this result for unit-disks in the plane has been known since 1968. Bollobás' proof uses the fact that for unit disks the perimeter is also monotonously non-decreasing along continuous expansions [3]. Perhaps surprisingly, this is not true if the disks in the collection have different radii. The critical term that spoils the monotonicity is contributed by the rotational motion of one disk about another. That contribution can be non-zero if exactly one of the two corners defined by the two circles belongs to the boundary of the union. Continuous expansions that decrease the perimeter are therefore possible, and one is shown in Figure 4.

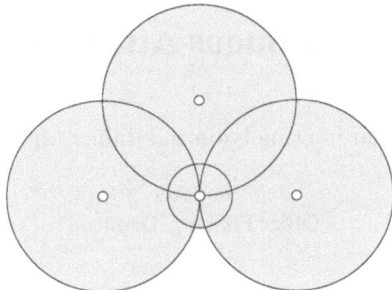

Figure 4: Moving the small disk vertically downward does not decrease any distances but does decrease the perimeter.

References

1. M. P. BENDSØE. *Optimization of Structural Topology, Shape, and Material.* Springer-Verlag, Berlin, Germany, 1995.
2. M. P. BENDSØE AND C. A. MOTA SOARES (EDS.) *Topology Design of Structures.* Kluewer, Dordrecht, The Netherlands, 1993.
3. B. BOLLOBÁS. Area of the union of disks. *Elem. Math.* **23** (1968), 60–61.
4. R. BRYANT, H. EDELSBRUNNER, P. KOEHL AND M. LEVITT. The area derivative of a space-filling diagram. Manuscript, 2001.
5. B. CSIKÓS. On the volume of the union of balls. *Discrete Comput. Geom.* **20** (1998), 449–461.
6. H. EDELSBRUNNER. The union of balls and its dual shape. *Discrete Comput. Geom.* **13** (1995), 415–440.
7. H. EDELSBRUNNER. Deformable smooth surface design. *Discrete Comput. Geom.* **21** (1999), 87–115.
8. H. EDELSBRUNNER AND P. FU. Measuring space filling diagrams and voids. Rept. UIUC-BI-MB-94-01, Beckman Inst., Univ. Illinois, Urbana, Illinois, 1994.
9. M. LEVITT AND A. WARSHEL. Computer simulation of protein folding. *Nature* **253** (1975), 694–698.
10. M. SPIVAK. *Calculus on Manifolds.* Addison-Wesley, Reading, Massachusetts, 1965.

A Time-Evolving Data Structure Scalable between Discrete and Continuous Attribute Modifications

Martin Danielsson and Rainer Müller

imc AG
Office Freiburg, Germany

Abstract. Time-evolving data structures deal with the temporal development of object sets describing in turn some kind of real-world phenomena. In the bitemporal case also objects having counterparts with an own predefined temporal component can be modelled. In this paper, we consider a subset of the problems usually covered by this context, having many real applications in which certain real-time constraints have to be met: synchronizability and random real-time access. We present a solution called the relational approach, which is based on a generalization of interval objects. By comparing this approach with the original simple transaction-based solution, we show its free scalability in the length of these interval objects, reducing the redundancy in data representation to a minimum.

1 Introduction and Motivation

Arbitrary complex contexts and relationships or application scenarios can be described exhaustively through a set of objects arranged in some kind of data structure. But data structures are forgettable or volatile in the sense that they usually do not describe more than a certain state or moment of a context. Most data structures include the current point of time forgetting the history, i.e., all modifications on the set of objects or on the objects themselves. However, the temporal evolvement of a set of objects can be useful or even necessary in certain scenarios in order to reconstruct or restore former states or versions, or even more, in order to replay or sift through what has happened with these objects in a certain period of time.

1.1 Application Scenarios

Real-life examples for such scenarios can obviously be very diverse. In order to provide a glimpse of the variety we mention a few selected ones: In human resource administration of a company's **ERP** system the employees, including all describing attributes such as name, address, salary, or SSN, are maintained as records in a database, or more generally, as objects in a data structure. However, a company's staff evolves through the years, new people are hired, others retire or leave the company. Employees get married, get children, move to a new house, get another position or a higher salary. For archiving purpose or, for instance, for tracking the salary development of a certain employee or the whole staff, it is obviously interesting and necessary to reconstruct or access "older objects" of an employee database.

R. Klein et al. (Eds.): Comp. Sci. in Perspective (Ottmann Festschrift), LNCS 2598, pp. 98–114, 2003.

Black boxes in aviation are another example. Today's **FDR**s (Flight Data Recorders) in aircraft cockpits record many different operating conditions during a flight by monitoring parameters such as altitude, airspeed and heading. In total, there are often more than several hundred different characteristics of the aircraft itself or of the surrounding flight conditions which can be stored by generating or by modifying the described objects in the FDR. With these objects retrieved from the FDR, a computer-animated video reconstruction of the flight can be generated. This animation and therewith the access and reconstruction of "older object versions" enable the investigating team to visualize, for example, the last moments of the flight before an accident.

In automotive design or construction several thousands of well-structured objects in a CAD system are often required to describe only parts of a car under development, such as the car's engine compartment. Tuning parameters in the CAD system during the design process usually affects numerous objects simultaneously. Therefore, it might be necessary for a designer to trace the development of a certain component through the whole process or to reconstruct older component versions due to failure in the current produced version or due to development policy changes.

Our last example considers presentations, talks or seminars held on the basis of electronically generated slides. These slides – usually produced and also presented with the help of common presentation systems – consist of several basic objects, such as text, simple geometric figures (rectangles, circles, polylines, etc.) or raster images. These objects have visual equivalents on the screen. They are modified along with the presentation, for instance, in order to become visible or to get moved. For a temporally asynchronous participation in such events or for learning purpose of participants, the access, reconstruction or more general: the replay of older object versions or version sets is required.

Data structures dealing with time-evolving data and providing time-dependent access to objects are usually subsumed under the category of **persistent**, **time-evolving** or **temporal** data structures. A more recent survey on this kind of data structure is provided by Salzberg & Tsotras[14]. Our interest and research in this area is considerably motivated by the last example mentioned above. The Chair of Thomas Ottmann at the Computer Science Institute of the University of Freiburg in Germany invested, beside the general engagement in the use of multimedia in education, a lot of time in how to realize the true system for presentation recording. The determination of what is true or reasonable from the user's point of view in such a system, has been distilled from several years of the chair's experience while investigating this area. Therefore, the goals of presentation recording from the teacher's and learner's perspective result in a few basic requirements that every system for presentation recording should reach. This can be summarized as follows: the recorded document should maintain – especially through the *conservation of the symbolic representation* of the presented data – the *quality and the live character* of the real event as authentic as possible while providing *flexible integration mechanisms* for the resulting document and *convenient navigation facilities* for the replay. This aspect is discussed in detail in [11].

Since 1995 the **Authoring on the Fly** (AOF) [10] system has been under development at Thomas Ottmann's Chair in Freiburg. AOF records audio, video, slides and corresponding graphical annotations in a highly automated way and produces documents in which the different streams are temporally synchronized. The AOF system has met the requirements, as described above, to a high extent [7]. An always important aspect in this discussion are arbitrary two- or three-dimensional animations which can visualize complex relationships or processes for learners. Since this task is always challenging – using only static learning material – the question how to integrate, present, annotate, and record animations in presentations has been a central part in the chair's research in presentation recording. A system for the production, presentation and recording of two-dimensional animations called JEDAS [8] has been realized and adopted to the AOF system [4]. As a general consequence the search for a suitable time-evolving data structure describing the objects on slides or in animations and their history has always been a key aspect in the research of both, presentation and animation recording. Several results from that research and the experience from their implementation in AOF or JEDAS have influenced and are still influencing the development of the commercial product Lecturnity[6] for presentation recording of our company *imc AG* in Germany.

In this paper we present a solution we call **relational approach** for the time-based access to time-evolving or temporal data structures or databases. In the following, the underlying problem of this relational approach will be described more precisely.

1.2   Problem Specification

The application for the problem described and covered in this paper are real-world scenarios in which *real-time* is a special issue in the sense that access, reconstruction, and maybe replay of object history should be realized with regard to real-time or life-time constraints. The FDR and the presentation recording cases above provide good examples for this category of real-world scenarios. In what follows, we specify the general characteristics of the time-evolving data describing some instances of our real-world scenarios. To complete the problem specification, we enumerate the basic requirements the data structure should meet for storing the time-evolving data. We therewith formalize the real-world scenario that underlies the problem considered in this paper.

We will now describe the problem stepwise according, to some degree, to common research domains and use the corresponding terminology. As in the domain of temporal, usually relational databases (e.g., [14]), we start with an initially empty set of objects we call the **object space**. These objects evolve over time where *time* is simply described through R^+. Objects generally consist of a time-invariant *key* and several time-variant *attributes*. Every object belongs to a *time interval*, in which it exists as part of the object space. According to the well-known *tuple-versioning model* [1], objects in our problem specification are therefore described through a tuple $(key, attributes, time interval)$. In contrast to this model and especially in contrast to the more elaborate *attribute-versioning model* [2] time intervals and time in general are not an integral part of the object itself and are only implicitly given. Our relational approach uses the time-invariant key as relation to provide objects with timing information. In [14] objects are called *alive* during their time interval whereas we call these objects **active**. In consequence, the object space at the point of time t, which are all active objects at this precise moment, is **the object space state of time** t. In the domain of temporal databases [14] this is often named logical database state of time t. Our objects can be grouped in *containers* and these containers in turn are objects themselves. The induced object hierarchy is completed by a *root container*, every object space must include at least one of them. Every object can only be included within one container. One of the mandatory attributes of an object, except for root containers, is the reference to its container. We therefore redefine an object as *active*, if and only if it is part of a container. A *record* in temporal databases (e.g., [13]) usually describes an object completely. In our problem specification an object is completely described by its attributes and its key. We call the attributes and the key the object's **properties**. Thus, all properties of all active objects at time *t* are sufficient to describe the object space state of time *t*.

We will now summarize what kind of modifications, or transactions in database terminology, on the object space are supported at which time and what is recorded. Objects can be added, can be deleted, and can be changed. Adding an object means to add it to a container of the object space. Deleting objects means to delete them from a container. Only root containers can directly be added to and deleted from the object space. Changing an object means to modify one of its attributes. Since membership of objects to a container is described through one of the object's attributes, changing is the only necessary modification operation on objects. As with object databases (e.g., [13]), objects in our problem specification are *compound* and *encapsulated*. Compound objects comprise attributes of arbitrary complex data types and not only those of simple data types. Objects encapsulate their describing data in the sense that they decide on their own which of their attributes are necessary to record and to reconstruct themselves. This implies that in our problem specification objects are able to deliver their own properties at every point of time and that they are also able to receive properties in order to reconstruct their older states. Thus, no instance must be able to understand or to interpret properties except the objects themselves.

Up to this point, our problem specification, to a certain degree, is similar to that usually investigated in the area of temporal databases or time-evolving data structures. We now introduce a special point in our problem specification in which a predefined process in our real-world scenario has to be described. So far, we have only considered object modifications which are static and unpredictable. In the temporal database context these objects are often called *plain objects* in contrast to *interval objects* having their own temporal component. An interval object combines a plain object with a time interval during which the object together with the specified attributes is active. A good example is a time-limited contract in the ERP scenario. Therefore, interval objects are continuous and, from the recording point of view, predictable. You can differentiate two notions of time: *transaction time* given by the static, unpredictable object modifications and *valid time* given by the continuous, predictable object modifications (cf. [14]). In order to describe the real-world scenarios covered in this paper with our problem specification, we extend the notion of interval objects to **transitions**. Transitions describe the modifications on one attribute of a certain object through a certain period of time. The form how these modifications are described within the transition is not fixed, as specified later in this paper. Examples for transitions can be commands in the FDR scenario passed to the aircraft's auto pilot system to describe, for instance, the aircraft's descending to a certain altitude in a certain time span. Another example might be moving geometric objects in a two- or three-dimensional animation, such as a ball flying through a room. A third example are text items of a text list on a slide continuously fading in item by item for presentation purposes. In the last example *color* might be the attribute of the text items modified by the transition over time. Obviously, transitions are a generalization of interval objects. In the animation context – one of the example scenarios of our problem specification – there is a conceptual intersection of our transitions with the ones used in the *path-transition paradigm*, first introduced by Stasko [15]. Here, transitions are often used to define the entities of an animation. In certain cases our transitions are equal to those used in this paradigm. In the database context, the problem underlying this paper belongs to the category of temporal databases covering *bitemporal time*, transaction time and valid time.

Requirements - Interesting Points The category of real-world scenarios considered in this paper reflects aspects of scenarios on a real-time basis. In these scenarios storing the history is always equivalent to recording. In most cases a reconstruction of older objects is motivated by a continuous real-time replay on the one hand and by a real-time access on the other. Both forms of reconstruction, real-time replay and real-time access, result from the fact that a recording of these scenarios is only motivated to enable an authentic replay of the reality, i.e., a replay in the same visual and auditive appearance from the user's point of view and especially in the same temporal flow as happened in reality. You can easily imagine why this is interesting regarding the FDR example. Due to the same constraints, real-time replay and access are required in the presentation recording scenario. [11] demonstrates in a more formal way how the requirements "real-time replay" and "real-time access" can be derived from the fundamental goals of presentation recording. Thus, real-time access and real-time replay are general requirements every reasonable system for presentation recording should meet.

To become more precise: For the following definition we assume that the only modification supported on the object space is adding and deleting objects. Even if an already active object is modified, a copy of this object with the modified attributes is added to the object space and the original one is deleted. This assumption results in the fact that the only modification made on active objects of the object space is adding new objects to and deleting active objects from containers. Thus, a new object space state is generated if and only if objects are added or deleted from active containers. We define the object space state directly derived from another state s as $succ(s)$. Per definition, no state exists between s and $succ(s)$ unequal to both states with regard to time. If we consider a sequence of object space states $(s_n)_{0 \leq n < k}$, we can define a function on this set of states that we call the **time base of** $(s_n)_n$: $\sigma : (s_n)_{0 \leq n < k} \longrightarrow R^+$. For every object space state $s \in (s_n)_n$, σ delivers the point of time $\sigma(s)$, when the state s is generated. We now define a sequence of object space states $(s_n)_n$ as **synchronizable** in relation to reality under the following conditions:

1. σ is strictly monotonic increasing
2. $s_n \neq s_{n+1}, \forall n : 0 \leq n \leq k-1$
3. $succ(s_n) = s_{n+1}, \forall n : 0 \leq n \leq k-1$

In other words, a synchronizable state sequence describes all modifications on the object space in an optimal granularity. Therefore, the number of states generated during the recording process in such a sequence is sufficient, but not larger than necessary to reconstruct what happened in reality. In summary: A synchronizable state sequence allows us to reconstruct exactly all events happening in a real-world scenario at the correct point of time with regard to real-time. The notion "synchronizable" is motivated by the fact that this feature enables a reconstruction and especially a replay of the object history temporally synchronized to other recordings at that period of time. Take, for instance, the audio recorded by the aircraft's voice recorder in the FDR example, or the lecturer's video captured in the presentation recording scenario.

If we consider a query or the access to a states sequence, the most important factor in our category of real-world scenarios is the absolute amount of time (in *ms*) required to find and reconstruct the queried state. A second requirement of our problem specification is therefore a very efficient access to arbitrary points of time in the states sequence, which we call **random real-time access**: $t_a(t) < \mu_{a,S} \in R_+ \, \forall t \in T$.

Here, $t_a(t)$ is the absolute time in *ms* required to find and reconstruct the state corresponding to time t. The threshold $\mu_{a,S}$ must be fixed depending on the considered real-world scenarios. For instance, regarding the replay in arbitrary speed (scrolling) of an animation used in a presentation or used to visualize the last moments of a flight, $20\,ms$ might be a good threshold for $\mu_{a,S}$ in order guarantee $25\,fps$ in a random sequence of access timestamps. Beside the scrolling in every speed through the history, this feature "random real-time access" enables the synchronization to arbitrary discrete time sources, i.e., time sources not delivering timestamps in an equidistant or monotonic increasing manner. Consider, for instance, the case where two different FDRs have to be evaluated in order to find the reason for the crash of two aircrafts. Finding the critical conditions for the crash might imply the synchronization of the states of one FDR to the states of the other with completely disjoint time bases.

This section can be summarized as follows: A solution for the problem specified above has to fulfil the following requirements: Every sequence of consecutive object space states described by the solution has to be

(a) Synchronizable, and has to provide (b) Random real-time access

In the following, we describe a simple approach to this problem, which, on the one hand, is a benchmark with regard to data volume and access time, since over years is has been the solution to record presentations used in the AOF system [11]. On the other hand, that simple solution is, under certain circumstances, an almost optimal solution for our problem.

1.3   Point of Origin: Event Queue and Object List

This approach to the problem specification covered in this paper consists of an initially empty list of objects and an initially empty queue of events. The key of the objects is implicitly given by the list identifiers they receive in the object list. The objects in the list are arranged according to their identifier in ascending order. The identifiers are assigned to the objects according to their time of insertion in the object space, i.e., every time a new object is added it is appended to the object list with the next consecutive identifier. Similar to the assumptions made in the last section, the modification of objects in the object space is realized by adding a copy with modified attributes and deleting the original object. Every time an object is added to or deleted from the object space, a new event is appended to the event queue. An *event* has the following format *(timestamp, number of active objects, list of active objects identifiers)*, where *timestamp* is the point of time when the event is generated. Thus, every event describes which objects are active in the time span between this event and the next event.

Obviously, the object space state sequence implicitly given by object list and event queue is synchronizable and provides random real-time access, since a binary search on the events is sufficient to reconstruct an object space state for a given point of time. Therefore, the combination of object list and event queue provides the easiest solution imaginable for the problem described in this paper. The drawback of this solution becomes evident when nearly continuous attribute modifications are described as in the case of animated objects. For instance, if an object with a visual equivalent is modified in only one of its attributes 25 times a second, all modifications result in a copy of this object and in a new event in the queue, listing all active objects, which is highly redundant.

In the following section, we will present the *relational approach* as our solution to the problem specified in this paper and describe how we take advantage of the inherent redundancy in the data structure above.

2   The Relational Approach

2.1   Continuous Transitions

The data structure has to cope with two different types of modifications of object attributes: *continuous* and *discrete changes* of objects. What a discrete change is, is clear. Furthermore, we want to formalize the notion of a continuous transition. A continuous change of an object makes use of a parameterized function which alters one or more attributes of an object for a specific time span in a continuous way. Mandatory parameters are the span of time in which the change is to occur (the *valid-time span* of the change), the start time of the change, and of course the object to affect (e.g., a function defining the throttle over time of an aircraft at lift-off, possible parameters may include the total weight of the aircraft and the wind strength and direction). Other parameters may be the type of attribute affected by the function (a specific function may be applied to different attributes of the same attribute type) or any other function modifier. Such a parameterized function with *fixed* parameter values is also called a *continuous object transition*. A transition is said to *end* or to be *closed* when the valid-time span of the transition has expired. Later on, we will also define a new transition type for discrete changes.

2.2   Object Space State Recording

In order to record the object space state over a specific time span in real time, we have to make clear what we need to keep track of. It is necessary to record both continuous and discrete changes. We will start with the first part, the problem for continuous transitions, and then reduce the problem of the discrete changes to objects to the continuous case.

Transition Recording. During recording, there is a *set of active transitions* at each point of time which has to be recorded in some way. In effect, this is a valid-time query for intervals. We will make the following assumptions: Each transition can be referred to by an ID, and the set of active transitions consists of such references. An easy way to record the set of active transitions is to simply make a snapshot of the IDs in the set each time the set changes. This obviously makes the reconstruction of the set of active transitions for a specific point of time very easy: it is a simple binary search. This solves the first part of our problem: reconstructing the set of active transitions.

However, even if only continuous and not discrete changes are allowed on objects, this is not enough to efficiently reconstruct the complete object space state, as the following example shows (see Fig. 1): During recording, T_1 is added at time t_1, T_2 at time t_2, T_1 ends at t_3, and T_2 ends at t_4. T_1 affects attribute A of an object o_1, and T_2 affects an attribute B of an object o_2. All other attributes of the two objects are not affected. We assume that all attribute values are set to predefined values before we try to reconstruct the object space state. Now let us take $t_0 \in (t_3, t_4)$. We can easily reconstruct the set of active transitions, which is $\{T_2\}$ here. T_2 would be applied to o_2 and thus o_2's attribute B would be updated correctly. However, the attribute A of o_1 would still remain unchanged to the predefined state. This is obviously not always correct: The attribute A must have the value defined by T_1 to time t_3.

A straight-forward solution to this problem would be to scan all transitions for end points prior to t_0, but this may be very time-consuming for a large t. Further, the access time for a reconstruction of the object space state is not the same for all t, and this is not desirable.

A more efficient solution is to make a snapshot of the object space state each time a transition ends (and thus is removed from the set of active transitions). Such an object space state snapshot consists of references to properties of all active objects. If an object's properties have not changed since the last snapshot, the last properties may (and should) be reused. If an object is being updated by a (continuous) transition and the only attributes that have changed since the last snapshot are the ones affected by the transition, the last object properties may also be reused: the attribute values would be overwritten at reconstruction anyway, as the transition already has to update the only attribute changed.

To make things more precise before returning to our example, we have to define some list-based data structures:

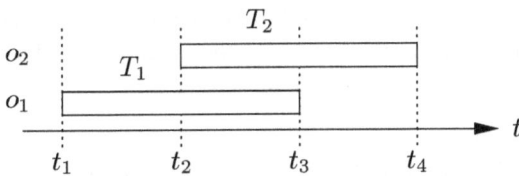

Figure 1. Example for transition recording

- **T-List.** The *transition list*, the *T-List*, contains all used transitions. All references to transitions imply a reference to this list.
- **P-List.** The *property list*, *P-List*, is a list of properties created by the objects during recording. A reference to properties always means a reference to an item in this list.
- **TAS.** The stream of snapshots of the set of active transitions is called the *transition access stream*, in short: *TAS*.
- **PAS.** Likewise, the stream into which snapshots of object properties are inserted is called the *properties access stream*. The first entry in the PAS is by definition a snapshot of object space state in its initial state.

Now let us return to the example. What happens during recording? The PAS and TAS are created on the fly and are empty at the beginning, i.e. before the recording starts. In the example, a transition T_1 is added immediately at the beginning of the recording, so the first entry into the TAS will be $\{T_1\}$ at t_1. By definition, the first entry into the PAS consists of the initial properties of the objects o_1 and o_2: $\{p_1^{(0)}, p_2^{(0)}\}$. As previously noted, only references to the T- and P-List are used in the TAS and PAS.

At time t_2, a second transition T_2 is added to the set of active transitions; this forces a new entry into the TAS: $\{T_1, T_2\}$. We do not need a PAS entry here: no transitions have ended yet. This changes at time t_3: T_1 ends. We now need both a new TAS entry ($\{T_2\}$) and a PAS entry. The object o_1 has changed and must create a new properties object $p_1^{(1)}$ to be referred to in the new PAS entry. Even though o_2 has also changed (through T_2), no new properties object is necessary: the only attributes that have changed are the ones being updated by the transition T_2. The PAS entry at t_3 thus looks like this: $\{p_1^{(1)}, p_2^{(0)}\}$. At time t_4 lastly, an empty TAS entry $\{\}$ is inserted. Now o_2 has changed and has to create a new properties object $p_2^{(1)}$. Since transition T_2 has ended, we also need a new PAS entry, $\{p_1^{(1)}, p_2^{(1)}\}$.

Correct reconstruction of the object space state is now straight-forward with the information collected during the recording: two binary searches give the correct TAS respectively PAS entries. First, all object properties in the PAS entry are applied to their respective objects. Then the transitions in the TAS entry are applied to their respective objects, updating the attribute values as appropriate.

Adding and Deleting Objects in the Object Space. Up to now, we have ignored how objects are added to and deleted from the object space. As mentioned in the introduction, an object is said to be active, i.e. part of the object space, if it belongs to a root container. When objects are added to the object space, simply a PAS entry is created. This PAS entry then contains properties for the newly added objects. Likewise, a PAS entry is forced when objects are deleted from the object space. This is the only case in which a PAS entry can occur without a TAS entry.

Recording Discrete Changes of Objects. We are still missing a solution for the second problem: how to record discrete changes in object attributes (except for changes of the active state, see above). In this part, we will show how to reduce this to the problem

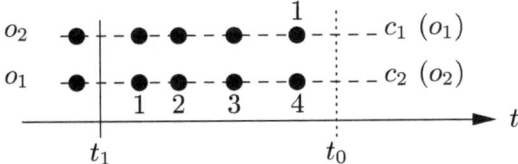

Figure 2. Example of complete transitions

solved above by introducing a new type of transitions: *complete transitions* (in contrast to *continuous* transitions).

A complete transition contains (time, attribute value) pairs for a specific attribute of a specific object. Thus, multiple complete transitions may exist simultaneously for a single object, but only one per attribute. We assume that an object knows about its complete transitions. If a discrete change is applied to an object at time t, the object checks for a previously existing complete transition of the changed attribute. If found, the new (time, attribute value) pair is added to a list of such pairs. If not found, a new complete transition of the requested type is created, the (time, value) pair is added, and then the newly created complete transition is inserted into the set of active transitions as described above. This obviously requires a TAS entry.

Types of Changes. Before stating the reconstruction algorithm for discrete object changes, we need to distinguish two types of attribute changes: *overwriting changes* and *modifying changes*. An overwriting change of an attribute value does not depend on the previous state of the changed attribute(s) (e.g., setting an employee's salary to $2000). In contrast, a modifying change alters an attribute's value on the basis of its previous value (e.g., giving an employee a $100 raise): in order to apply the change, the previous state has to be known.

During recording, this does not make any difference: the changes are simply collected in complete transitions of appropriate types. Modifying changes are collected in complete transitions *with history*, and overwriting changes are collected into complete transitions *without history*.

As noted before, complete transitions are treated quite as the continuous transitions, with the difference that complete transitions are only used for recording and not for describing purposes. When reconstructing the object space state for a time t_0, our algorithm returns a PAS and TAS entry (note that the PAS entry time stamp may differ from the TAS entry time stamp); but now complete transitions may also exist in the returned set of active transitions. Updating the object attributes is straight-forward for continuous transitions. Complete transitions need the time stamp of the PAS entry for attribute updating: all attribute changes prior to this time stamp need not (and must not) be applied to the object as they are already contained in the corresponding properties object. In Fig. 2 the c_1 line represents a complete transition without history (i.e., of overwriting changes) to object o_1. Assume we want to reconstruct the object space state to time t_0. As changes overwrite each other, we only need to find the time-value pair with the greatest time stamp smaller than or equal to t_0 and apply that change to the corresponding object o_1. In the second case, c_2 is a complete transition with history. We know that after the first step of the reconstruction algorithm, o_2 has the properties it had at time t_1 (they were contained in the PAS entry). We now only have to apply all the changes up to time t_0 in the correct order for reconstructing the correct object properties.

3   Analysis and Experimental Results

As we have seen, the reconstruction of an object space state for time t_0 is effected in a few different steps: (1) Finding the correct TAS and PAS entries in the corresponding streams, (2) applying the properties in the found PAS entry to the objects, and finally, (3) updating the objects according to the (continuous and complete) transitions in the TAS entry. We will now give a short worst-case run-time analysis for these steps.

Let us call the size of the TAS N_T, and let N_P be the size of the PAS. Step (1) thus takes $O(\log N_T + \log N_P)$ time.

The maximum number of objects over time in the object space is called N_o. If setting an object's properties is assumed to take a constant amount of time, step (2) can be done in $O(N_o)$ time.

In order to analyze step (3), we need to differentiate between the two transition types. We assume that updating an object through a continuous transition takes $O(1)$ time; this is reasonable as the update is a function evaluation and setting the object's attribute to the calculated value. If only continuous transitions have been used in the recording process, step (3) thus takes $O(M_T)$ time, M_T giving the maximum number of transitions used simultaneously during the recording process.

We now turn to the complete transitions. In the description of the recording process, it has not yet been defined how complete transitions end. Theoretically, they may contain an arbitrary number of (time, value) pairs. This is obviously not desirable in order to keep the access time for the object state reconstruction under a defined threshold. We therefore introduce a constant M_c which denotes the maximum number of (time, value) pairs in a complete transition. If the number of pairs has been reached in a complete transition during the recording, the transition is closed and removed from the set of active transitions. The next change of the same type thus induces the creation of a new complete transition of that type. Update time for a single complete transition without history is then $O(\log M_c + 1)$ (binary search plus an attribute update), and for a complete transition with history it is $O(\log M_c + M_c) = O(M_c)$. As M_c is a constant, we receive the following run-time analysis for the total reconstruction process:

$$O(\log N_T + \log N_P + N_o + M_T)$$

Thus, we expect the reconstruction process to be linear in the maximum number of objects and in the maximum number of simultaneous transitions (which is normally also limited in the number of objects). We do not expect the sizes of TAS or PAS to be important as they only affect the run-time logarithmically.

3.1   Synchronizability

The synchronizability (see section 1.2) of the TAS and the PAS is quite obvious, if we assume that both types of transitions are synchronizable: Entries are only added to TAS or PAS if changes to either the set of active transitions occur or if objects are added or deleted in the object space. As we have shown, the collected information also suffices to reconstruct the set of active transitions and the objects' properties for a specific point of time. What remains now is simply to show that the transitions, both complete and continuous, are synchronizable, too.

This can be seen easily in the case of the complete transitions. They contain (time, value) pairs, and the time stamps are the exact time stamps when the attribute change of the object occurred. So, the time base is evidently optimally granular (cf. section 1). The transitions do not contain any other information, but still enough to replay the changes to the objects at reconstruction of the object space state.

It cannot be shown directly that continuous transitions are synchronizable, though it is clear that they are, due to the following informal argument: A continuous transition is able to reconstruct a unique attribute value for a specific point of time, disregarding whether the object space state is currently being recorded or reconstructed. This guarantees that no information is lost during recording and no superfluous information is recorded. Combined with the well-known start time of the transitions which is recorded in the TAS, this shows that continuous transitions are synchronizable, too.

We may then conclude that the combination of TAS, PAS and both transition types are synchronizable and thus, our relational approach to object space state recording is synchronizable itself.

3.2   Experimental Results

The synchronizability of the data structure is a formal property, but the absolute *access time* of the data structure can only be calculated experimentally. We have seen in section 3 that the run-time for a single reconstruction of the object space state is linear in the maximum number of active objects and the maximum number of simultaneous transitions, but the run-time constants have not yet been estimated. For our animation recording example which required real-time access to the presentation recording for random visual scrolling [9], the access time has to be smaller than a given media threshold $\mu_{a,s}$ (see section 1.2).

With an example implementation of the data structure applied to animation recording (JEDAS [4]) we have made various experiments with differing transition setups and different amounts of objects. All tests ran over a time span of 5 Minutes (300 seconds). *Test 1* applied continuous transitions to two out of n objects. The transitions were 10 seconds long, and immediately after they ended, two new transitions were added to the set of active transitions so that the two objects were always being updated by transitions. All other $n - 2$ objects were not affected by any changes at all. *Test 2* is similar to Test 1 with the difference that the transition lengths are set to 500 ms and that the two transitions are not launched simultaneously: at t_0 the first transition is launched, then at $t_0 + 250$ ms, the second one is launched, and after one second the cycle restarts. This obviously boosts the number of TAS entries to four per second. The last test, *Test 3*, is an extension of Test 2: additionally, all other objects are updated by long, continuous background transitions for the complete time span of the test so that each TAS entry has $n, n - 1$ or $n - 2$ transition references.

These tests were made on a low-end 200 MHz computer and rendered the results presented in table 1. The tests all show the expected results: the average and maximum access times scale in all three examples linearly with the number of objects used in the test. The uncompressed bit rate values also show the expected results: test 1 has a very low bit rate which scales linearly with the number of objects. The obvious non-zero y interception is due to the fact that the number of transitions used during the recording is not affected by the number of objects. This phenomenon occurs in all the tests using a large number of small transitions on a fixed number of objects. Test 2 significantly enlarges the bit rate. This is due to the fact that TAS entries are inserted into the TAS four times per second. Test 3 lastly increases the bit rate once more. This is caused by the fact that here each TAS entry contains at least $n - 2$ references, whereas in test 2, one entry contains 2 references at most.

Test	1			2			3		
Stream									
Number of objects	25	50	100	25	50	100	25	50	100
Number of transitions	2	2	2	2	2	2	2+23	2+48	2+98
Transition length (ms)	10000	10000	10000	500	500	500	23:300000	48: 300000	98: 300000
							2: 500	2: 500	2: 500
Transition offsets (ms)	0	0	0	250	250	250	23: 0	48: 0	98: 0
							2: 250	2: 250	2: 250
Access time									
Average (ms)	6,1	11,1	20,1	6,2	11,3	20,0	7,6	14,1	23,5
Maximum (ms)	7,0	12,0	24,0	7,0	13,0	25,0	9,0	17,0	24,0
Bit rate uncompressed (kbit/s)	0,95	1,45	2,10	9,1	12,4	16,8	14,1	23,3	38,4
Bit rate compressed (kbit/s)	0,21	0,28	0,39	1,62	1,73	1,92	1,78	1,90	2,23

Table 1. Experimental results: access times and bit rates

By applying a Lempel-Ziv compression algorithm [5] to the streams, it was possible to lower the bit rates even more (see Tab. 1, compressed bit rates). The compression rates were very high, due to the following fact: In many cases, TAS and PAS entries are very similar (e.g., in test 3 the TAS entries always differ in exactly one reference only). This also shows that the TAS and PAS still contain redundancies in some cases.

The tests show that the relational approach fulfils the requirement of random real-time access defined in section 1, even for large numbers of objects (for presentation recording purposes) and on low-end computers.

4   Comparison to Event Queue and Object List

In this section we will show the relationship between the object list/event queue data structure (OL/EQ) and the relational approach we presented in section 2.

In contrast to the relational approach, OL/EQ does not directly support any continuous modifications to objects in the object space. The only changes allowed are discrete changes. This was one of the main aspects as the relational approach was developed. The only way to simulate continuous changes to objects with OL/EQ is to update the objects very frequently (to *discretize* a continuous transition). This obviously boosts the bit rate of such a recording, because to each discrete time a new object copy has to be created with the modified attribute. If we start from the other direction, with a long continuous transition in the relational approach, it is clear that you can split a transition into several short transitions, each one covering a change for a specific amount of time. Iterating this transition splitting leads to such small parts that each change only reflects the amount of time used in one step in the discretization of a transition in the OL/EQ approach. You may say that we have continuous transitions of length 0. Since the transitions which are added to the set of active transitions have valid-time spans of 0, they are immediately deleted again from that set. They update the objects just once. We thus have a still empty TAS entry, and a PAS entry (transitions have ended) which contains references to properties, and to new properties of those objects which have been affected by the transitions. Clearly, the PAS corresponds to the event queue in OL/EQ, and the P-List to the object list in this extreme situation. The two approaches correspond, and both produce a very large data volume.

The above shows a situation in which OL/EQ performs quite badly. Though, if we look at the area of presentation recording (without continuous changes to objects), the OL/EQ approach performs very well: if only static slides have to be displayed, the OL/EQ is almost optimal with regard to the problem constraints and requirements described in section 1.2. For each new slide the new objects are inserted into the object list, the corresponding IDs are added to the event queue (cf. Fig. 3) and the old objects are deleted from the event queue. Our relational approach shows a similar behavior: each time an object is added to or deleted from the object space, a PAS entry is created. We do not need any transitions or TAS entries, a PAS entry suffices. Obviously, here the two approaches are identical, and thus both nearly optimal in this case.

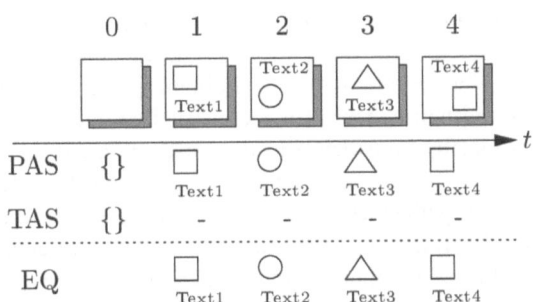

Figure 3. Presentation recording example: Comparing OL/EQ to the relational approach

5   Research Context

Almost all approaches dealing with similar problems as described in this paper fall into the category of temporal or time-evolving databases or persistent data structures (cf. [14] for a recent substantial survey). Compared to the relational approach these solutions have a completely different focus directed to the area of information systems and databases. Hence, real-time characteristics, such as providing a synchronizable stream or providing random real-time access, which is guaranteed by our approach, are, to that degree, no concern in this area and especially not covered by these solutions. Additionally, in the category of real-world scenarios underlying this paper, absolute access times decide whether a solution is appropriate or useless. A theoretical run-time analysis is definitely helpful to measure slight modifications on the same data structure but is not sufficient to assess a data structure as appropriate for our problem. In addition to the usual analysis style applied in the area of temporal databases, absolute response times measured in a well-chosen test environment are needed to estimate the real-time behavior of a data structure. The last aspect, reducing the set of comparable approaches, is the fact that our scenario belongs to the bitemporal case where objects have their own temporal component as real-world equivalent. On top of that, we need transitions as a generalized form of the interval objects for both, recording and representation purpose and this is not covered by the approaches of the bitemporal case. For example, we mention [3] as an early approach dealing with functions for time-varying attributes related in a broader sense to our transitions. [12] provides a good survey for the existing approaches in the bitemporal area. Partial persistence in data structures – see [6] as an example for the early forms – tries to overcome the same barriers from a graph-based point of view. However, here graph history is the first aim. Real-time modelling is not part of the persistence idea and the temporal access of these solutions is therefore mostly insufficient considered from the real-time perspective.

6   Conclusion

In this paper we introduced a solution called the relational approach for a problem specification belonging to the area of temporal databases or time-evolving data structures. In the real-world scenarios covered with this problem specification, the scenario's real-time behavior is an important concern for reconstruction and replay in addition to the bitemporal problems usually considered in this research area. In contrast to all other approaches our solution is directly focused on this real-time characteristic and supports a synchronizable replay together with a random real-time access to the data.

In addition to the theoretical run-time analysis the experimental results for access times show that our solution has enough achievement potential to deal with all real-time constraints ever valid in these scenarios, such as arbitrary forms of highly efficient access, temporally synchronized replay, as well as reconstruction runs through the stored data independent of direction and speed. We introduced different types of transitions as a generalization of interval objects used in the valid-time case of temporal databases. The evaluation of the memory requirements showed that these transitions play the central role in our relational approach regarding the produced data volumes. If continuous attribute modifications are described through arbitrary short transitions in the relational approach, structure and behavior of this approach are very similar to the combination of object list and event queue (OL/EQ). This modelling of transitions in the relational approach – artificially produced and not very probable – shows, on the one hand, the very close conceptual relationship of this new approach to the established OL/EQ. On the other hand, this highlights one of the strongest weaknesses of OL/EQ and how we took advantage of this fact in the relational approach, among other things, by means of transitions. In the case – more realistic than the artificial modelling mentioned above – in which we simply consider slides, displayed and removed without any further modifications, the relational approach is identical to OL/EQ. But, OL/EQ is nearly optimal for this special case it was originally designed for. Thus, the relational approach is obviously a generalization of an optimal, discrete solution and is freely scalable between discrete and continuous modifications. This motivates the title of our paper.

Future work lies, for instance, in removing the small redundancies in the access stream by using delta compression or a tree-based modelling. We also plan to describe object insertions and deletions themselves by transitions. Another goal is to apply an event-based approach while recording the reconstruction and replay itself in the relational approach.

References

1. J. A. Bubenko. The temporal dimension in information modelling. In *Technical Report RC-6187-26479, IBM T. J. Watson Research Center*, Nov. 1976.
2. J. Clifford and A. U. Tansel. On an algebra for historical relational database: Two views. In *Proc. of the ACM SIGMOD International Conference on Management of Data*, 1985.
3. J. Clifford and D. S. Warren. Formal Semantics for Time in Databases. *ACM Transactions on Database Systems*, 6(2):114–154, 1983.
4. M. Danielsson. Migration JEDAS-AOF. Master's thesis, Computer Science Institute, University of Freiburg, Germany, Jan. 2002.
5. P. Deutsch. Deflate compressed data format specification version 1.3. Request for Comment, RFC 1951, IETF, May 1996.
6. J. R. Driscoll, N. Sarnak, D. Sleator, and R. E. Tarjan. Making Data Structures Persistent. *Journal of Computer and System Science*, 38:86–124, 1989.
7. W. Hürst and R. Mueller. A Synchronization Model for Recorded Presentations and its Relevance for Information Retrieval. In *Proc. of ACM Multimedia '99*, Orlando, Fl., Oct. 1999.
8. T. Lauer, R. Mueller, and T. Ottmann. Animations for Teaching Purposes, Now and Tomorrow. *Special Issue on "Future of Computer Science Symposium (FOCSS)" of Journal of Universal Computer Science (J.UCS)*, 7(5), June 2001.
9. R. Mueller. *Wahlfreier Zugriff in Präsentationsaufzeichnungen am Beispiel integrierter Applikationen.* PhD thesis, Computer Science Institute, University of Freiburg, 2000.
10. R. Mueller and T. Ottmann. The "Authoring on the Fly" System for Automated Recording and Replay of (Tele)presentations. *Special Issue on "Multimedia Authoring and Presentation Techniques" of ACM/Springer Multimedia Systems Journal*, 8(3), May 2000.
11. R. Mueller and T. Ottmann. *Electronic Note-Taking*, chapter Technologies. Handbook on Information Technologies for Education and Training, Series: International Handbook on Information Systems, Editor: Adelsberger, H. H., Collis, B.,Pawlowski, J. M. Springer-Verlag, 2002.

12. G. Ozsoyoglu and R. T. Snodgrass. Temporal and real-time databases: A survey. *Knowledge and Data Engineering*, 7(4):513–532, 1995.
13. N. Pissinou, K. Makki, and Y. Yesha. On Temporal Modelling in the Context of Object Databases. *ACM Computing Surveys*, 22(3):8–15, Sept. 1993.
14. B. Salzberg and V. J. Tsotras. Comparison of Access Methods for Time-Evolving Data. *ACM Computing Surveys*, 31(2):158–221, June 1999.
15. J. Stasko. The Path-Transition Paradigm: A Practical Methodology for Adding Animation To Program Interfaces. *Journal of Visual Languages and Computing*, 3(1):213–236, Sept. 1990.

Fast Merging and Sorting on a Partitioned Optical Passive Stars Network

Amitava Datta and Subbiah Soundaralakshmi

Department of Computer Science & Software Engineering
University of Western Australia
Perth, WA 6009, Australia
{datta,laxmi}@csse.uwa.edu.au

Abstract. We present fast algorithms for merging and sorting of data on a multi-processor system connected through a *Partitioned Optical Passive Stars (POPS)* network. In a $POPS(d,g)$ network there are $n = dg$ processors and they are divided into g groups of d processors each. There is an optical passive star (OPS) coupler between every pair of groups. Each OPS coupler can receive an optical signal from any one of its source nodes and broadcast the signal to all the destination nodes. The time needed to perform this receive and broadcast is referred to as a *slot* and the complexity of an algorithm using the POPS network is measured in terms of number of slots it uses. Our sorting algorithm is more efficient compared to a simulated hypercube sorting algorithm on the POPS.

1 Introduction

The *Partitioned Optical Passive Stars (POPS)* network was proposed in [3; 7; 8; 10] as an optical interconnection network for a multiprocessor system. The POPS network uses multiple optical passive star (OPS) couplers to construct a flexible interconnection topology. Each OPS coupler can receive an optical signal from any one of its source nodes and broadcast the received signal to all of its destination nodes. The time needed to perform this receive and broadcast is referred to as a *slot*.

Berthomé and Ferreira [1] have shown that POPS networks can be modeled by directed stack-complete graphs with loops. This is used to obtain optimal embeddings of rings and de Bruijn graphs into POPS networks. Berthomé *et al.* [2] have also shown how to embed tori into POPS networks. Gravenstreter and Melhem [7] have demonstrated how to embed rings and tori into POPS networks.

Sahni [13] has shown simulations of hypercubes and mesh-connected computers using POPS networks. Sahni [13] has also presented algorithms for several fundamental operations like data sum, prefix sum, rank, adjacent sum, consecutive sum, concentrate, distribute and generalize. Though it is possible to solve these problems by using existing algorithms on hypercubes, the algorithms presented by Sahni [13] improve upon the complexities of the simulated hypercube algorithms. In another paper, Sahni [14] has presented fast algorithms for matrix multiplication, data permutations and BPC permutations on the POPS network. One of the main results in the paper by Sahni [13] is the simulation of an SIMD hypercube by a $POPS(d,g)$ network. Sahni [13] has shown

R. Klein et al. (Eds.): Comp. Sci. in Perspective (Ottmann Festschrift), LNCS 2598, pp. 115–127, 2003.

that an n processor $POPS(d,g)$ can simulate every move of an n processor SIMD hypercube using one slot when $d = 1$ and $2\lceil d/g \rceil$ slots when $d > 1$. It only makes sense to design specific algorithms for a $POPS(d,g)$ network for the case when $d > 1$. For the case $d = 1$, the POPS network is a completely connected network and it is easier to simulate the corresponding hypercube algorithm. Any algorithm designed for a $POPS(d,g)$ network should perform better than the corresponding simulated algorithm on the hypercube. Datta and Soundaralakshmi [5] have improved the algorithms for data sum and prefix sum for a POPS network with large group size, i.e., when $d > \sqrt{n}$. In this case, the number of groups in the $POPS(d,g)$ is small and consequently, there are more than $\sqrt{n}$ processors in each group. Datta and Soundaralakshmi [5] argue that this will be the case in most practical situations as the number of couplers will be much less than the number of processors. Other related work on the POPS network include an off-line permutation routing algorithm by Mei and Rizzi [11] and an on-line permutation routing algorithm by Datta and Soundaralakshmi [6] for a POPS network with large group size. Excellent review articles on the POPS networks are by Gravenstreter *et al.* [8], by Melhem *et al.* [10] and by Sahni [15].

In this paper, we present a fast algorithm for merging two sorted sequences on the POPS network. Then we use the merging algorithm to design a fast sorting algorithm on the POPS network.

The rest of this paper is organized as follows. In Section 2, we discuss some preliminaries. We present our merging algorithm in Section 3. We present the sorting algorithm in Section 4. Finally, we conclude with some remarks in Section 5.

2   Preliminaries

A $POPS(d,g)$ network partitions the n processors into g groups of size d each and optical passive stars (OPS) couplers are used to connect such a network. A coupler is used for connecting every pair of groups. Hence, overall g^2 couplers are needed. Each processor must have g optical transmitters, one transmitter each for transmitting to the g OPSs for which it is a source node. Further, each processors should have g optical receivers, for receiving data from each of the g couplers. Each OPS in a $POPS(d,g)$ network has degree d. In one slot, each OPS can receive a message from any one of its source nodes and broadcast the message to all of its destination nodes. However, in one slot, a processor can receive a message from only one of the OPSs for which it is a destination node. Melhem *et al.* [10] observe that faster all-to-all broadcasts can be implemented by allowing a processor to receive different messages from different OPSs in the same slot. However, in this paper, we assume that only one message can be received by a processor in one slot. A ten-processor computer connected via a $POPS(5,2)$ network is shown in Figure 1.

The g groups in a POPS network are numbered from 1 to g. A pair of groups is connected by an optical passive star (OPS) coupler. For coupler $c(i,j)$, the source nodes are the processors in group j and the destination nodes are the processors in group i, $1 \le i, j \le g$.

The most important advantage of a POPS network is that its diameter is 1. A message can be sent from processor i to processor j $(i \ne j)$ in one slot. To send a message

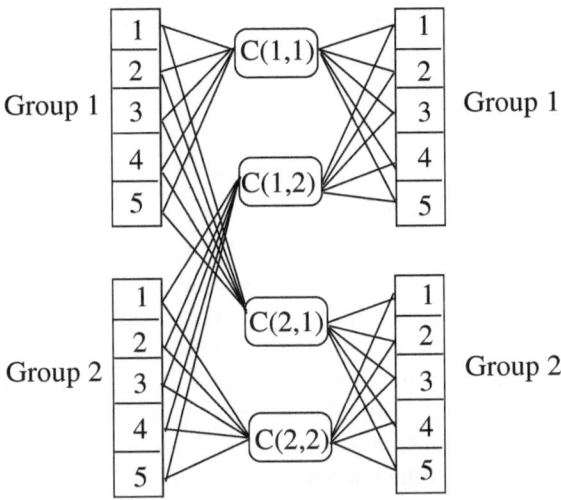

Figure 1. A 10-processor computer connected via a *POPS*(5,2) network. There are 2 groups of 5 processors each. Each group is shown twice to emphasize that it is both the source and the destination for some coupler.

from a processor in the i-th group to a processor in the j-th group, the processor in the i-th group uses the coupler $c(j,i)$, i.e., the coupler between the i-th group and the j-th group, with the i-th group as the source and the j-th group as the destination. The processor in the i-th group first sends the message to $c(j,i)$ and then $c(j,i)$ sends the message to the processor in the j-th group. Similarly, one-to-all broadcasts also can be implemented in one slot.

We refer to the following result obtained by Sahni [13].

Theorem 1. *[13] An n processor POPS(d,g) can simulate every move of an n processor SIMD hypercube using one slot when $d = 1$ and using $2\lceil d/g \rceil$ slots when $d > 1$.*

One consequence of Theorem 1 is that we can simulate an existing hypercube algorithm on a POPS network. Probably the most practical deterministic hypercube sorting algorithm is the bitonic sorting algorithm described by Ranka and Sahni [12]. This algorithm runs in $O(\log^2 n)$ time on a hypercube with n processors. There is a more efficient algorithm by Cypher and Plaxton [4; 9] with running time $O(\log n \log \log n)$ on an n processors hypercube. However, the algorithm by Cypher and Plaxton [4] is extremely complex and has a very large hidden constant. In most cases, the $O(\log^2 n)$ algorithm is implemented for its simplicity and ease of implementation. In this paper, we use the $O(\log^2 n)$ time algorithm given by Ranka and Sahni [12]. We immediately get the following theorem as a consequence of the hypercube simulation algorithm on a POPS network as given in Theorem 1.

Theorem 2. *It is possible to design a sorting algorithm on an n processor POPS(d,g) network that takes $O(2\lceil d/g \rceil \log^2 n)$ slots.*

In this paper, we design a custom algorithm for sorting on a $POPS(d,g)$ network and use the complexity in Theorem 2 as a benchmark to compare the performance of our custom sorting algorithm. First, we start with a merging algorithm in the next section.

3   Merging Using a POPS Network

The task of the merging algorithm is to merge two sorted sequences A and B of length K each into a merged sorted sequence of length $2K$. We assume that the sequences A and B are given initially in a $POPS(d,g)$ with $dg = 2K$ processors. Each processor holds one element. For simplicity, we assume that d divides K, in other words, all the elements of a sequence of length K can be stored in the processors of $\frac{K}{d}$ groups, one element per processor. We use the notation G_i to indicate the i-th group in $POPS(d,g)$.

We denote the elements of the sequence A by $a_1, a_2, \ldots, a_K$. Similarly, the elements of the sequence B are denoted by $b_1, b_2, \ldots, b_K$. For simplicity, we assume that all elements in the sequences A and B are distinct. However, our algorithm can be extended easily for the case when there are repeated elements. We denote by $rank(a_i : A)$, the rank of the element $a_i \in A$ in the sequence A. In other words, $rank(a_i : A)$ denotes the position of the element a_i in the sequence A. By $rank(a_i : B)$, we denote the number of elements in B that are less than a_i. The rank of the element a_i in the merged sequence $A \cup B$ is denoted by $rank(a_i : A \cup B)$. It is easy to see that $rank(a_i : A \cup B) = rank(a_i : A) + rank(a_i : B)$. Once we know $rank(a_i : A \cup B)$ for each element $a_i \in A$ and $rank(b_i : A \cup B)$ for each element $b_i \in B$, we know their ranks in the merged sequence and hence, we have merged the two sequences A and B. Since both A and B are sorted sequences, $rank(a_i : A)$ and $rank(b_i : B)$ for each a_i and b_i are already known. Hence, our main task is to compute $rank(a_i : B)$ and $rank(b_i : A)$. We discuss below how to compute $rank(b_i : A)$ for each element $b_i \in B$. The other computation is similar. We denote by $rank(B : A)$ the ranking of every element of B in A.

Our algorithm for computing $rank(b_i : A)$, for $1 \le i \le K$ is based on a parallel binary search strategy. In each step of the binary search, we determine the ranks of some of the elements of the sequence B in the sequence A. There are two phases in this binary search. In the first phase, we compute the ranks of the last elements from each group of the POPS that holds the sequence B. Once the ranks of all the last elements are known, we compute the rank of the other elements in the group. Recall that, K/d is the number of groups of the POPS holding elements from B. In the following discussion, by A (resp. B), we mean both the sequence A (resp. B) as well as the subset of processors holding the sequence A (resp. B). The meaning will be clear from the context.

3.1   First Phase

In the first phase of the binary search, we start with the last element of group number $\lfloor K/2d \rfloor$. We denote this element as $B_{\lfloor K/2d \rfloor}$. We broadcast $B_{\lfloor K/2d \rfloor}$ to all the processors in A. We assume that each processor in A knows the element stored in its neighbor. Hence, one processor (say, p) in A finds that $B_{\lfloor K/2d \rfloor}$ is in between its own and its

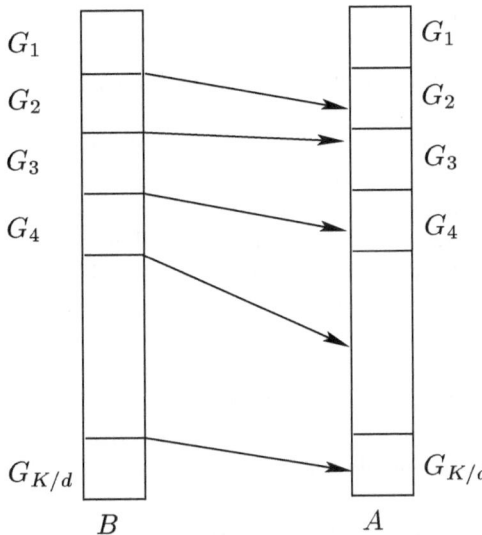

Figure 2. Illustration for the search in Section 3.1. Each of the sequences A and B has $\frac{K}{d}$ groups. The arrows indicate the ranks in B for the last element from each group in A.

neighbor's elements. Hence, $rank(B_{\lfloor K/2d \rfloor} : A)$ is p, the serial number of the processor p.

After this computation , the sequence B has two subsets of groups, groups 1 to $\lfloor K/2d \rfloor - 1$ and groups $\lfloor K/2d \rfloor + 1$ to K/d. Since both the sequences A and B are sorted, last elements of all the groups 1 to $\lfloor K/2d \rfloor - 1$ will be ranked among the elements of A from a_1 to $rank(B_{\lfloor K/2d \rfloor} : A)$. Similarly, the last elements of all the groups $\lfloor K/2d \rfloor + 1$ to K/d will be ranked among the elements $rank(B_{\lfloor K/2d \rfloor + 1} : A)$ to a_K. Hence, we have decomposed the ranking problem into two independent subproblems which can be solved in parallel. We do the binary search recursively in these two parts of the sequences A and B.

At the end of this parallel binary search, we have determined the ranks of the last elements from all the K/d groups. The number of elements ranked increases by a factor of 2 at every step. Each step requires two slots, one slot for the broadcast discussed above and one slot for getting back the rank in the originating processor. Hence, the total number of slots required is $2 \log(K/d) = 2 \log g$. We broadcast only one element from each group in this phase. Hence, we use one of the couplers associated with each group and there is no coupler conflict.

We consider the details of the second phase in two parts. In the next section we discuss the case when $g \geq d$, in other words, when $g \geq \sqrt{K}$ and $d \leq \sqrt{K}$ i.e., when the number of groups is larger than the number of processors in each group.

3.2   Second Phase when $g \geq d$

In the second phase, we rank all the elements within each group. We denote the K/d groups holding the elements from B as $G_i, 1 \leq i \leq K/d$ and the j-th element in G_i as $G_{i,j}, 1 \leq j \leq d$. Consider the k-th and $(k+1)$-th groups G_k and $G_{k+1}, 1 \leq k \leq (K/d-1)$. We know $rank(G_{k,d} : A)$ and $rank(G_{k+1,d} : A)$ from the computation in the first phase. All the elements from $G_{k+1,1}$ to $G_{k+1,d-1}$ must be ranked among the elements $rank(G_{k,d} : A)$ to $rank(G_{k+1,d} : A)$. Note that, these are the elements in G_{k+1}. We do a binary search to do this ranking. The binary search starts with ranking the element $G_{k+1,d/2}$ among the elements $rank(G_{k,d} : A)$ to $rank(G_{k+1,d} : A)$ by broadcasting $G_{k+1,d/2}$ to these processors in one slot.

The first task is to inform $G_{k+1,d/2}$, the indices of the two processors holding $rank(G_{k,d} : A)$ and $rank(G_{k+1,d} : A)$. Note that $G_{k,d}$ knows $rank(G_{k,d} : A)$ and $G_{k+1,d}$ knows $rank(G_{k+1,d} : A)$. Hence, $G_{k,d}$ and $G_{k+1,d}$ can broadcast $rank(G_{k,d} : A)$ and $rank(G_{k+1,d} : A)$ one after another among the processors in G_{k+1}. This computation takes two slots.

Since $g \geq d$, d or more couplers are associated with G_{k+1}. These couplers are $c(1, k+1), c(2, k+1), \ldots, c(g, k+1)$. Recall that the second (resp. first) index of the coupler gives the source (resp. destination) group index. The elements between $rank(G_{k,d} : A)$ and $rank(G_{k+1,d} : A)$ may be distributed in several groups and we denote these groups by G_{dest}. We use the following 2-slot routing to broadcast $G_{k+1,d/2}$ to all the processors in between $rank(G_{k,d} : A)$ and $rank(G_{k+1,d} : A)$.

$$G_{k+1,d/2} \rightarrow c(d/2, k+1) \rightarrow G_{d/2, k+1} \rightarrow c(dest, d/2) \rightarrow G_{dest}$$

Note that the notation $c(dest, d/2)$ indicates all the couplers that connect $G_{d/2}$ with the groups in G_{dest}. Once we know $rank(G_{k+1,d/2} : A)$, we have two independent subproblems as before. We continue this process and at each step, the number of elements to be ranked is double that of the previous step. The ranking stops when all the elements in the group are ranked in the sequence A. Unlike the first phase, we need to broadcast multiple elements from the same group and we need to avoid coupler conflicts. It is easy to see that there is no coupler conflict, since each element of a group has a coupler associated with it due to the fact that $g \geq \sqrt{K}$. Hence, all the broadcasts are done by different couplers. Further, no processor receives more than one element since each broadcast has a clearly separate group of destination processors. Hence, all the subproblems in each step of the binary search can be done in parallel without any coupler conflict and no processor receives more than one element. Each step requires 4 slots and the binary search terminates after $\log d$ steps, since there are d processors in each group. Hence, the total number of slots required is $4 \log d$.

The total number of slots required for phases 1 and 2 are $2 \log g + 4 \log d$. Note that we have only discussed the ranking of B in A, i.e., $rank(B : A)$. However, we have to compute $rank(A : B)$ as well for the sorting and this takes another $2 \log g + 4 \log d$ slots. Finally, the elements has to be sent to the correct processors according to their ranks. Since each group has $g \geq d$ couplers associated with it, we can send all the d elements in a group to their final destinations in 1 slot by using all these couplers. Hence, the overall slot requirement is $4 \log g + 8 \log d + 1$.

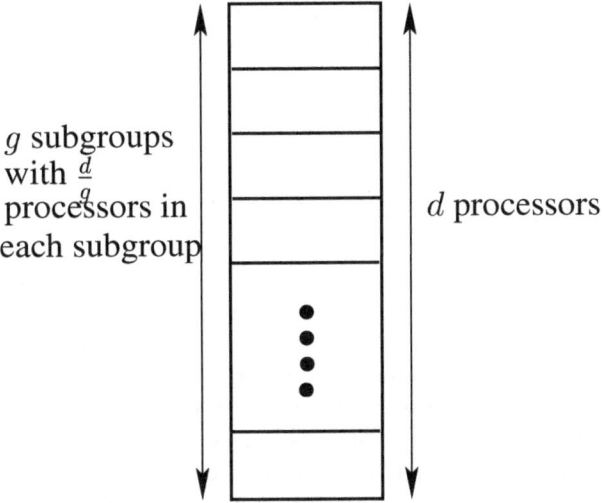

Figure 3. Illustration for Section 3.3. Each group of d processors is divided into g subgroups with $\frac{d}{g}$ processors in each subgroup.

3.3 Second Phase when $g < d$

In this case, the number of groups $g = \frac{K}{d}$ is less than the number of processors in each group. For the search in the second phase, the strategy in Section 3.2 does not work directly since the number of couplers associated with each group is smaller than the number of processors in the group. Moreover, for a processor $G_{i,j}$, the processor $G_{j,i}$ may not exist if $j > \frac{K}{d}$. Instead, we perform the search in the second phase in the following way.

We divide the d processors in G_i into g subgroups, each with $\frac{d}{g}$ processors. We have g couplers associated with G_i and there are g subgroups. Hence, the situation is similar to that in Section 3.2 and we can do a binary search with the first element from each subgroup. Once this binary search is complete in $2\log g$ slots, we have ranked all the first elements from each subgroup. However, the elements inside each subgroup are unranked. Since we have one coupler for each subgroup, we can sequentially rank all the $\frac{d}{g}$ elements in a subgroup using this coupler. The ranking of each element takes 4 slots, 2 slots to broadcast the element to all the processors in the subgroup and 2 slots to get back the rank to the processor holding the element. Hence, all the elements can be ranked in $\frac{4d}{g}$ slots. This ranking is done in parallel in all the subgroups.

For this case, The total slots required in phases 1 and 2, for computing $rank(A : B)$ and $rank(B : A)$, is $6\log g + 8\frac{d}{g}$. Since each group has d elements and g couplers associated with it, all the elements can be sent to their final processors in sorted order in $\frac{d}{g}$ slots and this can be done in parallel in all the groups. Hence, the total slot requirement for this case is $6\log g + \frac{9d}{g}$.

We can summarize the complexities of our merging algorithm in the following lemma.

Lemma 1. *Two sorted sequences of length K each can be merged into a single sorted sequence on a POPS(d,g), $K = dg$ with the following complexities :*

- *in $2\log g + 8\log d + 1$ slots if $g \geq d$*
- *in $6\log g + \frac{9d}{g}$ slots if $g < d$.*

4   Sorting Using a POPS Network

Our sorting algorithm is based on the sequential merge-sort algorithm. The merging algorithm in the previous section is the key tool used for the sorting algorithm. We assume that there are n numbers to be sorted on a $POPS(d,g)$ network, $dg = n$. Each processor initially has one element. The output of the algorithm is an ascending order of these n elements, one element per processor. The sequential merge-sort algorithm starts with n unsorted elements and builds sorted subsets of elements by pairwise merging smaller sorted subsets. In the first step, only pairs of single elements are merged. The complete sorted set is produced after $\log n$ such merging steps. In a $POPS(d,g)$, we cannot use this approach directly due to coupler restrictions. Instead, we start with sorted groups as our initial sorted subsets. The pairwise merging starts with merging a pairs of groups each of which is sorted. At the end of the merging, we get the complete sorted set. Hence, to apply our merging algorithm from the previous section, we first need to sort the elements in each group of the $POPS(d,g)$. This sorting is done differently for the cases when $d \leq \sqrt{n}$ and $d > \sqrt{n}$.

4.1   Sorting Each Group

Case 1. $d \leq \sqrt{n}$

First, we consider the case when $d = \sqrt{n}$. The number of groups $g = d = \sqrt{n}$ in this case and there are g^2 couplers. The $POPS(d,g)$ is a $POPS(d,d)$ in this case.

Consider the i-th group G_i and the d processors $G_{i,j}, 1 \leq i \leq g, 1 \leq j \leq d$. Our aim is to sort in ascending order the elements held by these d processors. We use a merge-sort technique based on computing cross-ranks between two sorted sequences for merging them. Our strategy is the following. We start with merging pairs of elements. In general, we merge pairs of sequences P and Q, each of which is sorted and has k elements, $1 \leq 2k \leq d$. We rank each element of P in the sequence Q through a strategy similar to binary search, however, we conduct multiple searches in parallel. Note that all the elements of P and Q are in the same group G_i and $2k \leq d$.

Consider the ranking $rank(P : Q)$ i.e., the ranking of all the elements of P in Q. By P_i, we denote the i-th element of $P, 1 \leq i \leq k$. We start our ranking by computing $rank(P_{\lceil \frac{k}{2} \rceil} : Q)$, i.e., by ranking the middle element of P in Q. This ranking is done by broadcasting $P_{\lceil \frac{k}{2} \rceil}$ to all the processors holding elements of Q. Assume that the processor $G_{i,j}$ holds the element $P_{\lceil \frac{k}{2} \rceil}$ and the elements of the sequence Q are in the processors $G_{i,r}, \ldots, G_{i,s}$. Then the data movement for this broadcast will be :

$$G_{i,j} \rightarrow c(j,i) \rightarrow G_{j,i} \rightarrow c(i,j) \rightarrow G_{i,k}, k = r \ldots s$$

This data movement takes 2 slots. After this computation, the $rank(P_{\lceil \frac{k}{2} \rceil} : Q)$ is determined by the processors $G_{i,r}, \ldots, G_{i,s}$. Let $P_{\lceil \frac{k}{2} \rceil}$ be in between two elements $G_{i,x}$ and $G_{i,x+1}$ in G_i. In that case, all the elements $P_1, \ldots P_{\lceil \frac{k}{2} \rceil}$ in P will be ranked among the elements $G_{i,r}, \ldots, G_{i,x}$ and all the elements $P_{\lceil \frac{k}{2} \rceil} + 1, \ldots, P_k$ will be ranked among the elements $G_{i,x+1}, \ldots, G_{i,s}$. Hence, as a result of the binary search, we have ranked one element of P in Q and decomposed the ranking problem into two equal size subproblems each of which is half the size of the original problem. We now do the ranking for these two subproblems in parallel in the next step. This will rank two elements from P into Q and produce 4 equal size ranking problems. Hence, in general, the number of ranked elements doubles at every step. Thus all the elements in P will be ranked in Q within $\log k$ steps. The total slot requirement is $4 \log k$ as each step requires 4 slots. This analysis is similar to the analysis in Section 3.2.

Recall that for sorting $P \cup Q$, we need to determine $rank(Q : P)$. Hence, we need additional $4 \log k$ slots and overall $8 \log k$ slots. Once the ranking is known, each element has to be sent to the appropriate processor in G_i according to the rank of the element. This can be done in one slot as we show below.

First, we need to show that we have enough couplers to do all the broadcasts necessary for this ranking. Note that we are considering the case when $d = g = \sqrt{n}$. Hence, there are g couplers connected to G_i. When we need to broadcast an element $G_{i,j}, 1 \leq j \leq d$, we have a unique coupler $c(j,i)$ connecting G_i to G_j and another unique coupler $c(i,j)$ connecting G_j to G_i. Hence, all broadcasts necessary for the binary search can be done in parallel. Since there are $d = g$ processors in G_i, we will need at most g couplers for the broadcasts. The elements can be sent to the correct processors according to their ranks in 1 slot since there are at most g elements and we have g couplers.

We utilize this generic merging in the following way. We start with pairwise merging of adjacent elements, then pairwise merging of 2-element groups, 4-element groups etc. In other words, at every merging stage, each group of sorted elements is twice the size of the groups in the previous stage. Hence, we need $\log d$ stages of merging and each stage requires $8 \log d + 1$ slots, as we have explained above. Thus, overall we need $8 \log^2 d + \log d$ slots to sort all the elements in a group. We do this sorting in all the groups in parallel, as we have g groups. Each group has $d = g$ processors and g^2 couplers.

A similar approach can be used for the case when $d < \sqrt{n}$. Note that in this case $g > \sqrt{n}$ and $g^2 > n$, the number of couplers is more than the number of processors. Hence, the above approach can be used easily for sorting each group.

Case 2. $d > \sqrt{n}$

In this case, each group G_i contains $d > \sqrt{n}$ processors and each group is connected to $g < \sqrt{n}$ couplers. Hence, we cannot use the strategy of Case 1 as there are not enough couplers to support the parallel binary search. Instead, we divide the processors in G_i into g subgroups of $\frac{d}{g}$ processors in each subgroup. We represent these g subgroups by

$K_1, K_2, \ldots, K_g$. Note that, since G_i has g couplers associated with it, we can associate a coupler $c(j,i)$ with each such subgroup $K_j, 1 \leq j \leq g$.

We sort the elements in each of the g groups by simulating the hypercube sorting algorithm in the following way. First, we consider the sorting of the elements in a group G_i. We distribute the elements in the g subgroups K_g among the g groups of the overall $POPS(d,g)$. Consider a subgroup $K_j, 1 \leq j \leq g$ and its $\frac{d}{g}$ elements. We send these $\frac{d}{g}$ elements in K_j to the group G_j, one element per processor, sequentially using the coupler $c(j,i)$. This data transmission can be done in parallel as each subgroup has a coupler associated with it. Hence, this data transmission takes $\frac{d}{g}$ slots. Now, we have g groups with each group containing $\frac{d}{g}$ elements. Hence, this is a sorting problem in a $POPS(\frac{d}{g},g)$ and according to Lemma 2, we need $\frac{2d}{g^2} \log(\frac{d}{g} \times g) = \frac{2d}{g^2} \log d$ slots to sort these elements. Once they are sorted, we can bring back all the sorted elements to the processors in G_i in additional $\frac{d}{g}$ slots by using the g couplers associated with G_i in parallel. Hence, the total slot requirement for sorting the elements in G_i is $\frac{2d}{g} + \frac{2d}{g^2} \log d$ slots.

This process is repeated for all the g groups and in overall $2d + \frac{2d}{g} \log d$ slots, we can sort the elements in each group.

We can reduce the complexity of this sorting considerably in the following way. Note that we only need all the elements of each group sorted and the initial location of these elements is not important. Hence, there is no need to distribute the elements of G_i among the g groups. When sorting the elements for G_i, we rather simulate the hypercube sorting algorithm on the g groups, with each group holding $\frac{d}{g}$ elements. For example, when we sort the elements destined for G_1, we take the first subgroup of $\frac{d}{g}$ elements from each of the g groups and simulate the hypercube sorting algorithm on them. Finally, we bring all the elements in these g subgroups to the first group in sorted order. Similarly, for sorting the elements for the second group, we take the second subgroups from all the groups and so on. This reduces the slot complexity of sorting all the g groups to $d + \frac{2d}{g} \log d$ slots.

However, we can further improve this complexity. When the sorting for a particular group, say G_1, is complete, we do not immediately bring all the sorted elements to G_1. We keep the sorted elements in their original groups. At the end of the g rounds of sorting, the elements of G_i can be classified in the following way. There are g subgroups of $\frac{d}{g}$ elements in G_i. The first subgroup is sorted and destined for G_1, the second subgroup is sorted and destined for G_2 and in general, the j-th subgroup, $1 \leq j \leq g$ is sorted and destined for G_j. G_i has one coupler each connected to each of the g groups. We simultaneously transfer g elements from G_i to the other groups in each slot using these g couplers. This data transfer can be done in parallel for each of the groups since each group has g couplers connected to all other groups. Hence, the d elements in each group can be transferred in $\frac{d}{g}$ slots. Thus the total slot complexity of the overall sorting reduces to $\frac{d}{g} + \frac{2d}{g} \log d$ slots.

The complexity of sorting the elements in each group is stated in the following lemma.

Lemma 2. *Elements in each group in a POPS(d,g) can be sorted within the following complexities :*

- *in $8\log^2 d + \log d$ slots if $g \geq d$,*
- *in $\frac{d}{g} + \frac{2d}{g}\log d$ slots if $g < d$.*

4.2   The Sorting Algorithm

We now discuss our complete sorting algorithm. First we sort elements within each group by the method in Section 4.1. Next, the merging starts with pairwise merging of groups following the method in Section 3. Since there are g groups, after $\log g$ steps, we get all the elements sorted in the $POPS(d,g)$. Hence, the total cost for merging is the slots mentioned in Lemma 1 multiplied by $\log g$. When $g \geq d$, this complexity is $2\log^2 g + 8\log g\log d + \log g = 8\log n + 2\log^2 g + \log g$ slots, as $n = gd$. When $g < \sqrt{n}$, this complexity is $6\log^2 g + \frac{9d}{g}\log g$. The complexity for sorting is clearly the combination of the complexity for sorting each group and then the complexity of merging. We state the complexity in the following theorem. The proof follows from Lemma 1 and 2.

Theorem 3. *n elements given in a POPS(d,g) with $n = dg$, can be sorted within the following complexities :*

- *$8\log n + 8\log^2 d + 2\log^2 g + \log d + \log g$ slots if $g \geq \sqrt{n}$*
- *$\frac{2d}{g}\log n + \frac{7d}{g}\log g + \frac{d}{g} + 6\log^2 g$ slots if $g < \sqrt{n}$.*

5   Conclusion

We have presented a simple algorithm for merging and sorting on a POPS network. Our algorithm is more efficient compared to the simulated hypercube sorting algorithm as can be seen by comparing the complexity in Lemma 1 with the complexity in Theorem 3. First, we consider the case when $g \geq d$, i.e., the number of groups is equal or larger than the number of processors in each group. In this case, the complexity of simulating the hypercube sorting algorithm is $2\log^2 n$ slots ignoring the constants in the $O(\log^2 n)$ complexity of the bitonic sorting algorithm in [12]. On the other hand, the dominating terms in Theorem 3 are $8\log^2 d + 2\log^2 g < 10\log^2 g$. It is unrealistic that in a practical POPS network, there will be more couplers than processors. Hence, we can assume that the group size will be at most $\sqrt{n}$. The complexity reduces to $2.5\log^2 n$ in that case. Though this is higher than $2\log^2 n$, we expect that a custom algorithm will be much more efficient since simulations usually incur a lot of overheads which we have not considered in Lemma 1.

However, the main strength of our algorithm is its low complexity for the case when $d > g$. The number of couplers in a realistic POPS network will be much smaller than the number of processors as noted by Gravenstreter and Melhem [7] and Datta and Soundaralakshmi [5]. In this case, the dominant term in our algorithm is $\frac{2d}{g}\log n$ which is an improvement of a factor of $\log n$ over the simulated hypercube sorting algorithm.

Acknowledgments : We would like to thank Prof. Thomas Ottmann for his continuing support and encouragement for the last decade. His words have always been inspirations for us. He is the most wonderful and inspirational colleague and friend for us. We dedicate this article to him on his 60th birthday.

The first author's work is partially supported by Western Australian Interactive Virtual Environments Centre (IVEC) and Australian Partnership in Advanced Computing (APAC).

References

1. P. Berthomé and A. Ferreira, "Improved embeddings in POPS networks through stack-graph models", *Proc. Third International Workshop on Massively Parallel Processing Using Optical Interconnections*, pp. 130-135, 1996.
2. P. Berthomé, J. Cohen and A. Ferreira, "Embedding tori in partitioned optical passive stars networks", *Proc. Fourth International Colloquium on Structural Information and Communication Complexity (Sirocco '97)*, pp. 40-52, 1997.
3. D. Chiarulli, S. Levitan, R. Melhem, J. Teza and G. Gravenstreter, "Partitioned optical passive star (POPS) multiprocessor interconnection networks with distributed control", *Journal of Lightwave Technology*, **14** (7), pp. 1901-1612, 1996.
4. R. Cypher and G. Plaxton, "Deterministic sorting in nearly logarithmic time on the hypercube and related computers", *Proc. 22nd Annual ACM Symposium on Theory of Computing*, pp. 193-203, 1990.
5. A. Datta and S. Soundaralakshmi, "Basic operations on a partitioned optical passive stars network with large group size", *Proc. 2002 International Conference on Computational Science*, Amsterdam, Lecture Notes in Computer Science, Springer-Verlag , Vol. 2329, pp. 306-315, 2002.
6. A. Datta and S. Soundaralakshmi, "Summation and routing on a partitioned optical passive stars network with large group size", Manuscript, 2002.
7. G. Gravenstreter and R. Melhem, "Realizing common communication patterns in partitioned optical passive stars (POPS) networks", *IEEE Trans. Computers*, **47** (9), pp. 998-1013, 1998.
8. G. Gravenstreter, R. Melhem, D. Chiarulli, S. Levitan and J. Teza, "The partitioned optical passive stars (POPS) topology", *Proc. Ninth International Parallel Processing Symposium*, IEEE Computer Society, pp. 4-10, 1995.
9. F. T. Leighton, *Introduction to Parallel Algorithms and Architectures : Arrays-Trees-Hypercubes*, Morgan-Kaufman, San Mateo, 1992.
10. R. Melhem, G. Gravenstreter, D. Chiarulli and S. Levitan, "The communication capabilities of partitioned optical passive star networks", *Parallel Computing Using Optical Interconnections*, K. Li, Y.Pan and S. Zheng (Eds), Kluwer Academic Publishers, pp. 77-98, 1998.
11. A. Mei and R. Rizzi, "Routing permutations in partitioned optical passive stars networks", *Proc. 16th International Parallel and Distributed Processing Symposium*, Fort Lauderdale, Florida, IEEE Computer Society, April 2002.
12. S. Ranka and S. Sahni, *Hypercube Algorithms with Applications to Image Processing and Pattern Recognition*, Springer-Verlag, New York, 1990.
13. S. Sahni, "The partitioned optical passive stars network : Simulations and fundamental operations", *IEEE Trans. Parallel and Distributed Systems*, **11** (7), pp. 739-748, 2000.
14. S. Sahni, "Matrix multiplication and data routing using a partitioned optical passive stars network", *IEEE Trans. Parallel and Distributed Systems*, **11** (7), pp. 720-728, 2000.

15. S. Sahni, "Models and algorithms for optical and optoelectronic parallel computers", *International Journal of Foundations of Computer Science*, **12** (3), pp. 249-264, 2001.
16. F. T. Leighton, *Introduction to Parallel Algorithms and Architectures : Arrays-Trees-Hypercubes*, Morgan-Kaufman, San Mateo, 1992.

Route Planning and Map Inference with Global Positioning Traces

Stefan Edelkamp and Stefan Schrödl

[1] Institut für Informatik
Georges-Köhler-Allee 51
D-79110 Freiburg
edelkamp@informatik.uni-freiburg.de
[2] DaimlerChrysler Research and Technology
1510 Page Mill Road
Palo Alto, CA 94303
schroedl@rtna.daimlerchrysler.com

Abstract. Navigation systems assist almost any kind of motion in the physical world including sailing, flying, hiking, driving and cycling. On the other hand, traces supplied by global positioning systems (GPS) can track actual time and absolute coordinates of the moving objects.

Consequently, this paper addresses efficient algorithms and data structures for the route planning problem based on GPS data; given a set of traces and a current location, infer a short(est) path to the destination.

The algorithm of Bentley and Ottmann is shown to transform geometric GPS information directly into a combinatorial weighted and directed graph structure, which in turn can be queried by applying classical and refined graph traversal algorithms like Dijkstras' single-source shortest path algorithm or A*.

For high-precision map inference especially in car navigation, algorithms for road segmentation, map matching and lane clustering are presented.

1 Introduction

Route planning is one of the most important application areas of computer science in general and graph search in particular. Current technology like hand-held computers, car navigation and GPS positioning systems ask for a suitable combination of mobile computing and course selection for moving objects.

In most cases, a possibly labeled weighted graph representation of all streets and crossings, called the *map*, is explicitly available. This contrasts other exploration problems like puzzle solving, theorem proving, or action planning, where the underlying problem graph is implicitly described by a set of rules.

Applying the standard solution of Dijkstra's algorithm for finding the single-source shortest path (SSSP) in weighted graphs from an initial node to a (set of) goal nodes faces several subtle problems inherent to route planning:

1. Most maps come on external storage devices and are by far larger than main memory capacity. This is especially true for on-board and hand-held computer systems.

R. Klein et al. (Eds.): Comp. Sci. in Perspective (Ottmann Festschrift), LNCS 2598, pp. 128–151, 2003.

2. Most available digital maps are expensive, since exhibiting and processing road information e.g. by surveying methods or by digitizing satellite images is very costly.
3. Maps are likely to be inaccurate and to contain systematic errors in the input sources or inference procedures.
4. It is costly to keep map information up-to-date, since road geometry continuously changes over time.
5. Maps only contain information on road classes and travel distances, which is often not sufficient to infer travel time. In rush hours or on bank holidays, the time needed for driving deviates significantly from the one assuming usual travel speed.
6. In some regions of the world digital maps are not available at all.

The paper is subdivided into two parts. In the first part, it addresses the process of map construction based on recorded data. In Section 2, we introduce some basic definitions. We present the *travel graph inference problem*, which turns out to be a derivate of the output sensitive sweep-line algorithm of Bentley and Ottmann. Subsequently, Section 3 describes an alternative statistical approach. In the second part, we provide solutions to accelerate SSSP computations for time or length optimal route planning in an existing accurate map based on Dijkstra's algorithm, namely A* with the Euclidean distance heuristic and refined implementation issues to deal with the problem of restricted main memory.

2   Travel Graph Construction

Low-end GPS data devices with accuracies of about 2-15 m and mobile data loggers (e.g. in form of palmtop devices) that store raw GPS data entries are nowadays easily accessible and widly distributed. To visualize data in addition to electronic road maps, recent software allows to include and calibrate maps from the Internet or other sources. Moreover, the adaption and visualization of topographical maps is no longer complicated, since high-quality maps and visualization frontends are provided at low price from organizations like *the Surveying Authorities of the States of the Federal Republic of Germany* with the TK50 CD series. Various 2D and 3D user interfaces with on-line and off-line tracking features assist the preparation and the reflection of trips.

In this section we consider the problem of generating a travel graph given a set of traces, that can be queried for shortest paths. For the sake of clarity, we assume that the received GPS data is accurate and that at each inferred crossing of traces, a vehicle can turn into the direction that another vehicle has taken.

With current technology of global positioning systems, the first assumption is almost fulfilled: on the low end, (differential) GPS yields an accuracy in the range of a few meters; high end positioning systems with integrated inertial systems can even achieve an accuracy in the range of centimeters.

The second assumption is at least feasible for hiking and biking in unknown terrain without bridges or tunnels. To avoid these complications especially for car navigation, we might distinguish valid from invalid crossings. Invalid crossing are ones with an intersection angle above a certain threshold and difference in velocity outside a certain interval. Fig. 1 provides a small example of a GPS trace that was collected on a bike on the campus of the computer science department in Freiburg.

```
# latitude, longitude, date (yyyymmdd), time (hhmmss)
```

```
48.0131754,7.8336987,20020906,160241
48.0131737,7.8336991,20020906,160242
48.0131720,7.8336986,20020906,160243
48.0131707,7.8336984,20020906,160244
48.0131716,7.8336978,20020906,160245
48.0131713,7.8336975,20020906,160246
```

Figure 1. Small GPS trace.

2.1   Notation

We begin with some formal definitions. *Points* in the plane are elements of $\mathbb{R} \times \mathbb{R}$, and *line segments* are pairs of points. A *timed point* $p = (x, y, t)$ has global coordinates x and y and additional time stamp t, where $t \in \mathbb{R}$ is the absolute time to be decoded in year, month, day, hour, minute, second and fractions of a second. A *timed line segment* is a pair of timed points. A *trace* T is a sequence of timed points $p_1 = (x_1, y_1, t_1), \ldots, p_n = (x_n, y_n, t_n)$ such that t_i, $1 \leq i \leq n$, is increasing. A *timed path* $P = s_1, \ldots, s_{n-1}$ is the associated sequence of timed line segments with $s_i = (p_i, p_{i+1})$, $1 \leq i < n$. The angle of consecutive line segements on a (timed) path and the velocity on timed line segments are immediate consequences of the above definitions.

The *trace graph* $G_T = (V, E, d, t)$ is a directed graph defined by $v \in V$ if its coordinates (x_v, y_v) are mentioned in T, $e = (u, v) \in E$ if the coordinates of u and v correspond to two successive timed points (x_u, y_u, t_u) and (x_v, y_v, t_v) in T, $d(e) = ||u - v||_2$, and $t(e) = t_v - t_u$, where $||u - v||_2$ denotes the Euclidean distance between (the coordinates of) u and v.

The *travel graph* $G'_T = (V', E', d, t)$ is a slight modification of G_T including its line segment intersections. More formally, let $s_i \cap s_j = r$ denote that s_i and s_j intersect in point p, and let $I = \{(r, i, j) \mid s_i \cap s_j = r\}$ be the set of all intersections, then $V' = V \cup \{r \mid (r, i, j) \in I\}$ and $E' = E \cup E_a \setminus E_d$, where $E_d = \{(s_i, s_j) \mid \exists r : (r, i, j) \in I\}$, and $E_a = \{(p, r), (r, q), (p', r), (r, q') \in V' \times V' \mid (r, i, j) \in I$ and $r = (s_i = (p, q) \cap s_j = (p', q'))\}$. Note that intersection points r have no time stamp. Once more, the new cost values for $e = (u, v) \in E' \setminus E$ are determined by $d(e) = ||u - v||_2$, and by $t(e) = t(e')d(e)/d(e')$ with respect to the original edge $e' \in E_d$. The latter definition of time assumes a uniform speed on every line segment, which is plausible on sufficiently small line segments.

The *travel graph* G'_D of a set D of traces $T_1, \ldots T_l$ is the travel graph of the union graph of the respective trace graphs $G_{T_1}, \ldots, G_{T_k}$, Where the union graph $G = (V, E)$ of two graphs $G_1 = (V_1, E_1)$ and $G_2 = (V_2, E_2)$ is defined as $V = V_1 \cup V_2$ and $E = E_1 \cup E_2$.

For the sake of simplicity, we assume that all crossings are in *general position*, so that not more than two line segments intersect in one point. This assumption is not a severe restriction, since all algorithms can be adapted to the more general case. We also might exclude matching endpoints from the computation, since we already know that two consecutive line segments intersect at the recorded data point. If a vehicle stops while the GPS recorder is running, zero-length sequences with strictly positive time delays are generated. Since zero-length segments cannot yield crossings, the problem of self loops might be dealt with ignoring these segments for travel graph generation and a re-inserting them afterwards to allow timed shortest path queries.

2.2   Algorithm of Bentley and Ottmann

The plane-sweep algorithm of Bentley and Ottmann [3] infers an undirected planar graph representation (the *arrangement*) of a set of segments in the plane and their intersections. The algorithm is one of the most innovative schemes both from a conceptual and from a algorithmical point of view.

From a conceptional point of view it combines the two research areas of *computational complexity* and *graph algorithms*. The basic principle of an imaginary *sweep-line* that stops on any interesting event is one of the most powerful technique in geometry e.g. to directly compute the Voronoi diagram on a set of n points in optimal time $O(n \log n)$, and is a design paradigm for solving many combinatorial problems like the minimum and maximum in a set of values in the optimal number of comparisons, or the maximum sub-array sum in linear time with respect to the number of elements.

From an algorithmical point of view the algorithm is a perfect example of the application of balanced trees to reduce the complexity of an algorithm. It is also the first *output-sensitive* algorithm, since its time complexity $O((n + k) \log n)$ is measured in both the input and the output length, due to the fact that n input segments may give rise to $k = O(n^2)$ intersections.

The core observation for route planning is that, given a set of traces D in form of a sequence of segments, the algorithm can easily be adapted to compute the corresponding travel graph G'_D. In difference to the original algorithm devised for computational geometry problems, the generated graph structure has to be directed. The direction of each edge e as well as its distance $d(e)$ and travel time $t(e)$ is determined by the two end nodes of the segment. This includes intersections: the newly generated edges inherit direction, distance and time from the original end points.

In Fig. 2 we depicted a snapshot of the animated execution of the algorithm in the client-server visualization Java frontend VEGA [15] *i*) on a line segment sample set and *ii*) on an extended trail according to Fig. 1. The sweep-line proceeds from left to right, with the completed graph to its left.

The algorithm utilizes two data structures: the *event queue* and the *status structure*. In the event queue the active points are maintained, ordered with respect to their x-coordinate. In the status structure the active set of segments with respect to the sweep line is stored in y-ordering. At each intersection the ordering of segments in the status structure may change. Fortunately, the ordering of segments that participate in the intersections simply reverses, allowing fast updates in the data structure. After new neighboring segments are found, their intersections are computed and inserted into the event queue. The abstract data structure needed for implementation are a priority queue for the event queue and a search tree with neighboring information for the status data structure. Using a standard heap for the former and a balance tree for the latter implementation yields an $O((n + k) \log n)$ time algorithm.

The lower bound of the problem's complexity is $\Omega(n \log n + k)$ and the first step to improve time performance was $O(n \log^2 n / \log \log n + k)$ [5]. The first $O(n \log n + k)$

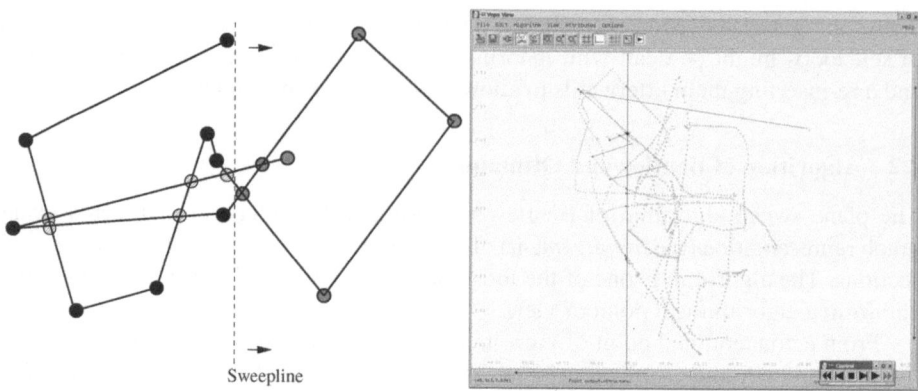

Sweepline

Figure 2. Visualization of the sweep-line algorithm of Bentley and Ottmann on a *i*) tiny and *ii*) small data set in the client-server visualization Java frontend VEGA.

algorithm [6] used $O(n + k)$ storage. The $O(n \log n + k)$ algorithm with $O(n)$ space is due to Balaban [2].

For trace graphs the $O((n + k) \log n)$ implementation is sufficiently fast in practice. As a small example trace file consider $n = 2^{16} = 65,536$ segment end points with $k = 2^8 = 256$ intersections. Then $n^2 = 2^{32} = 4,294,967,296$, while $(n + k) \log n = (2^{16} + 2^8) \cdot 16 = 1,052,672$ and $n \log n + k = 1,048,832$.

3   Statistical Map Inference

Even without map information, on-line routing information is still available, e.g. a driving assistance system could suggest *you are 2 meters off to the right of the best route*, or *you have not followed the suggested route, I will recompute the shortest path from the new position*, or *turn left in about 100 meter in a resulting angle of about 60 degrees*.

Nevertheless, route planning in trace graphs may have some limitations in the presentation of the inferred route to a human, since abstract maps compact information and help to adapt positioning data to the real-world.

In this section, an alternative approach to travel graphs is presented. It concentrates on the map inference and adaptation problem for car navigation only, probably the most important application area for GPS routing. One rationale in this domain is the following. Even in one lane, we might have several traces that overlap, so that the number of line segment intersections k can increase considerably. Take $m/2$ parallel traces on this lane that intersect another parallel $m/2$ traces on the lane a single lane in a small angle, then we expect up to $\Theta(m^2)$ intersections in the worst case.

We give an overview of a system that automatically generates digital road maps that are significantly more precise and contain descriptions of lane structure, including number of lanes and their locations, and also detailed intersection structure. Our approach is a statistical one: we combine 'lots of bad' GPS data from a fleet of vehicles, as opposed to 'few but highly accurate' data obtained from dedicated surveying vehicles operated by specially trained personnel. Automated processing can be much less expensive. The same is true for the price of DGPS systems; within the next few years, most new vehicles will likely have at least one DGPS receiver, and wireless technology is rapidly advancing to provide the communication infrastructure. The result will be more accurate, cheaper, up-to-date maps.

3.1   Steps in the Map Refinement Process

Currently commercially available digital maps are usually represented as graphs, where the nodes represent intersections and the edges are unbroken road *segments* that connect the intersections. Each segment has a unique identifier and additional associated attributes, such as a number of *shape points* roughly approximating its geometry, the road type (e.g., highway, on/off-ramp, city street, etc), speed information, etc. Generally, no information about the number of lanes is provided. The usual representation for a a two-way road is by means of a single segment. In the following, however, we depart from this convention and view segments as unidirectional links, essentially splitting those roads in two segments of opposite direction. This will facilitate the generation of the precise geometry.

The task of map refinement is simplified by decomposing it into a number of successive, dependent processing steps. Traces are divided into subsections that correspond to the road segments as described above, and the geometry of each individual segment is inferred separately. Each segment, in turn, comprises a subgraph structure capturing its lanes, which might include splits and merges. We assume that the lanes of a segment are mostly parallel. In contrast to commercial maps, we view an intersection as a structured region, rather than a point. These regions limit the segments at points where the traces diverge and consist of unconstrained trajectories connecting individual lanes in adjacent segments.

The overall map refinement approach can be outlined as follows.

1. Collect raw DGPS data (traces) from vehicles as they drive along the roads. Currently, commercially available DGPS receivers output positions (given as longitude/latitude/altitude coordinates with respect to a reference ellipsoid) at a regular frequency between 0.1 and 1 Hz.
   Optionally, if available, measurements gathered for the purpose of electronic safety systems (anti-lock brakes or electronic stability program), such as wheel speeds and accelerometers, can be integrated into the positioning system through a *Kalman filter* [14]. In this case, the step 2 (filtering or smoothing) can be accomplished in the same procedure.
2. Filter and resample the traces to reduce the impact of DGPS noise and outliers. If, unlike in the case of the Kalman filter, no error estimates are available, some of the errors can be detected by additional indicators provided by the receiver, relating to satellite geometry and availability of the differential signal; others (e.g., so-called *multipath* errors) only from additional plausibility tests, e.g., maximum acceleration according to a vehicle model. Resampling is used to balance out the bias of traces recorded at high sampling rates or at low speed. Details of the preprocessing are beyond the scope of the current paper and can be found in a textbook such as [23].

3. Partition the raw traces into sequences of segments by *matching* them to an initial base map. This might be a commercial digital map, such as that of Navigation Technologies, Inc. [22]. Section 3.2 presents an alternative algorithm for inferring the network structure from scratch, from a set of traces alone.

   Since in our case we are not constrained to a real-time scenario, it is useful to consider the context of sample points when matching them to the base map, rather than one point at a time. We implemented a map matching module that is based on a modified best-first path search algorithm based on the Dijkstra-scheme [9], where the matching process compares the DGPS points to the map shape points and generates a cost that is a function of their positional distance and difference in heading. The output is a file which lists, for each trace, the traveled segment IDs, along with the starting time and duration on the segment, for the sequence with minimum total cost (a detailed description of map matching is beyond the scope of this paper).

4. For each segment, generate a *road centerline* capturing the accurate geometry that will serve as a reference line for the lanes, once they are found. Our spline fitting technique will be described in Section 3.3.

5. Within each segment, cluster the perpendicular offsets of sample points from the road centerline to identify *lane* number and locations (cf. Section 3.4).

3.2   Map Segmentation

In the first step of the map refinement process, traces are decomposed into a sequence of sections corresponding to road segments. To this end, an initial base map is needed for map matching. This can either be a commercially available map, such as that of Navigation Technologies, Inc. [22]; or, we can infer the connectivity through a spatial clustering algorithm, as will be described shortly.

These two approaches both have their respective advantages and disadvantages. The dependence on a commercial input map has the drawback that, due to its inaccuracies (Navigation Technologies advertises an accuracy of 15 meters), traces sometimes are incorrectly assigned to a nearby segment. In fact, we experienced this problem especially in the case of highway on-ramps, which can be close to the main lanes and have similar direction.

A further disadvantage is that roads missing in the input map cannot be learned at all. It is impossible to process regions if no previous map exists or the map is too coarse, thus omitting some roads.

On the other hand, using a commercial map as the initial baseline associates additional attributes with the segments, such as road classes, street names, posted speeds, house numbers, etc. Some of these could be inferred from traces by related algorithms on the basis of average speeds, lane numbers, etc. Pribe and Rogers [25] describe an approach to learning traffic controls from GPS traces. An approach to travel time prediction is presented in [13]. However, obviously not all of this information can be independently recovered. Moreover, with the commercial map used for segmentation, the refined map will be more compatible and comparable with applications based on existing databases.

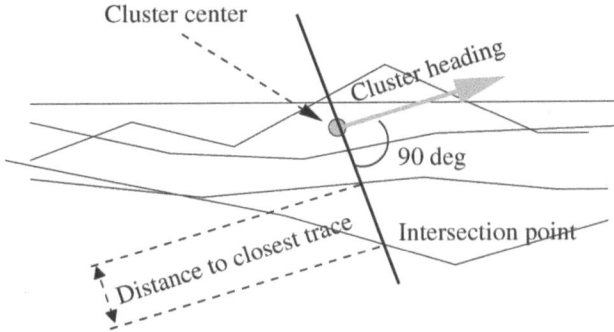

Figure 3. Distance between candidate trace and cluster seed

Road Segment Clustering In this section, we outline an approach to *inferring* road segments from a set of traces simultaneously. Our algorithm can be regarded as a *spatial clustering* procedure. This class of algorithms is often applied to recognition problems in image processing (See e.g. [10] for an example of road finding in aerial images). In our case, the main questions to answer are to identify common segments used by several traces, and to locate the branching points (intersections). A procedure should be used that exploits the contiguity information and temporal order of the trace points in order to determine the connectivity graph. We divide it into three stages: cluster seed location, seed aggregation into segments, and segment intersection identification.

Cluster Seed Location Cluster seed location means finding a number of sample points on different traces belonging to the same road. Assume we have already identified a number of trace points belonging to the same cluster; from these, a mean values for position and heading is derived. In view of the later refinement step described in Section 3.3, we can view such a cluster center as one point of the *road centerline*.

Based on the assumption of lane parallelism, we measure the distance between traces by computing their intersection point with a line through the cluster center that runs perpendicular to the cluster heading; this is equivalent to finding those points on the traces whose projection onto the tangent of the cluster coincides with the cluster location, see Fig. 3.

Our similarity measure between a new candidate trace and an existing cluster is based both on its minimum distance to other member traces belonging to the cluster, computed as described above; and on the difference in heading. If both of these indicators are below suitable thresholds (call them θ and δ, respectively) for two sample points, they are deemed to belong to the same road segment.

The maximum heading difference δ should be chosen to account for the accuracy of the data, such as to exclude sample points significantly above a high quantile of the error distribution. If such an estimate is unavailable, but a number of traces have already been found to belong to the cluster, the standard deviation of these members can give a clue. In general we found that the algorithm is not very sensitive to varation of δ.

The choice of θ introduces a trade-off between two kinds of segmentation errors: if it is too small, wide lanes will be regarded as different roads; in the opposite case,

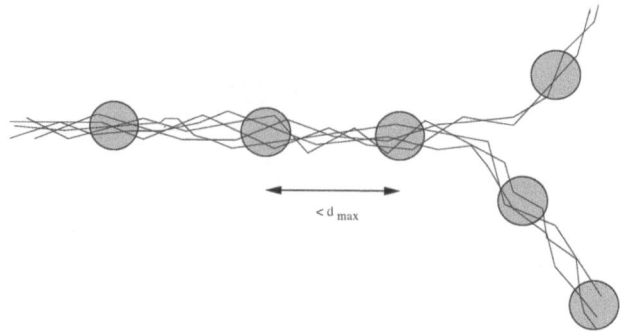

Figure 4. Example of traces with cluster seeds

nearby roads would be identified. In any case, the probablility that the GPS error exceeds the difference between the distance to the nearest road and the lane width is a lower bound for the segmentation error. A suitable value depends on the expected GPS errors, characteristics of the map (e.g., the relative frequencies of four-way intersections vs. freeway ramps), and also on the relative risks associated with both tyes of errors which are ultimately determined by the final application. As a conservative lower bound, θ should be at least larger than the maximum lane width, plus a tolerance (estimated standard deviation) for driving off the center of the lane, plus a considerable fraction of an estimated standard deviation of the GPS error. Empirically, we found the results with values in the range of 10–20 meters to be satisfying and sufficiently stable.

Using this similarity measure, the algorithm now proceeds in a fashion similar to the k-means algorithm [19]. First, we initialize the cluster with some random trace point. At each step, we add the closest point on any of the traces not already in the cluster, unless θ or δ is exceeded. Then, we recompute the average position and heading of the cluster center. Due to these changes, it can sometimes occur that trace points previously contained in the cluster do no longer satisfy the conditions for being on the same road; in this case they are removed. This process is repeated, until no more points can be added.

In this manner, we repeatedly generate cluster seeds at different locations, until each trace point has at least one of them within reach of a maximum distance threshold d_{max}. This threshold should be in an order of magnitude such that we ensure not to miss any intersection (say, e.g., 50 meters). A simple greedy strategy would follow each trace and add a new cluster seed at regular intervals of length d_{max} when needed. An example section of traces, together with the generated cluster centers, are shown in Fig. 4.

Segment Merging The next step is to *merge* those ones of the previously obtained cluster centers that belong to the same road. Based on the connectivity of the traces, two such clusters C_1 and C_2 can be characterized in that (1) w.l.o.g. C_1 precedes C_2, i.e., all the traces belonging to C_1 subsequently pass through C_2, and (2) all the the traces belonging to C_2 originate from C_1. All possible adjacent clusters satisfying this condition are merged. A resulting maximum chain of clusters $C_1, C_2, \ldots, C_n$ is called a *segment*, and C_1 and C_n are called the *boundary clusters* of the segment.

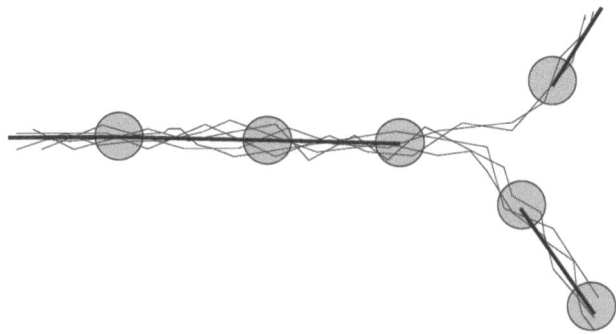

Figure 5. Merged cluster seeds result in segments

At the current segmentation stage of the map refinement process, only a crude geometric representation of the segment is sufficient; its precise shape will be derived later in the road centerline generation step (Section 3.3). Hence, as an approximation, adjacent cluster centers can either be joined by straight lines, polynomials, or one representative trace part (in our algorithms, we chose the latter possibility). In Fig. 5, the merged segments are connected with lines.

Intersections The only remaining problem is now to represent intersections. To capture the extent of an intersection more precisely, we first try to advance the boundary clusters in the direction of the split- or merge zone. This can be done by selecting a point from each member trace at the same (short) travel distance away from the respective sample point belonging to the cluster, and then again testing for contiguity as described above. We extend the segment iteratively in small increments, until the test fails.

The set of adjacent segments of an intersection is determined by (1) selecting all outgoing segments of one of the member boundary clusters; (2) collecting all incoming segments of the segments found in (1); and iterating these steps until completion. Each adjacent segment should be joined to each other segment for which connecting traces exist.

We utilize the concept of a *snake* borrowed from the domain of image processing, i.e., a contour model that is fit to (noisy) sample points. In our case, a simple star-shaped contour suffices, with the end points held fixed at the boundary cluster centers of the adjacent segments. Conceptually, each sample points exert an attracting force on it closest edge. Without any prior information on the shape of the intersection, we can define the 'energy' to be the sum of the squared distances between each sample point and the closest point on any of the edges, and then iteratively move the center point in an EM-style fashion in order to minimize this measure. The dotted lines in Fig. 6 correspond to the resulting snake for our example.

Dealing with Noisy Data Gaps in the GPS receiver signal can be an error source for the road clustering algorithm. Due to obstructions, it is not unusual to find gaps in the data that span a minute. As a result, interpolation between distant points is not reliable.

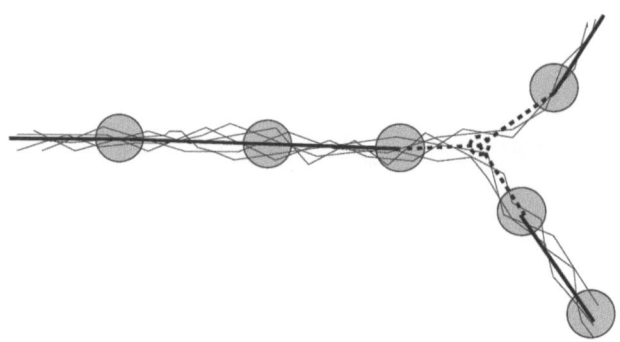

Figure 6. Traces, segments, and intersection contour model (dotted)

As mentioned above, checking for parallel traces crucially depends on *heading* information. For certain types of positioning systems used to collect the data, the heading might have been determined from the direction of differences between successive sample points. In this case, individual outliers, and also lower speeds, can lead to even larger errors in direction.

Therefore, in the first stage of our segmentation inference algorithm, filtering is performed by disregarding trace segments in cluster seeds that have a gap within a distance of d_{max}, or fall outside a 95 percent interval in the heading or lateral offset from the cluster center. Cluster centers are recomputed only from the remaining traces, and only they contribute to the subsequent steps of merging and intersection location with adjacent segments.

Another issue concerns the start and end parts of traces. Considering them in the map segmentation could introduce segment boundaries at each parking lot entrance. To avoid a too detailed breakup, we have to disregard initial and final trace sections. Different heuristics can be used; currently we apply a combined minimum trip length/speed threshold.

3.3 Road Centerline Generation

We now turn our attention to the refinement of individual segments. The *road centerline* is a geometric construct whose purpose is to capture the road geometry. The road centerline can be thought of as a weighted average trace, hence subject to the relative lane occupancies, and not necessarily a trajectory any single vehicle would ever follow. We assume, however, the lanes to be parallel to the road centerline, but at a (constant) perpendicular offset. For the subsequent lane clustering, the road centerline helps to cancel out the effects of curved roads.

For illustration, Fig. 7 shows a section of a segment in our test area. The indicated sample points stem from different traces. Clearly, by comparison, the shape points of the respective NavTech segment exhibit a systematic error. The centerline derived from the sample points is also shown.

It is useful to represent our curves in *parametric form*, i.e., as a vector of coordinate variables $C(u) = (x, y, z)(u)$ which is a function of an independent parameter u, for

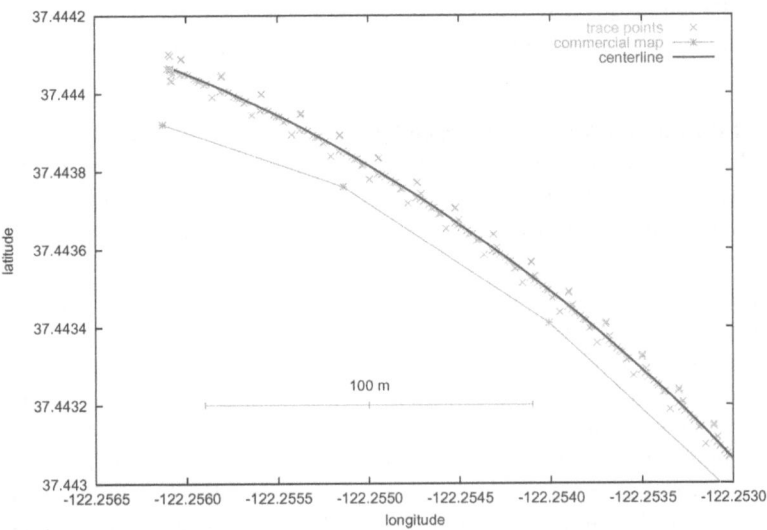

Figure 7. Segment part: NavTech map (bottom), trace points (crosses), and computed centerline

$0 \leq u \leq 1$. The centerline is generated from a set of sample points using a *weighted least squares fit*. More precisely, assume that $Q_0, \ldots, Q_{m-1}$ are the m data points given, $w_0, \ldots, w_{m-1}$, are associated weights (dependent on an error estimate), and $\bar{u}_0, \ldots, \bar{u}_{m-1'}$ their respective parameter values. The task can be formulated as finding a parametric curve $C(u)$ from a class of functions S such that the Q_k are approximated in the weighted least square sense, i.e.

$$s := \sum_{k=0}^{m-1} w_k \cdot \| Q_k - C(\bar{u}_k) \|^2$$

in a minimum with respect to S, where $\| \cdot \|$ denotes the usual Euclidean distance (2-norm). Optionally, in order to guarantee continuity across segments, the algorithm can easily be generalized to take into account derivatives; if heading information is available, we can use coordinate transformation to arrive at the desired derivative vectors.

The class S of approximating functions is composed of rational *B-Splines*, i.e., piecewise defined polynomials with continuity conditions at the joining knots (for details, see [24; 27]). For the requirement of continuous curvature, the degree of the polynomial has to be at least three.

If each sample point is marked with an estimate of the measurement error (standard deviation), which is usually available from the receiver or a Kalman filter, then we can use its inverse to weight the point, since we want more accurate points to contribute more to the overall shape.

The least squares procedure [24] expects the *number of control points n* as input, the choice of which turns out to be critical. The control points define the shape of the spline, while not necessarily lying on the spline themselves. We will return to the issue of selecting an adequate number of control points in Section 3.3.

Choice of Parameter Values for Trace Points For each sample point Q_k, a parameter value $\bar{u}_k$ has to be chosen. This parameter vector affects the shape and parameterization of the spline. If we were given a single trace as input, we could apply the widely used *chord length* method as follows. Let d be the total chord length $d = \sum_{k=1}^{m-1} |Q_k - Q_{k-1}|$. Then set $\bar{u}_0 = 0$, $\bar{u}_{m-1} = 1$, and $\bar{u}_k = \bar{u}_{k-1} + \frac{|Q_k - Q_{k-1}|}{d}$ for $k = 1, \ldots, m-2$. This gives a good parameterization, in the sense that it approximates a *uniform* parameterization proportional to the arc length.

For a set of k distinct traces, we have to impose a common ordering on the combined set of points. To this end, we utilize an initial rough approximation, e.g., the polyline of shape points from the original NavTech map segment s; if no such map segment is available, one of the traces can serve as a rough baseline for projection. Each sample point Q_k is *projected* onto s, by finding the closest interpolated point on s and choosing $\bar{u}_k$ to be the chord length (cumulative length along this segment) up to the projected point, divided by the overall length of s. It is easy to see that for the special case of a single trace identical to s, this procedure coincides with the chord length method.

Choice of the Number of Control Points The number of control points n is crucial in the calculation of the centerline; for a cubic spline, it can be chosen freely in the valid range $[4, m-1]$. Fig. 8 shows the centerline for one segment, computed with three different parameters n.

Note that a low number of control points may not capture the shape of the centerline sufficiently well ($n = 4$); on the other hand, too many degrees of freedom causes the result to "fit the error". Observe how the spacing of sample points influences the spline for the case $n = 20$.

From the latter observation, we can derive an upper bound on the number of control points: it should not exceed the average number of sample points per trace, multiplied by a small factor, e.g., $2 * m/k$.

While the appropriate number of control points can be easily estimated by human inspection, its formalization is not trivial. We empirically found that two measures are useful in the evaluation.

The first one is related to the *goodness of fit*. Averaging the absolute offsets of the sample points from the spline is a feasible approach for single-lane roads, but otherwise depends on the number and relative occupancies of lanes, and we do not expect this offset to be zero even in the ideal case. Intuitively, the centerline is supposed to stay roughly in the middle between all traces; i.e., if we project all sample points on the centerline, and imagine a fixed-length window moving along the centerline, then the average offset of all sample points whose projections fall into this window should be near to zero. Thus, we define the *approximation error* ε_{fit} as the average of these offsets over all windows.

The second measure checks for overfitting. As illustrated in Fig. 8, using a large number of control points renders the centerline "wiggly", i.e., tends to increase the

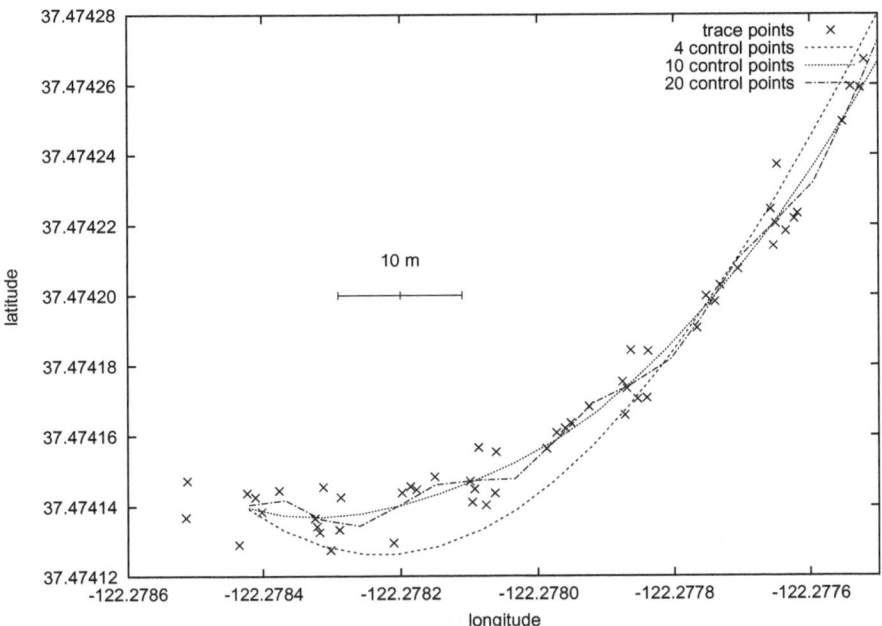

Figure 8. Trace points and centerlines computed with varying number of control points

curvature and makes it change direction frequently. However, according to construction guidelines, roads are designed to be piecewise segments of either straight lines or circles, with clothoids between as transitions. These geometric concepts constrain the curvature to be *piecewise linear*. As a consequence, the second derivative of the curvature is supposed to be zero nearly everywhere, with the exception of the segment boundaries where it might be singular. Thus, we evaluate the curvature of the spline at constant intervals and numerically calculate the second derivative. The average of these values is the *curvature error* ε_{curv}.

Fig. 9 plots the respective values of ε_{fit} and ε_{curv} for the case of Fig. 8 as a function of the number of control points. There is a tradeoff between ε_{fit} and ε_{curv}; while the former tends to decrease rapidly, the latter increases. However, both values are not completely monotonic.

Searching the space of possible solutions exhaustively can be expensive, since a complete spline fit has to be calculated in each step. To save computation time, the current approach heuristically picks the largest valid number of control points for which ε_{curv} lies below an acceptable threshold.

3.4   Lane Finding

After computing the approximate geometric shape of a road in the form of the road centerline, the aim of the next processing step is to infer the number and positions

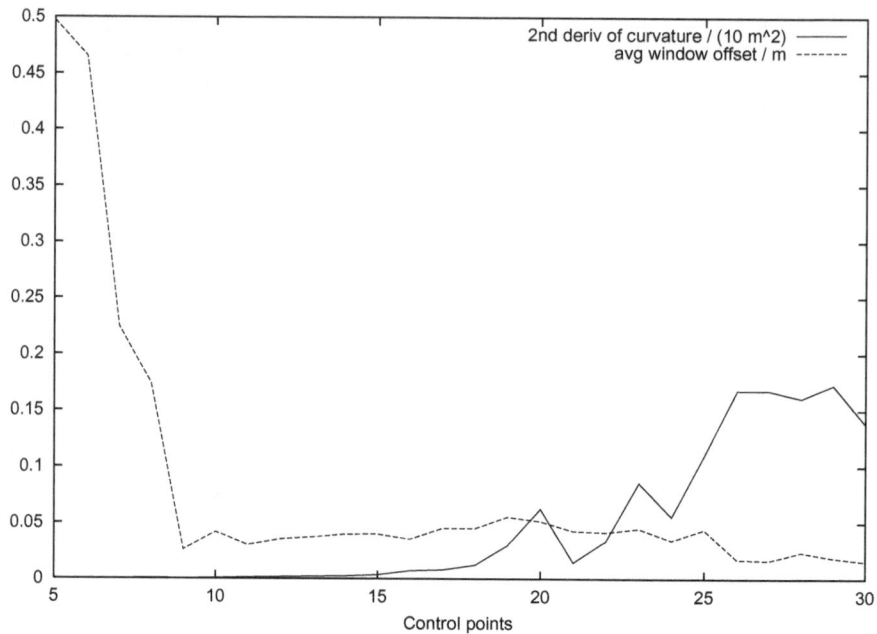

Figure 9. Error Measures ε_{fit} and ε_{curv} vs Number of Control Points

of its *lanes*. The task is simplified by canceling out road curvature by the following transformation. Each trace point P is *projected* onto the centerline for the segment, i.e., its nearest interpolated point P' on the map is determined. Again, the arc length from the first centerline point up to P' is the *distance along the segment*; the distance between P and P' is referred to as its *offset*. An example of the transformed data is shown in Fig. 10.

Intuitively, clustering means assigning n data points in a d-dimensional space to k clusters such that some distance measure within a cluster (i.e., either between pairs of data belonging to the same cluster, or to a cluster center) is minimized (and is maximized between different clusters). For the problem of lane finding, we are considering points in a plane representing the flattened face of the earth, so the Euclidean distance measure is appropriate.

Since clustering in high-dimensional spaces is computationally expensive, methods like the *k-means* algorithm use a hill-climbing approach to find a (local) minimum solution. Initially, k cluster centers are selected, and two phases are iteratively carried out until cluster assignment converges. The first phase assigns all points to their nearest cluster center. The second phase recomputes the cluster center based on the respective constituent points (e.g., by averaging) [19].

Segments with Parallel Lanes If we make the assumption that lanes are parallel over the entire segment, the clustering is essentially one-dimensional, taking only into account the offset from the road centerline. In our previous approach [26], a hierarchical

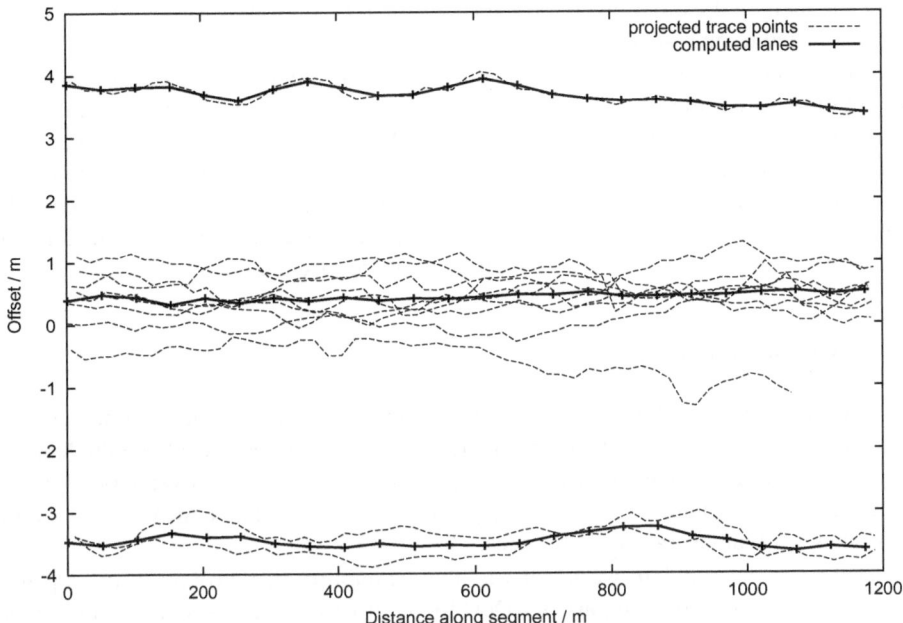

Figure 10. Traces projected on centerline (dashed), computed lane centers (solid)

agglomerative clustering algorithm (*agglom*) was used that terminated when the two closest clusters were more than a given distance apart (which represented the maximum width of a lane). However, this algorithm requires $O(n^3)$ computation time. More recently, we have found that it is possible to explicitly compute the *optimal* solution in $O(k \cdot n^2)$ time and $O(n)$ space using a dynamic programming approach, where n denotes the number of sample points, and k denotes the maximum number of clusters [27].

Segments with Lane Splits and Merges   We have previously assumed the ideal case of a constant number of lanes at constant offsets from the road centerline along the whole segment. However, lane widths may gradually change; in fact, they usually get wider near an intersection. Moreover, new lanes can start at any point within a segment (e.g., turn lanes), and lanes can merge (e.g., on-ramps). Subsequently, we present two algorithms that can accommodate the additional complexity introduced by lane merges and splits.

One-dimensional Windowing with Matching   One solution to this problem is to augment the one-dimensional algorithm with a windowing approach. We divide the segment into successive windows with centers at constant intervals along the segment. To minimize discontinuities, we use a Gaussian convolution to generate the windows. Each window is clustered separately as described above, i.e., the clustering essentially remains one-dimensional. If the number of lanes in two adjacent windows remains the same, we can associate them in the order given. For lane splits and merges, however, lanes in adjacent windows need to be *matched*.

To match lanes across window boundaries, we consider each trace individually (the information as to which trace each data point belongs to has not been used in the centerline generation, nor the lane clustering). Following the trajectory of a given trace through successive windows, we can classify its points according to the computed lanes. Accumulating these counts over all traces yields a matrix of *transition frequencies* for any pair of lanes from two successive windows. Each lane in a window is matched to that lane in the next window with maximum transition frequency.

3.5   Experimental Results

We have now completed our account of the individual steps in the map refinement process. In this section, we report on experiments we ran in order to evaluate the learning rate of our map refinement process. Our test area in Palo Alto, CA, covered 66 segments with a combined length of approximately 20 km of urban and freeway roads of up to four lanes, with an average of 2.44 lanes.

One principal problem we faced was the availability of a ground truth map with lane-level accuracy for comparison. Development of algorithms to integrate vision-based lane tracker information is currently under way. Unfortunatly, however, these systems have errors on their own and therefore cannot be used to gauge the accuracy of the pure position-based approach described in this article. Therefore, we reverted to the following procedure. We used a high-end real-time kinematic carrier phase DGPS system to generate a base map with few traces [32]. According to the announced accuracy of the system of about 5 cm, and after visual inspection of the map, we decided to define the obtained map as our baseline. Specifically, the input consisted of 23 traces at different sampling rates between 0.25 and 1 Hz.

Subsequently, we artificially created more traces of lower quality by adding varying amounts of gaussian noise to each individual sample position ($\sigma = 0.5 \ldots 2$ m) of copies of the original traces. For each combination of error level and training size, we generated a map and evaluated its accuracy.

Fig. 11 shows the resulting error in the *number of lanes*, i.e., the proportion of instances where the number of lanes in the learned map differs from the number of lanes in the base line map at the same position. Obviously, for a noise level in the range of more than half a lane width, it becomes increasingly difficult to distinguish different clusters of road centerline offsets due to overlap. Therefore, the accuracy of the map for the input noise level of $\sigma = 2$ m is significantly higher than that of the lower ones. However, based on the total spread of the traces, the number of lanes can still be estimated. For $n = 253$ traces, their error is below 10 percent for all of them. These remaining differences arise mainly from splits and merges, where in absence of the knowledge of lane markings it is hard to determine the exact branch points, whose position can heavily depend on single lane changes in traces.

Fig. 12 plots the mean absolute difference of the offsets of corresponding lanes between the learned map and the base map, as a function of the training size (number of traces). Again, the case $\sigma = 2$ m needs significantly more training data to converge. For

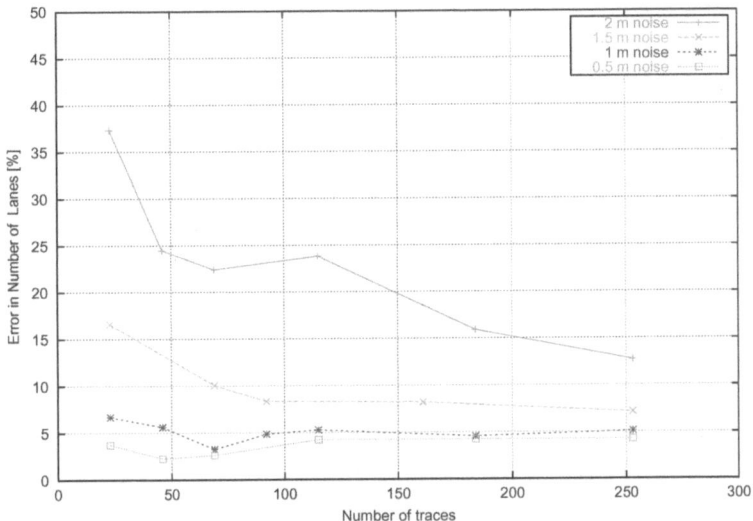

Figure 11. Error in determining the number of lane clusters

$\sigma <= 1.5$ m, the lane offset error decreases rapidly; it is smaller than 15 centimeters after $n = 92$ traces, and thus in the range of the base map accuracy.

4   Searching the Map Graph

Let us now turn our attention to the map usage in on-line routing applications. In some sense, both maps and travel graphs can be viewed as embeddings of weighted general graphs. Optimal paths can be searched with respect to accumulated shortest time t or distance d or any combination of them. We might assume a linear combination for a total weight function $w(u,v) = \lambda \cdot t(u,v) + (1 - \lambda) \cdot d(u,v)$ with parameter $\lambda \in \mathbb{R}$ and $0 \leq \lambda \leq 1$.

4.1   Algorithm of Dijkstra

Given a weighted graph $G = (V, E, w)$, $|V| = n$, $|E| = e$, the shortest path between two nodes can be efficiently computed by Dijkstra's single source shortest path (SSSP) algorithm [9].

Table 1 shows a implementation of Dijkstra's algorithm for implicitly given graphs that maintains a visited list *Closed* in form of a hash table and a priority queue of the nodes to be expanded, ordered with respect to increasing merits f.

The run time of Dijkstra's algorithm depends on the priority queue data structure. The original implementation of Dijkstra runs in $O(n^2)$, the standard textbook algorithm in $O((n + e) \log n)$ [7], and utilizing Fibonacci-heaps we get $O(e + n \log n)$ [12]. If the weights are small, buckets are preferable. In a Dial the i-th bucket contains all elements with a f-value equal to i [8]. Dials yields $O(e + n \cdot C)$ time for SSSP, with

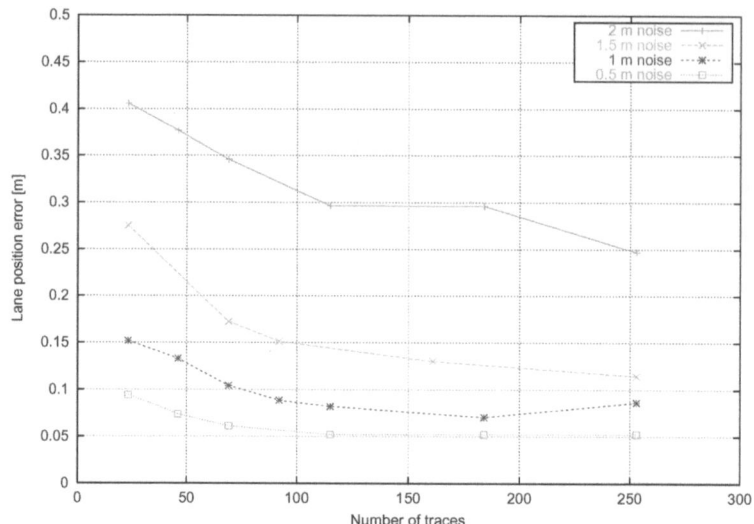

Figure 12. Average error in lane offsets

$C \leftarrow max_{(u,v) \in E}\{w(u,v)\}$. Two-Level Buckets have top level and bottom level of length $\lceil \sqrt{C+1} \rceil + 1$, yielding the run time $O(e + n\sqrt{C})$. An implementation with *radix heaps* uses buckets of sizes $1, 1, 2, 4, 8, \ldots$ and imply $O(e + n\log C)$ run time, two-level heap improve the bound to $O(e + n\log C/\log\log C)$ and a hybrid with Fibonacci heaps yields $O(e + n\sqrt{\log C})$ [1]. This algorithm is almost linear in practice, since when assuming 32 bit integers we have $\lceil \sqrt{\log C} \rceil \leq 6$. The currently best result are *component trees:* with $O(n+e)$ time for undirected SSSP on a random access machine of word length w with integer edge weights in $[0..2^w - 1]$ [31]. However, the algorithm is quite involved and likely not to be practical.

4.2 Planar Graphs

Travel graphs have many additional features. First of all, the number of edges is likely to be small. In the trail graph the number of edges equals the number of nodes minus 1, and for l trails $T_1, \ldots, T_l$ we have $|T_1|, \ldots, |T_l| - l$ edges in total. By introducing k intersections the number of edges increases by $2k$ only. Even if intersections coincide travel graphs are still planar, and by Eulers formula the number of edges is bounded by at most three times the number of nodes. Recall, that for the case of planar graphs linear time algorithms base on graph separators and directly lead to network flow algorithms of the same complexity [17].

If some intersections were rejected by the algorithms to allow non-intersecting crossing like bridges and tunnels, the graph would loose some of its graph theoretical properties. In difference to general graphs, however, we can mark omitted crossings to improve run time and storage overhead.

Dijkstra: **A***

$Open \leftarrow \{(s,0)\}$ $Open \leftarrow \{(s,h(s))\}$
$Closed \leftarrow \{\}$
while $(Open \neq \emptyset)$
  $u \leftarrow Deletemin(Open)$
  $Insert(Closed, u)$
  if $(goal(u))$ **return** u
  for all v **in** $\Gamma(u)$
    $f'(v) \leftarrow f(u) + w(u,v)$ $+h(v) - h(u)$
    if $(Search(Open, v))$
      if $(f'(v) < f(v))$
        $DecreaseKey(Open(v, f'(v)))$
    else if not $(Search(Closed, v))$
      $Insert(Open, (v, f'(v)))$

Table 1. Implementation of Dijkstra's SSSP algorithm vs. A*.

4.3   Frontier Search

Frontier search [18] contributes to the observation that the newly generated nodes in any graph search algorithm form a connected horizon to the set of expanded nodes, which is omitted to save memory.

The technique refers to Hirschberg's linear space divide-and-conquer algorithm for computing maximal common sequences [16]. In other words, frontier search reduces a $(d+1)$-dimensional memorization problem into a d-dimensional one. It divides into three phases. In the first phase, a goal t with optimal cost f^* is searched. In the second phase the search is re-invoked with bound $f^*/2$; and by maintaining shortest paths to the resulting fringe the intermediate state i from s to t is detected. In the last phase the algorithm is recursively called for the two subproblems from s to i, and from i to t.

4.4   Heuristic Search

Heuristic search includes an additional node evaluation function h into the search. The estimate h, also called heuristic, approximates the shortest path distance from the current node to one of the goal nodes. A heuristic is *admissible* if it provides a lower bound to the shortest path distance and it is *consistent*, if $w(u,v) + h(v) - h(u) \geq 0$. Consistent estimates are admissible.

Table 1 also shows the small changes in the implementation of A* for consistent estimates with respect to Dijkstra's SSSP algorithm. In the priority queue *Open* of generated and not expanded nodes, the f-values are tentative, while in set *Closed* the f-values are settled. On every path from to the initial state to a goal node the accumulated heuristic values telescope, and if any goal node has estimate zero, the f values of each encountered goal node in both algorithms are the same. Since in Dijkstra's SSSP al-

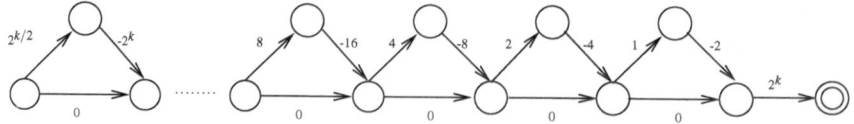

Figure 13. A graph with an exponential number of re-openings

gorithm the f-value of all expanded nodes match their graph theoretical shortest path value we conclude that for consistent estimates, A* is complete and optimal.

Optimal solving the SSSP problem for admissible estimates and negative values of $w(u,v) + h(v) - h(u)$ leads to re-openings of nodes: already expanded nodes in *Closed* are pushed back into the search frontier *Open*. If we consider $w(u,v) + h(v) - h(u)$ as the new edge costs, Fig. 13 gives an example for such a re-weighted graphs that leads to exponentially many re-openings. The second last node is re-opened for every path with weight $\{1, 2, \ldots, 2^k - 1\}$. Recall that if h is consistent, no reopening will be necessary at all.

In route planning the Euclidean distance of two nodes is a heuristic estimate defined as $h(u) = \min_{g \in G} ||g - u||_2$ for the set of goal nodes G is both admissible and consistent Admissibility is granted, since no path on any road map can be shorter than the flight distance, while consistency follows from the triangle inequality for shortest path. For edge $e = (u, v)$ we have $\min_{g \in G} ||g - v||_2 \leq \min_{g \in G} ||g - u||_2 + w(u, v)$. Since nodes closer to the goal are more attractive, A* is likely to find the optimum faster. Another benefit from this point of view is that all above advanced data structures for node maintenance in the priority as well as space saving strategies like frontier search can be applied to A*.

5   Related Work

Routing schemes often run on external maps and external maps call for refined memory maintenance. Recall that external algorithms are ranked according to *sorting complexity* $O(sort(n))$, i.e., the number of external block accesses (I/Os) necessary to sort n numbers, and according to *scanning complexity* $O(scan(n))$, i.e., the number of I/Os to read N numbers. The usual assumption is that N is much larger than B, the block size. Scanning complexity equals $O(n/B)$ in a single disk model. On planar graphs, SSSP runs in $O(sort(n))$ I/Os, where n is the number of vertices. As for the internal case the algorithms apply graph separation techniques [30]. For general BFS at most $O(\sqrt{n} \cdot scan(n + e) + sort(n + e))$ I/Os [20] are needed, where e is the number of edges in the graph. Currently there is no $o(n)$ algorithm for external SSSP. On the other hand, $O(n)$ I/Os are by far too much in route planning practice.

Fortunately, one can utilize the spatial structure of a map to guide the secondary mapping strategy with respect to the graph embedding. The work of [11] provides the new search algorithm and suitable data structures in order to minimize page faults by a local reordering of the sequence of expansions. Algorithm *Localized A** introduces an operation *deleteSome* instead of strict *deleteMin* into the A* algorithm. Nodes corresponding to an active page are preferred. When maintaining an bound α on obtained solution lengths until *Open* becomes empty the algorithm can be shown to be complete and optimal. The loop invariant is that there is always a node on the optimal solution path with correctly estimated accumulated cost. The authors prove the correctness and completeness of the approach and evaluate it in a real-world scenario of searching a large road map in a commercial route planning system.

In many fields of application, shortest path finding problems in very large graphs arise. Scenarios where large numbers of on-line queries for shortest paths have to be processed in real-time appear for example in traffic information systems. In such systems, the techniques considered to speed up the shortest path computation are usually based on pre-computed information. One approach proposed often in this context is a space reduction, where pre-computed shortest paths are replaced by single edges with weight equal to the length of the corresponding shortest path. The work of [29] gives a first systematic experimental study of such a space reduction approach. The authors introduce the concept of multi-level graph decomposition. For one specific application scenario from the field of timetable information in public transport, the work gives a detailed analysis and experimental evaluation of shortest path computations based on multi-level graph decomposition.

In the scenario of a central information server in the realm of public railroad transport on wide area networks a system has to process a large number of on-line queries for optimal travel connections in real time. The pilot study of [28] focuses on travel time as the only optimization criterion, in which various speed-up techniques for Dijkstra's algorithm were analyzed empirically.

Speed-up techniques that exploit given node coordinates have proven useful for shortest-path computations in transportation networks and geographic information systems. To facilitate the use of such techniques when coordinates are missing from some, or even all, of the nodes in a network [4] generate artificial coordinates using methods from graph drawing. Experiments on a large set of train timetables indicate that the speed-up achieved with coordinates from network drawings is close to that achieved with the actual coordinates.

6   Conclusions

We have seen a large spectrum of efficient algorithms to tackle different aspects of the route planning problem based on a given set of global positioning traces.

For trail graph inference the algorithm of Bentley and Ottmann has been modified and shown to be almost as efficient as the fastest shortest path algorithms. Even though this solves the basic route planning problem, different enhanced aspects are still open. We indicate low memory consumption, localized internal computation, and fast on-line performance as the most challenging ones.

Map inference and map matching up to lane accuracy suite better as a human-computer interface, but the algorithmic questions include many statistical operations and are non-trivial for perfect control. On the other hand, map inference based on GPS information saves much money especially to structure unknown and continuously changing terrains.

Low-end devices will improve GPS accuracy especially by using additional iner-
tia information of the moving object, supplied e.g. by a tachometer, an altimeter, or a
compass. For (3D) map generation and navigation other sensor data like sonar and laser
(scans) can be combined with GPS e.g. for outdoor navigation of autonomous robots
and in order to close uncaught loops.

The controversery if the GPS routing problem is more a geometrical one (in which
case the algorithm of Bentley/Ottmann applies) or a statistical one (in which clustering
algorithms are needed) is still open. At the moment we expect statistical methods to
yield better and faster results due to their data reduction and refinement aspects and
we expect that a geometrical approach will not suffice to appropriately deal with a large
and especially noisy data set. We have already seen over-fitting anomalies in the statistic
analyses. Nevertheless, a lot more research is needed to clarify the the quest of a proper
static analysis of GPS data, which in turn will have a large impact in the design and
efficiency of the search algorithms.

We expect that in the near future, the combination of positioning and precision map
technology will give rise to a range of new vehicle safety and convenience applications,
ranging from warning, advice, up to automated control.

Acknowledgments   S. Edelkamp is supported by DFG in the projects *heuristic search*,
and *directed model checking*. Thanks to Shahid Jabbar for visualizing the GPS trace.

References

1. R. K. Ahuja, K. Mehlhorn, J. B. Orbin, and R. E. Tarjan. Faster algorithms for the shortest
   path problem. *Journal of the ACM*, pages 213–223, 1990.
2. I. J. Balaban. An optimal algorithm for finding segment intersection. In *ACM Symposium on
   Computational Geometry*, pages 339–364, 1995.
3. J. L. Bentley and T. A. Ottmann. Algorithms for reporting and counting geometric intersections. *Transactions on Computing*, 28:643–647, 1979.
4. U. Brandes, F. Schulz, D. Wagner, and T. Willhalm. Travel planning with self-made maps.
   In *Workshop on Algorithm Engineering and Experiments (ALENEX)*, 2001.
5. B. Chazelle. Reporting and counting segment intersections. *Computing System Science*,
   32:200–212, 1986.
6. B. Chazelle and H. Edelsbrunner. An optimal algorithm for intersecting lines in the plane.
   Journal of the ACM, 39:1–54, 1992.
7. T. H. Cormen, C. E. Leiserson, and R. L. Rivest. *Introduction to Algorithms*. The MIT Press,
   1990.
8. R. B. Dial. Shortest-path forest with topological ordering. *Communication of the ACM*,
   12(11):632–633, 1969.
9. E. W. Dijkstra. A note on two problems in connection with graphs. *Numerische Mathematik*,
   1:269–271, 1959.
10. P. Doucette, P. Agouris, A. Stefanidis, and M. Musavi. Self-organized clustering for road
    extraction in classified imagery. *Journal of Photogrammetry and Remote Sensing*, 55(5-
    6):347–358, March 2001.
11. S. Edelkamp and S. Schroedl. Localizing A*. In *National Conference on Artificial Intelli-
    gence (AAAI)*, pages 885–890, 2000.
12. M. L. Fredman and R. E. Tarjan. Fibonacci heaps and their uses in improved network opti-
    mization algorithm. *Journal of the ACM*, 34(3):596–615, 1987.

13. S. Handley, P. Langley, and F. Rauscher. Learning ot predict the duration of an automobile trip. In *Knowledge Discovery and Data Mining (KDD)*, pages 219–223, 1998.
14. A. C. Harvey. *Forecasting, Structural Time Series Models and the Kalman Filter*. Cambridge University Press, 1990.
15. C. A. Hipke. *Verteilte Visualisierung von Geometrischen Algorithmen*. PhD thesis, University of Freiburg, 2000.
16. D. S. Hirschberg. A linear space algorithm for computing common subsequences. *Communications of the ACM*, 18(6):341–343, 1975.
17. P. Klein, S. Rao, M. Rauch, and S. Subramanian. Faster shortest-path algorithms for planar graphs. *Special Issue of Journal of Computer and System Sciences on selected papers of STOC 1994*, 55(1):3–23, 1997.
18. R. E. Korf and W. Zhang. Divide-and-conquer frontier search applied to optimal sequence allignment. In *National Conference on Artificial Intelligence (AAAI)*, pages 910–916, 2000.
19. J. B. MacQueen. Some methods for classification and analysis of multivariate observations. In *Symposium on Math, Statistics, and Probability*, volume 1, pages 281–297, 1967.
20. K. Mehlhorn and U. Meyer. External-memory breadth-first search with sublinear I/O. In *European Symposium on Algorithms (ESA)*, 2002.
21. G. W. Milligan and M. C. Cooper. An examination of procedures for determining the number of clusters in a data set. *Psychometrika*, 50(1):159–179, 1985.
22. C. Navigation Technologies, Sunnyvale. Software developer's toolkit, 5.7.4 solaris edition, 1996.
23. B. W. Parkinson, J. J. Spilker, P. Axelrad, and P. Enge. *Global Positioning System: Theory and Applications*. American Institute of Aeronautics and Astronautics, 1996.
24. L. Piegl and W. Tiller. *The nurbs book*. Springer, 1997.
25. C. A. Pribe and S. O. Rogers. Learning to associate driver behavior with traffic controls. In *Proceedings of the 78th Annual Meeting of the Transportation Review Board*, Washington, DC, January 1999.
26. S. Rogers, P. Langley, and C. Wilson. Mining GPS data to augment road models. In *Knowledge Discovery and Data Mining (KDD)*, pages 104–113, 1999.
27. S. Schroedl, S. Rogers, and C. Wilson. Map refinement from GPS traces. Technical Report RTC 6/2000, DaimlerChrysler Research and Technology North America, Palo Alto, CA, 2000.
28. F. Schulz, D. Wagner, and K. Weihe. Dijkstra's algorithm on-line: An empirical case study from public railroad transport. *Journal of Experimental Algorithmics*, 5(12), 2000.
29. F. Schulz, D. Wagner, and C. Zaroliagis. Using multi-level graphs for timetable information. In *Workshop on Algorithm Engineering and Experiments (ALENEX)*, 2002.
30. L. Thoma and N. Zeh. I/O-efficient algorithms for sparse graphs. In *Memory Hierarchies*, Lecture Notes in Computer Science. Springer, 2002. To appear.
31. M. Thorup. Undirected single-source shortest paths with positive integer weights in linear time. *Journal of the ACM*, 46:362–394, 1999.
32. J. Wang, S. Rogers, C. Wilson, and S. Schroedl. Evaluation of a blended DGPS/DR system for precision map refinement. In *Proceeedings of the ION Technical Meeting 2001*, Institute of Navigation, Long Beach, CA, 2001.

The Aircraft Sequencing Problem

Torsten Fahle, Rainer Feldmann, Silvia Götz, Sven Grothklags, and Burkhard Monien

Department of Computer Science, University of Paderborn,
Fürstenallee 11, 33102 Paderborn, Germany
{tef,obelix,sylvie,sven,bm}@uni-paderborn.de

Abstract. In this paper we present different exact and heuristic optimization methods for scheduling planes which want to land (and start) at an airport – the Aircraft Sequencing Problem (ASP). We compare two known integer programming formulations with four new exact and heuristic solution methods regarding quality, speed and flexibility.

1 Motivation

Worldwide air traffic has experienced a significant growth of 5 to 10 percent a year during the last decades and this trend is expected to continue for the next years despite the terror attacks on Sep. 11th, 2001. This growth has resulted in substantial increases in delay at nearly every major airport. However, environmental and geographic constraints limit the opportunities to increase system capacity by building new airports or adding new runways at existing airports. So one major aim must be to use the existing capacities as efficiently as possible to increase the throughput and work against the growing delays.

For nearly every major airport the Terminal Area with its runways is the limiting factor of the whole system. All arriving and departing planes which come/go from/to different directions come together and have to be merged to use the available runways while for safety reasons strict separation requirements have to be respected. These separation requirements greatly depend on the order in which the planes land.

A large and heavy plane, like the Boeing 747, generates a great deal of air turbulence (wake vortices) and a plane flying too close behind could lose its aerodynamic stability. For safety reasons therefore landing a Boeing 747 necessitates a relatively large time delay before other planes can land. On the other hand a light plane, like the Avro RJ85, generates little air turbulence and therefore landing such a plane necessitates only a relatively small time delay before other planes can land. Table 1 shows typical minimal separation requirements for landing planes depending on the weight class they belong to.

Planes taking off impose similar restrictions on successive operations. If a runway is used for both, taking off and landing, switching between starting and landing planes results in even higher separation requirements. Furthermore, if two starting planes are going to use the same air corridor later on additional separation times have to be respected because the separation requirements in air corridors are much larger than during take offs or landings.

R. Klein et al. (Eds.): Comp. Sci. in Perspective (Ottmann Festschrift), LNCS 2598, pp. 152–166, 2003.
© Springer-Verlag Berlin Heidelberg 2003

| Leading | Trailing plane | | |
plane	Light	Medium	Heavy
Light	3	3	3
Medium	5	4	4
Heavy	6	5	4

Table 1. Minimal landing separations (in nautical miles) between different plane types.

So the order in which planes land and start at an airport can have a great impact on the overall system throughput and thereby also the amount of delay to endure.

Besides the separation requirements each plane normally has to land resp. start during a predefined time window. For an arriving aircraft its earliest landing time is determined by its shortest way to the runway and its maximum airspeed. The latest time is given at least by the period the aircraft can be put into holding (circling) before it runs out of fuel. Obviously, for a landing aircraft this time window is a hard constraint.

For a starting aircraft its earliest starting time can be determined by the time it is ready to leave the gate and the taxiing time it needs to reach the runway. In theory there does not need to exist a latest starting time if one allows an arbitrary delay but in practice this delay is limited.

Besides these 'natural' time windows most starting planes in Europe must respect so called CFMU-slots, which are 15 minutes time windows during which they are allowed to take off. If they miss their slot, they have to apply for another CFMU-slot, which normally takes quite a long time, especially during peak hours. These slots are centrally given away by the EUROCONTROL Central Flow Management Unit (CFMU) in Brussels for whole western Europe to minimize air traffic congestion. Similar restrictions exist in the United States.

One difficult issue is the choice of an appropriate objective function to be able to measure the quality of different aircraft sequences. There are (at least) three different parties involved that partly have different and contrary opinions about how a 'good' aircraft sequence should look like.

Air Traffic Control (ATC) is the most important party as it is exclusively responsible for the safety of the landing and starting aircrafts and has to execute the actual sequencing. The main objectives of ATC are
  - maximizing the throughput of the runways
  - minimizing inbound delays, in short that means minimizing the time that *approaching* planes are in the air
  - adherence to CFMU-slots
  - minimizing the workload of the ATC controllers

Airline The main objectives of the airlines are
  - punctuality regarding the (printed) schedule[1]
  - maximizing connectivity between incoming and outgoing flights
  - respecting airline priorities within their own flights

---

[1] This kind of punctuality is (surprisingly) completely ignored by ATC at present.

Airport The main objectives of the airport are
- punctuality regarding the operative schedule[2]
- minimizing the need for gate changes due to delays

Looking at the above list one central aspect of a 'good' schedule is its punctuality. Though every party has their own definition of punctuality.

2   Introduction

In this paper we study the problem of efficiently scheduling the landing and starting times of planes at an airport. We call this problem the Aircraft Sequencing Problem (ASP). We present different exact and heuristic optimization methods for the ASP and compare them regarding quality, speed and flexibility. The results presented here are part of a study (Priority Based Collaborative Aircraft Scheduling) that was done together with Deutsche Flugsicherung (DFS) and Lufthansa Systems (LHS).

From the point of view of the ASP landing and starting planes don't differ in the way they are modeled. They share the same parameters and restrictions (predefined time window, separation requirements). If it is not necessary for other reasons to distinguish between landing and starting planes, we therefore for simplicity only speak about landing planes from now on.

Formally we are given a set of planes for which the optimal landing times on the runway have to be computed while respecting each plane's time window and the separation criteria specified for each pair of planes. This problem is a slightly unusual application of a machine–scheduling problem. It is a special type of machine–scheduling problem with sequence-dependent processing times and with earliness and tardiness penalties for jobs not completed on target. However, there is an important difference between the ASP and common types of machine–scheduling applications: In the ASP the separation constraints must not only be satisfied between immediately successive planes, but between all pairs of planes.

The problem was presented first by Beasley et al. [4]. Ernst et al. [7] developed a fast simplex–based lower-bounding method for the ASP and used it in a heuristic as well as in an exact branch–and–bound method. As in [4] and [7], we are only interested in modeling the *decision* problem here, that is we are only interested in deriving, for each plane, a landing time. Along with this, the ATC controller must also determine solutions to the *control* problem of how to land the planes. In addition, we concentrate on the static case of scheduling aircraft landings, where we have complete information on the planes to schedule. The dynamic case was dealt with by Beasley et al. [3].

There are many variants of the ASP and many approaches for solving it. Most previous work in this area used simulation models ([1]), queuing models ([9]) or heuristics ([6; 12]). A dynamic programming approach was considered by Psaraftis [10; 11] and Bianco et al. [5] adopted a job–shop scheduling algorithm. Beasley et al. [4] provided a detailed review of literature in this area.

---

[2] The operative schedule is installed approximately one to two hours before touch down resp. take off.

The existing solution techniques for the ASP–variant used in this paper are based on a Mixed Integer Programming formulation (see section 4.1). We wanted to know how good other optimization methods are suited for the ASP and developed new approaches that use Constraint Programming, model the ASP as a (boolean) Satisfiability problem or are Local Search heuristics.

The rest of the paper is structured as follows. In section 3 we formally define the Aircraft Sequencing Problem as used in this paper. Section 4 is dedicated to the implemented optimization methods for the ASP. Four exact and two heuristic optimization approaches are presented. In section 5 we present numerical results and compare the different optimization methods regarding solution quality, speed, flexibility and extensibility. Section 6 sums up.

3 Problem Formulation

We present a model for the Aircraft Sequencing Problem that corresponds to the one introduced in [4]. In this paper we only consider the simplest model with only one runway and no additional restrictions. In section 5 we discuss possible extensions for the different optimization methods presented in section 4.

The following notations are needed to determine the validity of schedules for the ASP. They define the time windows during which the planes must land and the separation requirements between pairs of planes.

P the set of planes $P = \{1, \ldots, n\}$, $|P| = n$
E_i the earliest landing time of plane $i \in P$
L_i the latest landing time of plane $i \in P$
S_{ij} the minimal separation time required between the planes i and j, if plane i lands before j, $S_{ij} \geq 0 \; \forall i, j \in P$, $S_{ii} = 0 \; \forall i \in P$

For some of the optimization methods presented next it is necessary that all times (E_i, L_i and S_{ij}) are integral. In practice that is obviously no restriction, so in this paper we assume that all times are integers.

A solution for the ASP can be defined by specifying the exact landing times of all planes.

t_i the landing time of plane $i \in P$

A given solution (defined by the t_is of all planes $i \in P$) is *valid* if the following two conditions hold:

$$\forall i \in P : E_i \leq t_i \leq L_i \tag{1}$$
$$\forall i, j \in P : \text{if } t_i \leq t_j \text{ then } t_i + S_{ij} \leq t_j \tag{2}$$

Condition 1 assures that every landing time lies within the specified time window of the corresponding plane. Condition 2 ensures that the minimal separation times between planes are adhered.

In this paper we use a cost function to evaluate our landing schedules, where the costs of any plane are linearly related to the deviation from its preferred landing time. By minimizing this cost function we are looking for on–time schedules. To calculate the cost function we need the following information:

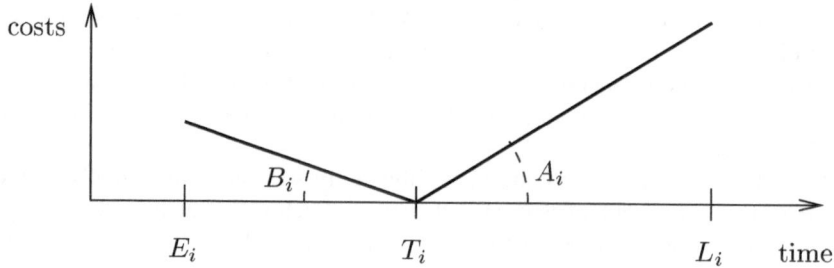

Figure 1. Run of the cost function for a single plane i.

T_i the target (preferred) landing time of plane $i \in P$

B_i the penalty cost (per time unit) for plane $i \in P$ when landing before its target time T_i

A_i the penalty cost (per time unit) for plane $i \in P$ when landing after its target time T_i

The total costs of a given solution can then be computed by

$$\sum_{i=1}^{n} (B_i \cdot \max\{0, T_i - t_i\} + A_i \cdot \max\{0, t_i - T_i\}) \tag{3}$$

and we are looking for a valid solution with minimal costs. This cost function, on one hand, models a major aspect of a 'good' solution, namely its punctuality. On the other hand it is simple enough (piecewise linear) so that it can be easily expressed by all of the investigated optimization methods. Figure 1 sketches the run of the cost function for a single plane.

4   Optimization Methods for the ASP

In this section we present the six optimization methods for the Aircraft Sequencing Problem that we implemented. There are two Integer Programming models, one that uses a continuous time representation (4.1) and one that discretizes time (4.2). Both models are taken from [4].

The rest of the optimization methods are new approaches we developed for the ASP. In section 4.3 we describe a Constraint Programming formulation. Section 4.4 shows how the ASP can be modeled as a Satisfiability problem. This formulation is only able to decide if a valid solution exists – it does not search for optimal solutions. Finally, in Section 4.5 we present two Local Search heuristics, a Hill Climbing and a Simulated Annealing algorithm.

4.1   Mixed Integer Program

For a (more or less) literal implementation of the ASP expressed by (1)-(3) as a Mixed Integer Program (MIP) we need the following variables:

t_i the landing time of plane $i \in P$
b_i the duration plane $i \in P$ lands before its target time T_i
a_i the duration plane $i \in P$ lands after its target time T_i
δ_{ij} 0: plane i lands before plane j; 1: otherwise

Then the ASP can be written as an MIP in the following way:

Minimize

$$\sum_{i=1}^{n} (B_i \cdot b_i + A_i \cdot a_i) \tag{4}$$

under

$$
\begin{aligned}
E_i \leq t_i \leq L_i && \forall i \in P && (5)\\
\delta_{ij} \in \{0,1\} && \forall i,j \in P, i \neq j && (6)\\
\delta_{ij} + \delta_{ji} = 1 && \forall i,j \in P, i < j && (7)\\
t_i + S_{ij} \leq t_j + M \cdot \delta_{ij} && \forall i,j \in P, i \neq j && (8)\\
t_i + b_i - a_i = T_i && \forall i \in P && (9)
\end{aligned}
$$

Condition (5) corresponds directly to (1) and ensures that all planes land during their specified time window.

To express the separation condition (2) in this MIP we need to know the order, in which the planes land. Conditions (6) and (7) take care of that by setting the boolean variable δ_{ij} to zero if plane i should land before j. Because of (7) either δ_{ij} or δ_{ji} is zero, meaning that either plane i lands before j or vice versa.

Depending on the order in which two planes i and j land either the separation constraint $t_i + S_{ij} \leq t_j$ or $t_j + S_{ji} \leq t_i$ must hold. Condition (8) puts only these constraints into action, for which the δ_{ij}-variables are zero. If the δ_{ij}-variable is one constraint (8) is made redundant by adding a big enough constant M to the right side. To strengthen the LP-relaxation of this MIP, M should actually be made as small as possible and dependent on the planes involved, e.g. $M = L_i + S_{ij} - E_j$ when used with variable δ_{ij}.

The cost function (4) must be modeled using the 'helper' variables b_i and a_i. Equation (9) ties together the actual landing time t_i and the deviations b_i and a_i. Note that in an optimal solution not both b_i and a_i will be non-zero[3], and so the deviation from the target time T_i is computed correctly.

4.2   Integer Program

One can obtain a different linear programming formulation for the ASP if time is discretized. We only need one type of (boolean) variables:

x_{it} 1: plane i lands at time t; 0: otherwise

---

[3] This is true if $B_i \geq 0$ and $A_i \geq 0$ holds, or, to be more precise, only if the cost function is concave, e.g. $B_i + A_i \geq 0$.

For simplicity we introduce C_{it} for the costs if plane i lands at time t, e.g. $C_{it} = B_i \cdot \max\{0, T_i - t\} + A_i \cdot \max\{0, t - T_i\}$; note that C_{it} is constant. Then the ASP can be formulated as the following Integer Program (IP):

Minimize

$$\sum_{i=1}^{n} \sum_{t=E_i}^{L_i} C_{it} \cdot x_{it} \tag{10}$$

under

$$x_{it} \in \{0, 1\} \qquad \forall i \in P, \forall t \in [E_i, L_i] \tag{11}$$

$$\sum_{t=E_i}^{L_i} x_{it} = 1 \qquad \forall i \in P \tag{12}$$

$$x_{it} + x_{ju} \leq 1 \qquad \forall i, j \in P, i \neq j, \forall t \in [E_i, L_i], \\ \forall u \in [t, t + S_{ij} - 1] \cap [E_j, L_j] \tag{13}$$

Conditions (11) and (12) ensure that every plane lands exactly at one time. The separation constraints are modeled by condition (13). It assures that, if plane i lands at time t, then plane j cannot land during the interval $[t, t + S_{ij} - 1]$ by forbidding that x_{it} and x_{ju} can both be one at the same time.

This model was presented in [4] but was not explored further because of disappointing computational experiences. One disadvantage of this model is its potential huge number of variables, which also causes a huge number of inequalities of type (13). We reformulated the model slightly and were able to reduce the number of inequalities by replacing condition (13) with

$$x_{it} + \sum_{u=\max\{t-S_{ji}+1, E_j\}}^{\min\{t+S_{ij}-1, L_j\}} x_{ju} \leq 1 \qquad \forall i, j \in P, i < j, \forall t \in [E_i, L_i] \tag{14}$$

Here we are exploiting the fact, that at most one of the x_{ju} can be 1 in a valid solution, so we can sum them up and the inequality must still be less than or equal to 1. This not only reduces the number of inequalities needed but also strengthens the LP relaxation.

4.3   Constraint Programming

Our Constraint Programming (CP) model for the ASP is similar to the MIP formulation. It uses the following variables with the specified domains:

t_i	$[E_i, L_i]$	the landing time of plane $i \in P$
$cost_i$	$[0, \infty]$	costs induced by plane $i \in P$
$succ_{ij}$	boolean	*true*: plane i lands before plane j; *false*: otherwise

Then the constraint set for the ASP can be written as follows:

$$\text{minimize} \sum_{i=1}^{n} cost_i \qquad (15)$$

$$cost_i = B_i \cdot \max\{0, T_i - t_i\} + A_i \cdot \max\{0, t_i - T_i\} \qquad \forall i \in P \qquad (16)$$

$$succ_{ij} \Longleftrightarrow \neg succ_{ji} \qquad \forall i, j \in P, i < j \qquad (17)$$

$$succ_{ij} \Longrightarrow t_i + S_{ij} \le t_j \qquad \forall i, j \in P, i \ne j \qquad (18)$$

Constraints (15) and (16) model the objective function of the ASP. The $succ_{ij}$ variables play the same role as the δ_{ij} variables in the MIP model and determine the order in which the planes should land. Constraints (17) and (18) correspond to conditions (7) and (8) of the MIP.

This constraint set is sufficient to model the ASP, but to exploit the power of CP to reduce the domains of the variables we have added some redundant constraints (for example a transitivity constraint: $succ_{ij} \wedge succ_{jk} \Longrightarrow succ_{ik}$). Furthermore we implemented a special fixing procedure for the t_i variables to guide CP to the optimal solution: Once the current time window of a plane i (domain of its t_i variable) cannot be affected by other planes, t_i is fixed to the time that causes the least possible costs for this plane.

4.4 SAT

The formulation of the ASP as a Satisfiability Problem (SAT) uses a similar approach as the IP formulation – it discretizes time. In contrast to all other optimization methods presented here the SAT formulation is only able to decide if there exists a valid solution for the ASP. It does not deliver an optimal solution, because it is hard to express something like 'costs' with a boolean expression only.

The SAT formulation uses the same set of boolean variables as the IP formulation with a slightly different interpretation:

x_{it} *true*: plane i lands at time t *or before*; *false*: otherwise

Choosing the variables in such a way allows us to construct the ASP out of clauses with only a *small constant* number of variables. This speeds up the used (Q)SAT solver ([8]) considerably. In a readable form the needed conditions (which are actually clauses) look like this:

$$x_{iL_i} = true \qquad \forall i \in P \qquad (19)$$

$$x_{it} \Longrightarrow x_{i,t+1} \qquad \forall i \in P, \forall t \in [E_i, L_i - 1] \qquad (20)$$

$$x_{it} \wedge \neg x_{i,t-1} \Longrightarrow x_{j,t-S_{ji}} \vee \neg x_{j,t+S_{ij}-1} \qquad \forall i, j \in P, i \ne j, \forall t \in [E_i, L_i] \qquad (21)$$

Clause (19) ensures that every plane has landed at least at its latest possible time L_i. Clause (20) ensures that if a plane is on the ground at time t it will stay on the ground until the end of its time window. Finally clause (21) models our separation restrictions: If plane i lands at time t ($x_{it} \wedge x_{i,t-1}$) then plane j either must have landed S_{ji} time units before $(x_{j,t-S_{ji}})$ or must not land during the next S_{ij} time units, e.g. it must not be on the ground after $S_{ij} - 1$ time units $(\neg x_{j,t+S_{ij}-1})^4$.

The complete theory that is given to our SAT solver looks like this:

$$\bigwedge_{\forall i \in P} x_{iL_i} \qquad \wedge \tag{22}$$

$$\bigwedge_{\forall i \in P, \forall t \in [E_i, L_i-1]} (\neg x_{it} \vee x_{i,t+1}) \qquad \wedge \tag{23}$$

$$\bigwedge_{\forall i,j \in P, i \neq j, \forall t \in [E_i,L_i]} (\neg x_{it} \vee x_{i,t-1} \vee x_{j,t-S_{ji}} \vee \neg x_{j,t+S_{ij}-1}) \tag{24}$$

Clauses (22)-(24) directly correspond to (19)-(21).

The SAT formulation is not able to look for optimal solutions but it can be easily extended to produce fault-tolerant solutions by using quantified boolean formulas. We did not actually implement this QSAT formulation, but experiences with other problems show that this should be an easy task.

4.5   Local Search

We developed two local search heuristics for the ASP, a Hill Climbing algorithm (HC) and a Simulated Annealing algorithm (SA). Both use the same solution representation, neighborhood structure and initial solution, which are the crucial problem dependent parts of any local search algorithm.

A solution is represented only by the order in which the planes should land, e.g. it is a permutation π of P. To be able to calculate the costs of such a solution we need a method to determine the actual landing times. This is done in a heuristic way:

- Plane π_1 lands as early as possible, e.g. $t_{\pi_1} = E_{\pi_1}$.
- All other planes follow as close as possible, e.g.

$$t_{\pi_i} = \max\{E_{\pi_i}, t_{\pi_1} + S_{\pi_1,\pi_i}, \dots, t_{\pi_{i-1}} + S_{\pi_{i-1},\pi_i}\}$$

If the separations S_{ij} fulfill the triangle inequality, e.g $S_{ij} + S_{jk} \geq S_{ik}$, one can determine the landing time of plane π_i just by looking at its predecessor:

$$t_{\pi_i} = \max\{E_{\pi_i}, t_{\pi_{i-1}} + S_{\pi_{i-1},\pi_i}\}$$

Using this heuristic all planes land as early as the permutation π permits. Nevertheless, the produced solution can be invalid, but only because a plane lands *after* its latest landing time. If there exists a valid solution for which the planes land in the order indicated by π, the heuristic will also construct a valid solution.

In favor of a simple neighborhood we allow all possible permutations π as solutions, even if they are illegal. Instead, we punish these illegal solutions with some high penalty costs.

---

[4] It can happen that some of the variables used in condition (21) actually do not exist. This has been done to avoid special cases and improve readability. A variable x_{it} with $t < E_i$ should be replaced by the constant *false* and a variable x_{it} with $t > L_i$ by the constant *true*.

We implemented two different types of neighborhood operations for the ASP. They generate a new permutation π' out of an existing permutation $\pi = (\pi_1, \ldots, \pi_n)$. They both need two additional parameters $k, l \in P$:

Swap The planes at position k and l ($k < l$) are exchanged, e.g.

$$\pi' = (\pi_1, \ldots, \pi_{k-1}, \pi_l, \pi_{k+1}, \ldots, \pi_{l-1}, \pi_k, \pi_{l+1}, \ldots, \pi_n)$$

Insert The plane at position k is taken out of π and inserted at position l ($k \neq l$), e.g.

$$\pi' = (\pi_1, \ldots, \pi_{k-1}, \pi_{k+1}, \ldots, \pi_l, \pi_k, \pi_{l+1}, \ldots, \pi_n), \text{ if } k < l$$
$$\pi' = (\pi_1, \ldots, \pi_{l-1}, \pi_k, \pi_l, \ldots, \pi_{k-1}, \pi_{k+1}, \ldots, \pi_n), \text{ if } k > l$$

Both neighborhood types are used concurrently by our local search algorithms.

Local search algorithms need an initial solution to start with. As our methods can deal with illegal solutions we don't need to spend much effort in constructing an initial solution. We choose an initial permutation σ, where the planes are ordered according to their earliest landing times, e.g. $\forall k, l \in P : k < l \Longrightarrow E_{\sigma_k} \leq E_{\sigma_l}$.[5]

Hill Climbing Our HC implementation is a randomized algorithm. The neighbors of the current solution are visited in a random order. Furthermore, we don't look for the best neighbor of the current solution but we switch to the first visited neighbor with costs less than the costs of the current solution.

Simulated Annealing We use a simple geometric cooling schedule for our SA algorithm. You need to specify the initial temperature T_0, the cooling factor cf and the acceptance ratio ar at which the algorithm should terminate. For all our experiments we used $T_0 = 10000$, $cf = 0.95$ and $ar = 0.005$.

5   Computational Results

We implemented the described optimization methods and ran tests on a number of problem instances. The programs were implemented in C/C++ under Linux using the gcc 2.95.3 compiler suite. The MIP and IP algorithm use ILOG Cplex 7.0 as linear programming solver and the CP algorithm uses ILOG Solver 5.0 as CP solver. The SAT implementation is based on a SAT solver developed at the University of Paderborn ([8]). All tests were carried out on a 933Mhz Pentium III workstation running RedHat Linux 7.2.

The problem instances that we used were taken from three different sources:

Beasley These are the eight ASP instances from J.E. Beasley's OR-Library, see [2] for details on obtaining the data.

---

[5] Experiments have shown that especially Simulated Annealing does not depend on the initial solution. We have done tests with random permutations and permutations that were ordered according to L_i and T_i. The solution quality and running times are nearly the same.

Pad These instances are self–generated by an ASP generator. The generator assures that the instances have a valid solution.

Fra These instances are based on the actual flight events at Frankfurt Airport on August 8th, 2000. For different periods of this day the landings and take offs were used as input data. We generated two sets of instances with (slightly) different separation criterias: short (good weather) and long separations (bad weather).

In tables 2, 3 and 4 the running times of the different optimization methods are shown. In the column named '#AC' the number of aircrafts of the corresponding problem instance is given. The following columns contain the running times in seconds of the optimization approaches. The running times of the two *randomized* heuristics HC and SA are actually the mean values of 10 runs each.

Empty entries indicate that the corresponding approach could not solve the problem within one day. As the exact methods could only solve the two smallest problems of the Fra–set table 3 only contains these two rows. In table 4 running times with an asterisk (*) indicate that the heuristic was not able to find a valid solution.

As expected the running times of both heuristics (HC and SA) are by far the smallest ones, especially on the larger instances. Besides HC is an order of magnitude faster than SA.

The SAT approach normally also finds *valid* solutions as fast as the heuristics, but cannot do any optimization. Furthermore it was not able to decide within one day whether the three problems Fra9-11, Fra11-13 and Fra13-15 have a valid solution or not. These are probably the most difficult instances but for Fra9-11 a valid solution was found by SA.

CP can only deliver solutions for the three smallest Beasley instances. It seems that there doesn't exist enough 'tight' constraints and therefore CP cannot exploit its major strength, its domain reduction techniques.

The MIP and IP show varying behavior on the problem instances. For the Beasley instances MIP is the faster solution method, for the rest of the instances IP wins. The reason for this lies in the fact that the time windows of the Beasley problems are much larger than the time windows of the Pad– and Fra–set. As the IP approach discretizes time, it produces Integer Programs with many variables for the Beasley instances which are costly to solve. But the LP–relaxation of the IP model is tighter than the relaxation of the MIP model and so instances with smaller time windows are solved faster by the IP approach.

As the heuristics HC and SA cannot guarantee to find optimal solutions the solution quality of these two approaches is of interest. In tables 5 and 6 the optimal solution value is shown besides the minimal, maximal and mean solution values (10 runs per instance) of the heuristics. Empty entries in the 'optimal' column mean that the optimal solution value is unknown. Empty values in the columns of the heuristics mean that they were unable to find valid solutions for the corresponding problem instance.

The solution quality of SA is always better than the solution quality of HC. Often SA is able to compute the optimal solution. Moreover SA produces solutions with a smaller variance in quality than HC.

The basic ASP presented in this paper can be extended naturally in different ways. The presented optimization methods vary in their capabilities to model such extensions:

instance	#AC	MIP	IP	CP	SAT	HC	SA
Beasley1	10	0.02	92.12	16.15	0.16	0.00	0.03
Beasley2	15	0.60	3970.65	306.17	0.38	0.01	0.10
Beasley3	20	0.10	181.73	11.34	0.75	0.01	0.18
Beasley4	20	22.41			0.78	0.01	0.19
Beasley5	20	105.62			0.73	0.01	0.18
Beasley6	30	0.00	13.56		0.00	0.02	0.05
Beasley7	44	1.30			0.13	0.04	0.96
Beasley8	50	2.49	841.78		5.23	0.15	1.74
Pad1	10	0.00	0.01		0.00	0.00	0.00
Pad2	20	0.03	0.05		0.00	0.01	0.14
Pad3	30	0.34	0.17		0.00	0.02	0.63
Pad4	40	50.26	1.70		0.00	0.06	1.39
Pad5	50	101.31	1.20		0.01	0.08	1.73
Pad6	60	2401.04	114.04		0.02	0.25	2.82
Pad8	80		622.73		0.03	0.54	5.99
Pad10	100		38654.65		0.04	1.09	11.65

Table 2. Running times (in seconds) of the different optimization methods on the Beasley and Pad instances.

instance	#AC	long separations			short separations		
		MIP	IP	CP	MIP	IP	CP
Fra9-9:30	29		663.04		446.59	22.94	
Fra9-10	53					77.98	

Table 3. Running times (in seconds) of the exact optimization methods on the Fra instances.

instance	#AC	long separations			short separations		
		SAT	HC	SA	SAT	HC	SA
Fra9-9:30	29	0.01	0.03	0.38	0.01	0.03	0.41
Fra9-10	53	0.02	0.17	1.71	0.03	0.17	1.98
Fra9-11	111		*(0.17)	8.18	0.06	1.54	13.39
Fra11-13	123		*(1.28)	*(8.42)	0.06	2.08	16.07
Fra13-15	112		*(0.92)	*(6.14)	0.07	1.57	12.86
Fra15-17	119	0.22	1.94	14.21	0.07	1.75	15.89
Fra17-19	111	0.17	1.42	10.20	0.06	1.41	10.94
Fra19-21	112	0.17	1.65	13.83	0.07	1.65	14.13

Table 4. Running times (in seconds) of the heuristic optimization methods on the Fra instances.

instance	optimal	Hill Climbing			Simulated Annealing		
		minimal	maximal	mean	minimal	maximal	mean
Beasley1	700	820	1520	960	820	820	820
Beasley2	1480	1710	1710	1710	1710	1710	1710
Beasley3	820	1350	1650	1440	1350	1350	1350
Beasley4	2520	2520	2520	2520	2520	2520	2520
Beasley5	3100	3100	3100	3100	3100	3100	3100
Beasley6	24442	24442	24442	24442	24442	24442	24442
Beasley7	1550	1550	1550	1550	1550	1550	1550
Beasley8	1950	3055	3055	3055	3055	3055	3055
Pad1	531.61	531.61	531.61	531.61	531.61	531.61	531.61
Pad2	995.93	1025.53	1025.53	1025.53	1025.53	1025.53	1025.53
Pad3	2418.12	2418.12	2418.12	2418.12	2418.12	2418.12	2418.12
Pad4	3345.43	3475.09	4330.71	3645.20	3345.43	3345.43	3345.43
Pad5	2424.68	2434.11	2434.11	2434.11	2434.11	2434.11	2434.11
Pad6	3675	3755	3825	3788.33	3755	3825	3766.50
Pad8	4820	4910	6000	5255.56	4820	5015	4900.00
Pad10	6605	6605	7045	6749.50	6605	6800	6677.50

Table 5. Solution quality of the heuristic optimization methods on the Beasley and Pad instances.

instance	optimal	Hill Climbing			Simulated Annealing		
		minimal	maximal	mean	minimal	maximal	mean
long separations							
Fra9-9:30	4200	4310	5140	4505	4200	4590	4317
Fra9-10		8980	10710	9720	8990	9350	9101
Fra9-11					35350	37260	36500
Fra11-13							
Fra13-15							
Fra15-17		63620	67240	64641	62680	66540	64374
Fra17-19		23060	25540	24670	22880	24410	23640
Fra19-21		26360	28630	27180	24690	28580	25418
short separations							
Fra9-9:30	2540	2540	2890	2662	2540	2570	2552
Fra9-10	4340	4400	4940	4492	4370	4410	4384
Fra9-11		12000	14480	12643	11650	12170	11842
Fra11-13		32840	36740	34883	31060	33330	31903
Fra13-15		19640	23060	21221	19530	20780	19923
Fra15-17		11150	13460	12595	11150	12130	11592
Fra17-19		8120	10100	9278	7800	8220	7972
Fra19-21		8260	8510	8450	8260	8760	8460

Table 6. Solution quality of the heuristic optimization methods on the Fra instances.

multiple runways All of the presented optimization methods can be extended to schedule the landings on more then one runway. In [4] this is shown for MIP and IP.

non–linear objective function As described in section 1 we probably need a more complex objective function to measure the quality of ASP solutions in practice. The MIP approach is restricted most in this respect because it can model linear objective functions only. CP, HC and SA hardly put any restrictions on the objective function.

dynamic ASP In practice the ASP is a dynamic problem where the problem data changes (slightly) from time to time: new planes appear, time windows and separation criteria change, etc. It is desirable for the ATC controllers that the updated schedules don't differ too much from previous schedules. Beasley describes a general penalty–based method for this in [3] that can be adopted for all optimization methods but SAT.

The presented optimization approaches should be divided into two categories: exact and heuristic methods. The exact methods (IP, MIP, CP and SAT) always find an optimal solution if one exists. The SAT approach is an exception, as it is only looks for valid rather than optimal solutions. The heuristic approaches (HC and SAT) cannot guarantee the quality of the calculated solutions and may even fail to find existing valid solutions. On the other hand their modeling capabilities are in no way inferior to the ones of CP.

CP very powerful modeling capabilities, but by far the slowest method

MIP fastest exact optimization method for instances with big time windows, difficulties in modeling non-linearities

IP fastest exact approach for instances with small time windows, capable to express non-linear cost function

SAT no optimization, very fast on all but the most difficult instances, with QSAT–extension: capable to produce fault–tolerant solutions

HC fastest optimization method, medium to good solution quality with medium variance

SA good solution quality with small variance, second fastest approach

Both heuristic optimization approaches (HC and SA) for the ASP work well. HC is faster, SA produces solutions with better quality and both are very flexible in modeling additional requirements. Integer Programming (IP) is the best compromise of the exact solution methods, at least if the predefined time windows of the planes don't get too large.

6   Conclusions

The major goal of this work was to compare six different exact and heuristic solution methods for the basic Aircraft Sequencing Problem and to determine how well they are suited for additional extensions that arise in practice.

Besides two known Mixed Integer Programming formulations four new exact and heuristic solution methods were developed: a Constrained Programming approach, a Satisfiability formulation, a Hill Climber and a Simulated Annealing algorithm. Computational results were presented for a number of test problems with up to 123 planes.

Furthermore we compared the implemented methods regarding their potential for extensions to the ASP model used in this paper.

It is of interest to see if the results that we describe in this paper can actually be confirmed for extended ASP models with multiple runways, additional constraints and more complex objective functions.

References

1. A. Andreussi, L. Bianco, and S. Ricciardelli. A simulation model for aircraft sequencing in the near terminal area. *European Journal of Operational Research*, 8:345–354, 1981.
2. J.E. Beasley. Obtaining test problems via internet. *Journal of Global Optimization*, 8:429–433, 1996.
3. J.E. Beasley, M. Krishnamoorthy, Y.M. Sharaiha, and D. Abramson. The displacement problem and dynamically scheduling aircraft landings. *to appear*.
4. J.E. Beasley, M. Krishnamoorthy, Y.M. Sharaiha, and D. Abramson. Scheduling aircraft landings – the static case. *Transportation Science*, 34:180–197, 2000.
5. L. Bianco, S. Ricciardelli, G. Rinaldi, and A. Sassano. Scheduling tasks with sequence–dependent processing times. *Naval Research Logistics*, 35:177–184, 1988.
6. R.G. Dear and Y.S. Sherif. The dynamic scheduling of aircraft in high density terminal areas. *Microelectronics and Reliability*, 29:743–749, 1989.
7. A.T. Ernst, M. Krishnamoorthy, and R.H. Storer. Heuristic and exact algorithms for scheduling aircraft landings. *Networks*, 34:229–241, 1999.
8. R. Feldmann, B. Monien, and S. Schamberger. A distributed algorithm to evaluate quantified boolean formulae. In *Proc. of the 17th National Conference on Artificial Intelligence (AAAI-2000)*, pages 285–290, 2000.
9. J. Milan. The flow management problem in air traffic control: A model of assigning priorities for landings at a congested airport. *Transportation, Planning and Technology*, 20:131–162, 1997.
10. H.N. Psaraftis. A dynamic programming approach to the aircraft sequencing problem. Research report R78-4, Flight Transportation Laboratory, MIT, 1978.
11. H.N. Psaraftis. A dynamic programming approach for sequencing groups of identical jobs. *Operations Research*, 28:1347–1359, 1980.
12. C.S. Venkatakrishnan, A. Barnett, and A.R. Odoni. Landings at Logan Airport: Describing and increasing airport capacity. *Transportation Science*, 27, 1993.

On the Computational Complexity of Hierarchical Radiosity

Robert Garmann[2] and Heinrich Müller[1]

[1] Informatik VII (Computer Graphics)
University of Dortmund
D-44221 Dortmund, Germany
mueller@cs.uni-dortmund.de
http://ls7-www.cs.uni-dortmund.de
[2] robert.garmann@gmx.de
http://www.weller-garmann.de

Abstract. The hierarchical radiosity algorithm is an efficient approach to simulation of light with the goal of photo-realistic image rendering. Hanrahan et. al. describe the initialization and the refinement of links between the scene's patches based upon a user-specified error parameter ε. They state that the number of links is linear in the number of elements if ε is assumed to be a constant. We present a result based upon the assumption that the geometry is constant and ε approaches 0 in a multigridding-procedure. Then the algorithm generates $L = \Theta(N^2)$ links where N denotes the number of elements generated by the algorithm.

1 Introduction

Radiosity is used in calculations of computer graphics which determine the diffuse interreflection of light in a three-dimensional environment for photo-realistic image synthesis. The task is to determine the radiosity function B by solving the radiosity equation

$$B(\mathbf{y}) = B_e(\mathbf{y}) + \rho(\mathbf{y}) \int_S G(\mathbf{x}, \mathbf{y}) V(\mathbf{x}, \mathbf{y}) B(\mathbf{x}) d\mathbf{x}. \tag{1}$$

$\mathbf{x}, \mathbf{y}$ denote surface points of the environment S. The radiosity B located at a point $\mathbf{y}$ is determined by integrating over the radiosity incident from all points $\mathbf{x}$ of the environment S, and weighting the result by a reflection coefficient ρ. $B_e(\mathbf{y})$ denotes the radiosity emitted by $\mathbf{y}$ if the point belongs to a light source. This work concerns radiosity in *flatland*, that is $S \subset \mathbb{R}^2$.

The transfer of energy from point $\mathbf{x}$ to point $\mathbf{y}$ is weighted by the terms G and V. $G(\mathbf{x}, \mathbf{y}) = \cos\theta_x \cos\theta_y/(2r_{xy})$ defines the purely geometric relationships in flatland. θ_x and θ_y are the angles between the edge $\overline{\mathbf{xy}}$ of length r_{xy} and the surface normals at $\mathbf{x}$ and $\mathbf{y}$. V is a visibility function which equals one if $\mathbf{x}$ and $\mathbf{y}$ are mutually visible, otherwise zero. See [1; 6] for a derivation of the radiosity equation from physics.

Sometimes the functions ρ, G and V are combined into one function

$$k(\mathbf{x}, \mathbf{y}) = \rho(\mathbf{y}) G(\mathbf{x}, \mathbf{y}) V(\mathbf{x}, \mathbf{y})$$

which is called the *kernel* of the radiosity equation.

R. Klein et al. (Eds.): Comp. Sci. in Perspective (Ottmann Festschrift), LNCS 2598, pp. 167–178, 2003.

Since the radiosity equation cannot be solved analytically in general, approximating algorithms have been developed. One of these algorithms is the hierarchical radiosity algorithm. The hierarchical radiosity algorithm, together with several modifications proposed since its appearance, is widely used because of its practical computational efficiency. In the original paper by Hanrahan et al. [3], a theoretical argument for the good computational behavior was given, too. In this contribution we refine and generalize the complexity analysis, with the surprising result that in a multigridding setting which is relevant in practice, the asymptotic worst-case behavior is an order worse than that given by Hanrahan et al. under a more restrictive assumption.

Section 2 is devoted to a summary of the hierarchical radiosity algorithm. In section 3 we give a proof of the computational complexity of the algorithm for a special simple two-dimensional scene. One basic aid for the proof is the decision on a tractable oracle function. In section 4 we complement the theoretical result by experimental investigations of several simple scenes, using an exact oracle function.

2   Hierarchical Radiosity

The hierarchical radiosity algorithm of Hanrahan et al. [3] is a procedure which solves the radiosity equation using a finite element approach. The radiosity equation is transformed into the linear system of equations

$$b_i = e_i + \sum_{j=1}^{n} k_{ij} b_j, \quad i = 1, \ldots, n, \tag{2}$$

where n denotes the number of surface elements into which the surface of the given scene is decomposed. The coefficients b_i, e_i and k_{ij} represent discrete approximations of B, B_e and k over constant basis functions corresponding to the elements.

All radiosity algorithms have roughly two components. These can be described by *setting up* the equations, i. e. computing the entries of the linear system, and *solving* the linear system. The latter typically invokes some iterative solution scheme.

Hierarchical radiosity considers the possible set of interactions between elements in a recursive enumeration scheme. An interaction between two elements is called a *link*. The algorithm has to insure that every transport, i. e. every surface interacting with other surfaces, is accounted for exactly once. This goal is achieved by applying the following procedure to every input surface with every other input surface as a second argument [5]:

```
ProjectKernel(Element i, Element j)
  error= Oracle(i,j);
  if (Acceptable(error) || RecursionLimit(i,j))
    link (i,j);
  else
    if (PreferredSubdivision(i,j) == i)
      ProjectKernel(LeftChild(i), j);
      ProjectKernel(RightChild(i), j);
```

```
else
    ProjectKernel(i, LeftChild(j));
    ProjectKernel(i, RightChild(j)).
```

First a function `Oracle` is called which estimates the error across a proposed interaction between elements `i` and `j`. If this estimated error satisfies the predicate `Acceptable`, a link is defined between the elements `i` and `j`, and the related coefficient of the kernel is calculated. Resource limitations may require to terminate the recursion even if the error is not acceptable yet. This predicate is evaluated by `RecursionLimit`. If the error is too high then the algorithm recurs by subdividing elements into two child elements. Typically it will find that the benefit in terms of error reduction is not equal for the two elements in question. For example, one element might be larger than the other one, and it will be more helpful to subdivide the larger one in order to reduce the overall error. This decision is taken by `PreferredSubdivision`, and a recursive call is initiated on the child interactions which arise from splitting one of the parent elements.

There are different ways to estimate the error involved with a link between elements `i` and `j`. Hanrahan et al. [3] calculate an approximate form-factor corresponding to the link, and use it to obtain an approximate upper bound on the transferred energy. Several other estimation techniques are possible [4].

The predicate `Acceptable` may itself become more and more stringent during the calculation, creating a fast but inaccurate solution first and using it as a starting point for successive solutions with less error. This technique is called *multigridding*.

3   Analysis for a Simple Scene

3.1   The Result

We are going to formulate a complexity result for the number of elements, N, and the number of links, L, which are generated by the hierarchical radiosity algorithm. In order to do so, we have to specify the behavior of the functions and predicates of the algorithm.

Let $\varepsilon > 0$. We accept a link if the error predicted by the oracle function is less than or equal to ε. The oracle function itself used in the analysis is the approximation

$$\frac{\max(A_x, A_y)}{r_{xy}}$$

of the form-factors between two elements, where A_x, A_y denote the sizes of the two elements involved, and where r_{xy} is the distance between the centers of the elements. The form-factors express the geometric relationship relevant for radiosity exchange between two elements [1; 6]. They are defined as an integral of $G(x,y)V(x,y)$ over the two elements, divided by the size of the radiosity-sending surface. We use this simple oracle because of clarity of the proof[1].

For the analysis we modify the algorithm slightly. Instead of splitting just one of the elements `i` and `j`, we now split both elements and then recur four times:

---

[1] The analysis may also be applied to the oracle $\left(\frac{\max(A_x, A_y)}{r_{xy}}\right)^2$, which is a stronger bound of the real error [5]. It can be easily shown that this does not affect the relationship between L and N.

```
ProjectKernel(Element i, Element j)
  error= Oracle(i,j);
  if (Acceptable(error) || RecursionLimit(i,j))
    link (i,j);
  else
    ProjectKernel(LeftChild(i), LeftChild(j));
    ProjectKernel(RightChild(i), LeftChild(j));
    ProjectKernel(LeftChild(i), RightChild(j));
    ProjectKernel(RightChild(i), RightChild(j));
```

This modification does not affect the relationship between the number of patches and links seriously. The modified algorithm does not generate more than twice the number of links established by the original algorithm. The number of elements differs by not more than a constant factor less than two.

Finally, in the analysis the predicate `RecursionLimit` is assumed to return `false` for arbitrary pairs of elements.

Using these modifications and specifications we will prove that

$$N = \Theta\left(g\varepsilon^{-1}\right),$$
$$L = \Theta\left(N\varepsilon^{-1}\right), \qquad (3)$$

where g is a parameter determined by the initial geometry of the scene.

We can interpret the terms for L and N of this result in different ways. Let us first assume that ε is fixed and g variable. Then

$$L = \Theta(N\varepsilon^{-1}) = \Theta(N),$$

that is, the number of links scales linearly with the number of elements. Hanrahan et al. [3] arrived at this result, because they assumed a constant error threshold.

An other assumption may be that the geometry term g is fixed and ε is varied, e. g. in a multigridding procedure. Then the number of elements becomes $N = \Theta(\varepsilon^{-1})$, or, vice versa, $\varepsilon^{-1} = \Theta(N)$, and we get

$$L = \Theta(N\varepsilon^{-1}) = \Theta(N^2),$$

i. e. the number of links scales quadratically with the number of elements.

Theorem 1. *Under the assumption that the initial geometry is fixed and the error parameter is variable, the hierarchical radiosity algorithm generates a quantity of links which grows quadratically with the number of elements.* ☐

The remainder of this section is devoted to the proof of equations (3).

3.2   The Scene

Our analysis concerns the simple scene depicted in Fig. 1. The scene consists of two directly facing parallel two-dimensional patches, each having a length of l units. The perpendicular distance of the two patches is d. An analysis of a general two-dimensional two-patch scene is described in [2].

Let A_x and A_y denote the lengths of two elements generated by the algorithm. Since both initial patches have an equal length l, the modified algorithm assures that $A_x = A_y$ if A_x and A_y are connected by a link. In the following, $A := A_x$ denotes the size of the elements and r the distance between their centers.

3.3 Number of Links

Figure 2 shows the subdivision hierarchies generated by the algorithm for each of the two initial patches. Double arrows are drawn between elements that are linked. Links occur at different levels of the hierarchies. All plotted links together are responsible for connecting a single point of one patch, indicated by a small dot, to all points of the other patch. Nearby points are accounted for by links on deep levels; distant points are represented by links on coarser levels. Every single element is connected to a small subset of elements of the other hierarchy. Figure 3 shows the region of neighbors of an element. Hanrahan et al. [3] stated that the length D of the region of neighbors is constant. We will derive a precise formula which describes how D depends on ε.

Consider any link $A_x \leftrightarrow A_y$ established by the algorithm. The link was introduced because the predicate Acceptable was satisfied for the link, thus

$$\frac{A}{r} \leq \varepsilon.$$

From this we get the following lower bound for the distance of an established link:

$$r \geq r_{min} := A\varepsilon^{-1}. \tag{4}$$

The link $A_x \leftrightarrow A_y$ may have been established by the algorithm only if its parent link has been refined. The parent link was refined, because it did not satisfy the predicate Acceptable:

$$\frac{2A}{r_{parent}} > \varepsilon.$$

From this we get

$$r_{parent} < 2A\varepsilon^{-1}.$$

Consider the relationship between r and r_{parent}. Figure 4 shows the link with distance r and the four possibilities where the parent link could be located. By the triangle inequality we know that

$$r - A \leq r_{parent} \leq r + A,$$

or equivalently

$$\exists \delta \in [-1, 1] : \quad r = r_{parent} + \delta \cdot A.$$

From this we get the following upper bound on the distance of an established link:

$$r < r_{max} := 2A\varepsilon^{-1} + \delta \cdot A. \tag{5}$$

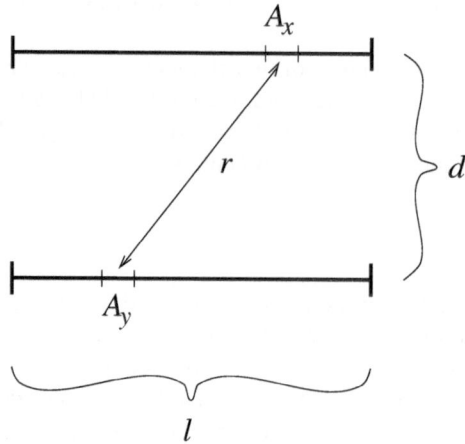

Figure 1. A simple scene.

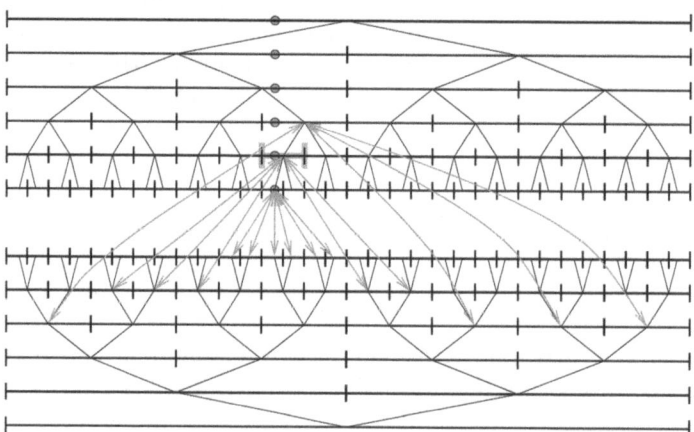

Figure 2. Links connect elements at different levels.

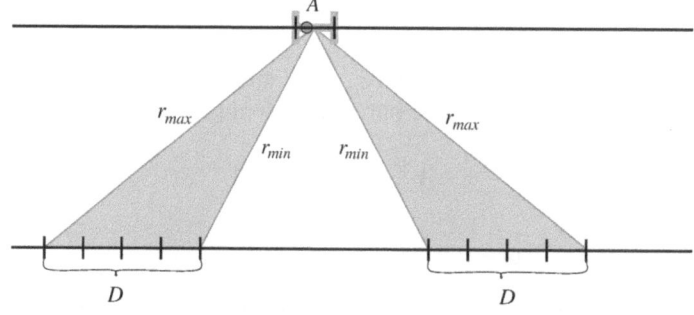

Figure 3. Region of neighbors of a single element.

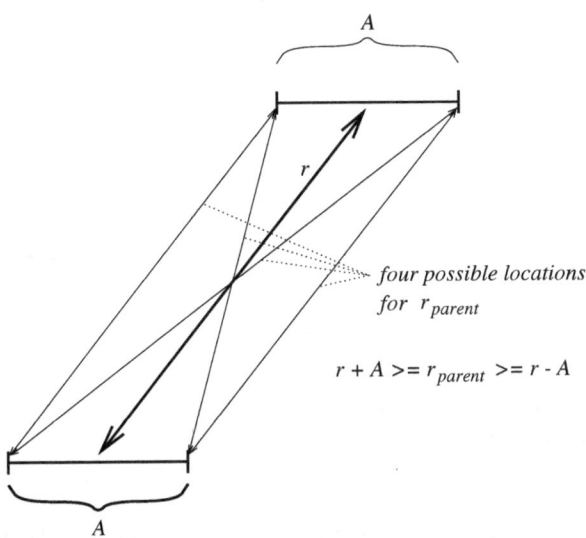

Figure 4. Relationship between the distance r of a link connecting two elements of size A and the distance r_{parent} of the parent link.

The length of the region of neighbors, D, can be bounded by the triangle inequality as follows (see Fig. 3):

$$r_{max} - r_{min} \leq D \leq r_{max} + r_{min}$$

where

$$r_{max} - r_{min} = A \cdot (\varepsilon^{-1} + \delta)$$
$$r_{max} + r_{min} = A \cdot (3\varepsilon^{-1} + \delta).$$

Since both bounds are of order $A \cdot \Theta(\varepsilon^{-1})$, we have $D = A \cdot \Theta(\varepsilon^{-1})$ as well.

So far we have derived a formula that describes how the length of the region of neighbors of a single element, D, depends on ε and the size of the element, A. The number of neighbors of the element is simply $\Theta\left(\frac{D}{A}\right)$, because the neighbors of the element have got the size A. The total number of links in the scene is the number of neighbors times the number of elements, which is

$$L = \Theta\left(N\frac{D}{A}\right) = \Theta\left(N\varepsilon^{-1}\right).$$

This proves the second equation of (3). $\square$

3.4   Number of Elements

The fact that the initial patches we are dealing with are in parallel assures that both initial patches are subdivided uniformly into leaves of equal size. Every leaf is linked

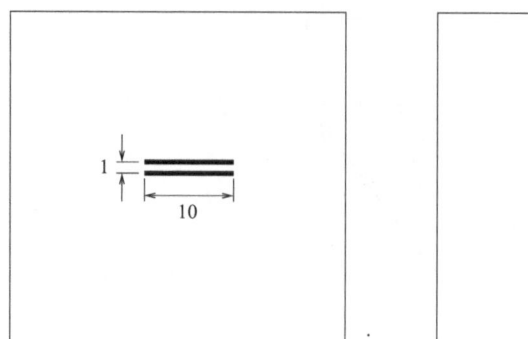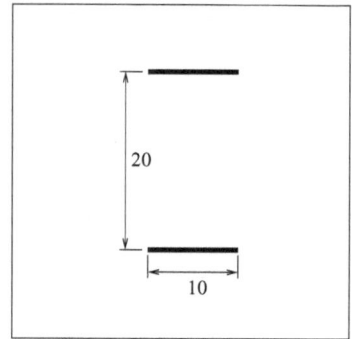

Figure 5. Two simple scenes. The left scene consists of two parallel patches with a small distance. In the right scene the distance is large.

to at least its directly facing counterpart, otherwise it would have been subdivided. Let A_{leaf} denote the size of the elements at the leaves. We show that

$$\frac{1}{2}d \cdot \varepsilon < A_{leaf} \le d \cdot \varepsilon.$$

The bounds can be derived from the value of the predicate `Acceptable` if it is applied to links that have got the minimum distance $r = d$. The upper bound arises from the fact that the predicate `Acceptable` is satisfied for those links which connect pairs of leaves with a minimum distance d between. The lower bound holds because the predicate `Acceptable` was not satisfied for links between two patches of size $2 \cdot A_{leaf}$.

Now we use the term $A_{leaf} = \Theta(d \cdot \varepsilon)$ for the size of the leaves in order to count the number of leaves, N_{leaf}, of the element hierarchy of a single initial patch. The number simply is

$$N_{leaf} = \frac{l}{A_{leaf}} = \Theta\left(\frac{l}{d}\varepsilon^{-1}\right).$$

Since we have only two initial patches, and since a complete binary tree has $2 \cdot N_{leaf} - 1$ nodes, we conclude

$$N = 2 \cdot (2 \cdot N_{leaf} - 1) = \Theta\left(\frac{l}{d}\varepsilon^{-1}\right).$$

By setting $g := \frac{l}{d}$ we have proven the first equation of (3).   □

4   Experiments with an Exact Oracle

A potential objection to the given proof is that the oracle function used yields a bad estimation of the error. For that reason, we have numerically analyzed the four simple scenes depicted in Fig. 5 and 6 with an *exact* oracle for the geometry error. Reducing the geometry error is a justified way to gain a small error in the resulting radiosity function

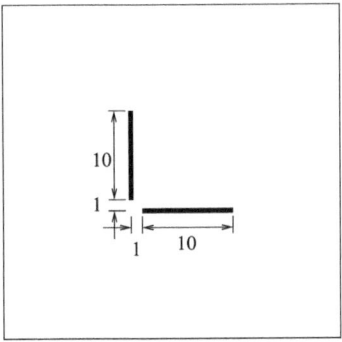 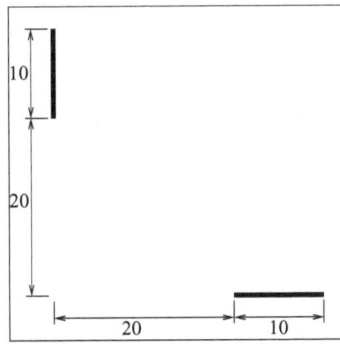

Figure 6. Two simple scenes with orthogonally located patches. In the left scene the distance is small, in the right scene the distance is large.

[5]. We have calculated the L^1-error of the geometry term G evaluated at the centers $\mathbf{x}_c$ and $\mathbf{y}_c$ of two elements:

$$\int_{A_x} \int_{A_y} \mid G(\mathbf{x}_c, \mathbf{y}_c) - G(\mathbf{x}, \mathbf{y}) \mid \mathrm{dydx}.$$

Actually the L^1-error was approximated using a 1024×1024 discretization of the geometry term G between the two initial patches. The RecursionLimit predicate was forced to always return false. Furthermore we used the original version of the algorithm where only one of two elements is subdivided if the error is not accepted. In the test the function PreferredSubdivision selected the larger element out of its parameters.

For a sequence of continuously decreasing ε-values the resulting elements and links have been calculated. Figures 7 and 8 show a diagram for each of the four scenes. Every resulting configuration is represented by a dot which has the number of links (vertical axis) and the number of elements (horizontal axis) as coordinates. Clearly, because of the discrete nature of the algorithm, in general there are several different values of L possible for the same N. This arises from the fact that the algorithm may refine links between patches which are already subdivided. Then the number of links increases while the number of patches is not varied. This reveals in "stacked" dots in the diagrams. Nevertheless, the least-squares-fits of quadratic polynomials shown in the diagrams manifest the theoretical result that the number of links depends quadratically on the number of elements.

Acknowledgements

The authors would like to thank Georg Pietrek, Peter Schröder, and Frank Weller for their helpful hints and remarks.

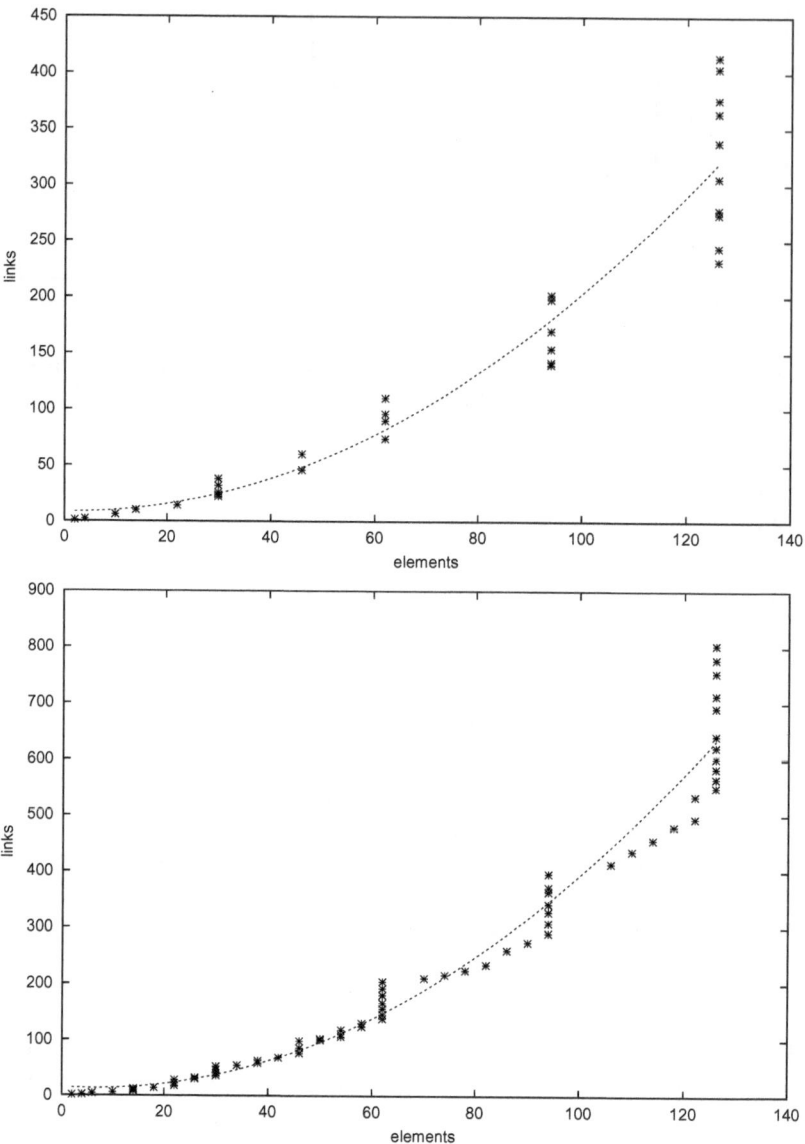

Figure 7. Empirical analysis for the scenes consisting of parallel patches. The top diagram shows results for the small distance scene, the bottom one for the large distance scene. Measurements have been performed for a sequence of continuously decreasing ε-values. Every dot means one measurement of the number of links (vertical axis) and the number of elements (horizontal axis). The continuous curves have been determined by a least-squares-fit of a quadratic polynomial.

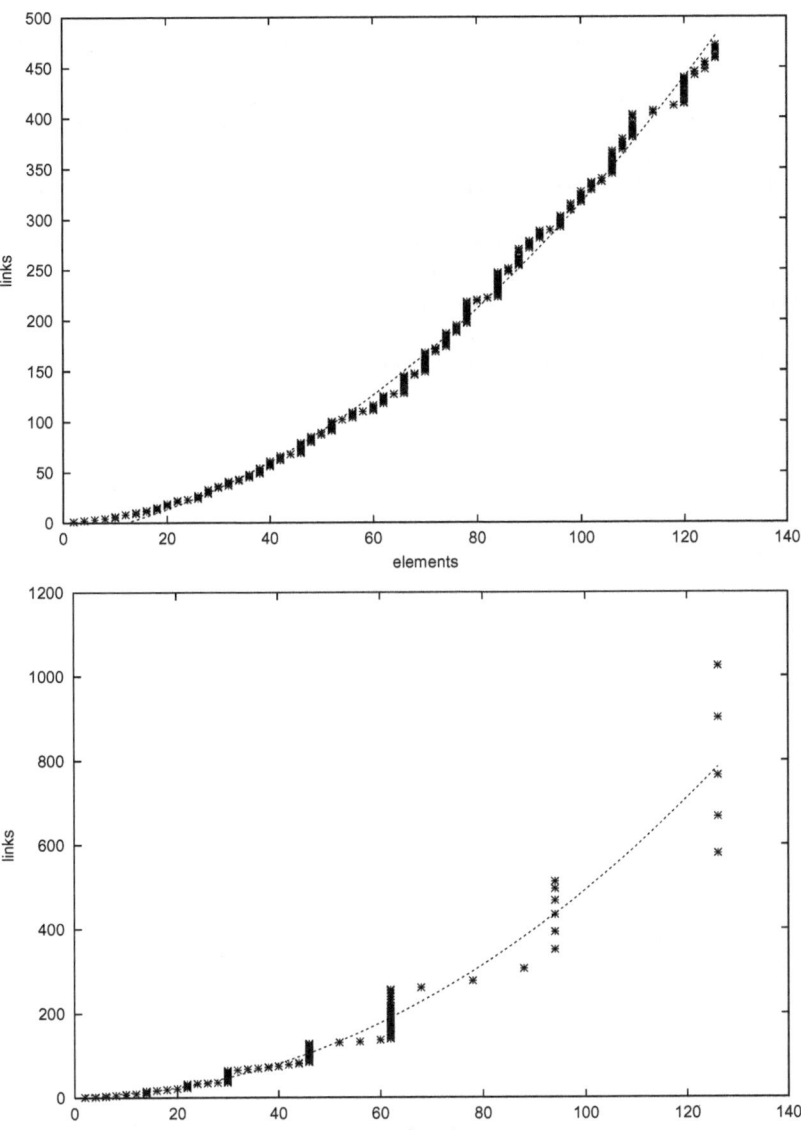

Figure 8. Empirical analysis for the scenes with two orthogonal patches. The top diagram shows results for the small distance scene, the bottom one for the large distance scene. The continuous curves have been determined by a least-squares-fit of a quadratic polynomial.

References

1. M. F. Cohen and J. R. Wallace. *Radiosity and Realistic Image synthesis*. Academic Press Professional, Cambridge, 1993.
2. R. Garmann. "Hierarchical radiosity - an analysis of computational complexity." Research Report 584/1995, Fachbereich Informatik, Universität Dortmund, Germany, D-44221 Dortmund, August 1995. Also available at http://ls7-www.informatik.uni-dortmund.de
3. P. Hanrahan, D. Salzman, and L. Aupperle. "A rapid hierarchical radiosity algorithm." *Computer Graphics*, vol. 25, pp. 197–206, July 1991.
4. D. Lischinski, B. Smits, and D. P. Greenberg. "Bounds and error estimates for radiosity." In *Proceedings of SIGGRAPH 94*, pp. 67–74, 1994.
5. P. Schröder. *Wavelet Algorithms for Illumination Computations*. PhD thesis, Princeton University, November 1994.
6. F. X. Sillion and C. Puech. *Radiosity & Global Illumination*. Morgan Kaufman, San Francisco, 1994.

Yes, Trees May Have Neurons

60 Varieties of Trees

Alois P. Heinz

University of Applied Sciences Heilbronn, Max-Planck-Str. 39, D-74081 Heilbronn,
heinz@fh-heilbronn.de

Abstract. Neural Trees are introduced. These descendants of decision trees are used to represent (approximations to) arbitrary continuous functions. They support efficient evaluation and the application of arithmetic operations, differentiation and definite integration.

1 Introduction

After Thomas Ottmann, whose 60th birthday we are celebrating, had been appointed to the new chair of algorithms and data structures in Freiburg, he gave his inaugural lecture about trees in computer science [1]. In this talk he presented a 48 item list with some of the most prominent trees. New varieties of trees and algorithms on trees have been developed since then, many of them by Thomas Ottmann or under his auspices. One of the newer varieties is the Neural Tree that will be described here in some detail.

A Neural Tree is like a decision tree whose inner nodes (decision nodes) have been replaced by *neurons*. The neurons serve as to soften the sharp decisions of the replaced nodes. The mapping represented by the tree thus becomes a continuous function. This allows a completely new set of operations being applied to the tree, like differentiation and integration. Other operations like sum, product, and cross product are permitted also, properties that can be used to show that Neural Trees are universal approximators. The evaluation algorithms for Neural Tree functions and derived functions will use the information stored in several paths from the root to a subset of the leafs depending on the input. Under certain conditions only a small fraction of all nodes is visited which makes the evaluations very efficient.

In the next section we give the definition of Neural Trees, followed by a short analysis of their modeling abilities in Sect. 3. We describe and analyze an efficient evaluation procedure for the Neural Tree function in Sect. 4. In Sect. 5 an efficient top-down procedure for the evaluation of the function's Jacobian matrix is given that minimizes the costly matrix operations. An evaluation procedure for derivatives of the Neural Tree function of arbitrary order using Taylor series expansions of a spatial decomposition is presented in Sect. 6. In Sect. 7 we investigate the minimal conditions that have to be satisfied to assure the existence of continuous derivatives of Neural Tree functions of order up to a given n. In Sect. 8 we present an algorithm for the exact and efficient evaluation of definite integrals of the Neural Tree function. We draw our conclusions in Sect. 9.

R. Klein et al. (Eds.): Comp. Sci. in Perspective (Ottmann Festschrift), LNCS 2598, pp. 179–190, 2003.
© Springer-Verlag Berlin Heidelberg 2003

2 Definition of Neural Trees

A Neural Tree T with input dimension d and output dimension c can be defined as a labeled binary tree. Each inner node k of T is called a neuron and is equipped with a normalized weight vector $w_k \in \mathbb{R}^d$, a threshold $c_k \in \mathbb{R}$ and a positive radius $R_k \in \mathbb{R}^+ \cup \{\infty\}$. Each leaf L is labeled with a vector $v_L \in \mathbb{R}^c$. A *decision function* $\varphi_k : \mathbb{R}^d \to [0,1]$ is associated with each neuron k,

$$\varphi_k(x) := \sigma((w_k \cdot x - c_k)/R_k) \quad , \tag{1}$$

where $\sigma : \mathbb{R} \to [0,1]$ is a continuously differentiable *activation function* with symmetry point in $(0,1/2)$ and with a nonnegative derivative $\Lambda \equiv \sigma'$. For any real function g, $\hat{g}(x)$ will be used in the sequel as an abbreviation for $1 - g(x)$. So it is correct to write $\sigma(x) = \hat{\sigma}(-x)$. We call σ a *lazy activation function* if additionally $\sigma(x) = 0$ for all x less than -1 and if a polynomial $s(x)$ exists with $\sigma(x) = s(x)$ for all x in the interval $[-1,0]$. The *evaluation* $v_T(x)$ of T with respect to a given input vector $x \in \mathbb{R}^d$ is defined as the evaluation of the root of T,

$$v_T(x) := v_{\text{root}(T)}(x) \quad , \tag{2}$$

where the evaluation of a leaf L is defined as the leaf vector,

$$v_L(x) := v_L \quad , \tag{3}$$

and the evaluation of any neuron k is recursively defined as

$$v_k(x) := v_{k_l}(x)\,\hat{\varphi}_k(x) + v_{k_r}(x)\,\varphi_k(x) \quad . \tag{4}$$

Here and in the following k_l and k_r are used to denote the left and right descendants of k, respectively. An alternative definition of $v_T(x)$ is often better suited for an efficient implementation. If for an arbitrary node q of T the *responsibility* of q with respect to x is defined as

$$\phi_q(x) := \begin{cases} 1 & \text{if } q = \text{root}(T) \\ \phi_k(x)\,\hat{\varphi}_k(x) & \text{if } q = k_l \\ \phi_k(x)\,\varphi_k(x) & \text{if } q = k_r \quad , \end{cases} \tag{5}$$

then $v_T(x)$ is given by a sum over all leafs L of T,

$$v_T(x) = \sum_{L \in T} v_L \phi_L(x) \quad . \tag{6}$$

Suppose the leafs[1] of T are $L_1, \ldots, L_\ell$. Then the *responsibility vector* $\Phi_T(x) \in \mathbb{R}^\ell$ of T with respect to x has $\phi_{L_m}(x)$ as its m-th component and the *leaf weight matrix* $\mathbf{W}_T \in \mathbb{R}^{c \times \ell}$ has v_{L_m} as its m-th column. Now it is easy to verify that

$$v_T(x) = \mathbf{W}_T \Phi_T(x) \quad . \tag{7}$$

---

[1] The number of leafs ℓ determines the *capacity* of the tree. It's easy to see that the number of examples of the mapping that could be represented *exactly* grows at least linearly with ℓ.

Note that $\mathbf{W}_T$ is independent of x and that the components of $\Phi_T(x)$ are all non-negative and sum up to unity. $\Phi_T(x)$ is usually a very sparse vector [3].

The next section will show that Neural Tree functions are complete with respect to a certain set of operations. The evaluation of a Neural Tree function and other derived functions can be done very efficiently. The time and space requirements are sublinear in the tree size if a lazy activation function is used, because then only a fraction of the tree has to be visited during an evaluation as a consequence of its hierarchical structure.

3   Modeling Abilities of Neural Trees

What kind of functions can be represented by Neural Trees? It is easy to see that constant functions and sigmoid functions may be represented. But any other continuous function $f : \mathbb{R}^d \to \mathbb{R}^c$ of interest can be approximated too. This follows from the fact that arithmetic is applicable also.

Lemma 1. *The sum of two Neural Tree functions is a Neural Tree function.*

Proof. Consider two Neural Trees T_1 and T_2 with compatible input and output dimensions. Then construct T_{1+2} as follows: Scale all leaf vectors of T_1 and T_2 by a factor of 2 and let both trees be subtrees of a new root k with infinitely large radius $R_k = \infty$. Thus $\varphi_k(x) = 1/2$ for all $x \in \mathbb{R}^d$. Then

$$v_{T_{1+2}}(x) = 2\mathbf{W}_{T_1}\frac{\Phi_{T_1}(x)}{2} + 2\mathbf{W}_{T_2}\frac{\Phi_{T_2}(x)}{2} = v_{T_1}(x) + v_{T_2}(x) \tag{8}$$

Lemma 2. *The (pointwise) product of two Neural Tree functions is a Neural Tree function.*

Proof. For two Neural Trees T_1 and T_2 with compatible input and output dimensions construct $T_{1\odot 2}$ as follows: Take T_1 and replace each leaf L_1 of T_1 by a copy of T_2 that has all leaf vectors component-wise multiplied by v_{L_1}. Then

$$v_{T_{1\odot 2}}(x) = \sum_{L_1 \in T_1, L_2 \in T_2} v_{L_1} \odot v_{L_2}\, \phi_{L_1}(x)\phi_{L_2}(x) \tag{9}$$

$$= \left(\sum_{L_1 \in T_1} v_{L_1}\phi_{L_1}(x)\right) \odot \left(\sum_{L_2 \in T_2} v_{L_2}\phi_{L_2}(x)\right) \tag{10}$$

$$= v_{T_1}(x) \odot v_{T_2}(x) \tag{11}$$

Theorem 1. *Any given continuous function $f : K \to \mathbb{R}$, $K \subset \mathbb{R}^n$ and compact, can be uniformly approximated by a sequence of Neural Tree functions to within a desired accuracy.*

Proof. The theorem is a direct consequence of the Stone-Weierstrass theorem [4] and follows from Lemmas 1 and 2, if we additionally consider that constants are Neural Tree functions and that there are Neural Tree functions that have different values for a given pair of distinct input points.

Theorem 2. *Let $f : K \to \mathbb{R}$, $K \subset \mathbb{R}^n$ and compact, be a given continuous function with continuous derivatives. Then Neural Tree functions are capable of simultaneous uniform approximation of f and its derivatives.*

Proof (Sketch). A first Neural Tree T with small radii can be built that decomposes K according to a n-dimensional grid. Each leaf L of T belongs to a small hyper-rectangle $H_L \subset K$ such that $\phi_L(x) > 0$ implies $x \in H_L$. Then each leaf L of T can be replaced by a Neural Tree T_L with one neuron and two leafs, whose function approximates to a hyperplane within H_L and has the same value and derivatives than f in the center x_{H_L} of H_L. Increasing the radii to half the size of the grid width smoothes the Neural Tree function. Decreasing the grid width (and thus increasing the height) leads to a better approximation.

Now one can imagine how Neural Trees can be constructed by expert designers using local decomposition or arithmetic operators like sum and product. The cross product operation is applicable as well. A Neural Tree will usually be constructed automatically by a topology enhancing and parameter modifying procedure (a so-called learning algorithm) using a given training set [5]. In this case the height can be restricted to a logarithmic function of the tree size.

4   Efficient Neural Tree Evaluation

The efficient evaluation of a modeled function as well as of derived functions is extremely important especially in control applications. The basis of the efficient evaluation is given in particular by the fact that $\sigma(x)$ is different from 0 or 1 only in the interval $(-1, 1)$ which allows effective lazy evaluation. The evaluation of a given neuron k makes use of both descendant nodes only if $\varphi_k(x) \in (0, 1)$. The evaluation algorithm is given in Fig. 1.

```
EVAL := proc (k: node, x: vector(d)): vector(c);
    if leaf (k) then return (v_k) fi;
    real φ := σ((w_k · x − c_k)/R_k);
    if φ = 0 then return (EVAL (k_ℓ, x))
        elif φ = 1 then return (EVAL (k_r, x))
        else return ((1 − φ) × EVAL (k_ℓ, x) + φ × EVAL (k_r, x))
    fi
end;
```

Figure 1. The procedure EVAL evaluates a Neural Tree function recursively. The algorithm takes advantage of lazy evaluation

An upper bound estimate of the average case runtime $A(N,P)$ (the number of nodes visited) as a function of the total number of tree nodes N and a probability parameter P ($0 < P < 1$) can be derived as follows. First, we assume that we know the upper bound[2] P such that for any neuron k and any side s ($s = l$: left or $s = r$: right) the conditional probability $P_{k,s}$ for the event that k_s has to be visited if k is visited is bounded by $(1 + P)/2$. Note that then the conditional probability for visiting both descendants of any node is upper bounded by P. Under the given assumptions the probability for any node in height h to be visited is less than or equal to $(1 + P)^h/2^h$. Since this is a non-increasing function of h, from all trees with N nodes including leafs those trees show worst-case behavior where all levels smaller than $H(N) = \lfloor \log_2(N + 1) \rfloor$ are filled with nodes and the remaining nodes are in level $H(N)$. Computing the sum of these nodes' probabilities yields a bound for number of nodes that have to be visited in the average-case:

$$A(N,P) \leq \frac{1}{P} \left[(1 + P)^{\log_2(N+1)} - 1 \right] \tag{12}$$

$$\leq \frac{1}{P} \sqrt[r]{N + 1} \text{ with } r = (\log_2(1 + P))^{-1} \in (1, \infty) \tag{13}$$

Thus, in the average case the runtime is in the order of some root of the tree size. If P is less than about 40%, for example, the average case evaluation runtime is proportional to less than the square root of the number of tree nodes[3]. It is easy to see from (12) that $A(N,P)$ is linear in N if $P = 1$ and logarithmic in N if P approaches 0.

5   Efficient Top-Down Jacobian Evaluation

Many applications require the on-line inversion [6] of some model function $f : \mathbb{R}^d \rightarrow \mathbb{R}^c$ such as the *inverse kinematics problem* in robotics [7] or the *inverse plant modeling* in control [8]. Numerical inversion techniques [9] most often involve iterated evaluations of the functions' Jacobian $\mathbf{J}^{(f)} : \mathbb{R}^d \rightarrow \mathbb{R}^{c \times d}$, whose (k,i)-th entry consists of the partial derivative of the k-th component of f with respect to the i-th component of its input.

The evaluation of the Jacobian is considered to be a very expensive operation since either d-dimensional row vectors have to be forward propagated [10] or c-dimensional column vectors have to be backward propagated [11;12;13] through each node of the function graph. Different mixed propagation modes [14;15;16] with slightly enhanced efficiency may be applicable, but even they have to inspect each single node and perform a number of operations with it.

Note that the recursive definition of the Neural Tree evaluation implies a flow of vector valued information from the bottom of the tree to the top. Applying any of the known *automatic differentiation* modes [17] to the evaluation procedure in Fig. 1 yields a program that computes the Jacobian and propagates $c \times d$-matrices bottom-up. The algorithm proposed here differs from this approach in that it performs in a top-down fashion avoiding bottom-up propagation of these larger intermediate results.

Using (7) the Jacobian of the Neural Tree function $\mathbf{v}_T$ can be described as

$$\mathbf{J}^{(v_T)} = \mathbf{W}_T \nabla_x \Phi_T(x) \ , \tag{14}$$

---

[2] P could be computed from a known distribution of input vectors.
[3] The probabilities observed in earlier experiments ranged from about 5% to 40%.

where the components of $\nabla_x \Phi_T(x)$ can be computed one after another according to the following recursive scheme

$$\nabla_x \phi_q(x) = \begin{cases} (0, \dots, 0) & \text{if } q = \text{root}(T) \\ \nabla_x \phi_k(x) \hat{\phi}_k(x) - \phi_k(x) \nabla_x \phi_k(x) & \text{if } q = k_l \\ \nabla_x \phi_k(x) \phi_k(x) + \phi_k(x) \nabla_x \phi_k(x) & \text{if } q = k_r \end{cases} \tag{15}$$

It's important to notice that the m-th row of $\nabla_x \Phi_T(x) \in \mathbb{R}^{\ell \times d}$ consists of only zeros if L_m belongs to the left subtree of an ancestor k with decision function $\phi_k(x) = 1$ or to the right subtree of k with $\phi_k(x) = 0$. In this case the responsibility of L_m is zero and there is no contribution from L_m to the Jacobian.

The complete recursive top-down Jacobian evaluation algorithm is now given in Fig. 2. It visits only nodes with non-disappearing responsibilities with respect to the ac-

```
TOP_DOWN_J :=
    proc (k: node, x: vector(d), φ: real, ∇φ: vector(d), J: matrix(c, d));
        local ∇φs: vector(d), n: integer, i: integer;
        if leaf (k) then for n from 1 to c do
            for i from 1 to d do J[n, i] += v_k[n] × ∇φ[i]  od od
        else y := (w_k · x − c_k)/R_k;
            φ := σ(y);
            if φ < 1 then ∇φs := ∇φ × (1 − φ) − φ × Λ(y)/R_k × w_k;
                        TOP_DOWN_J (k_l, x, φ × (1 − φ), ∇φs, J) fi;
            if φ > 0 then ∇φs := ∇φ × φ + φ × Λ(y)/R_k × w_k;
                        TOP_DOWN_J (k_r, x, φ × φ, ∇φs, J) fi
        fi
    end;
```

Figure 2. The recursive top-down procedure TOP_DOWN_J evaluates the Jacobian of a Neural Tree function at a given input point $x \in \mathbb{R}^d$. The global matrix $\mathbf{J}$ has to be initialized with zeros before the procedure is called with the following arguments: the root of the tree, (a reference to) the vector x, the root's responsibility ($= 1$) and its gradient (a d-dimensional vector of zeros), and (a reference to) $\mathbf{J}$. Non-contributing subtrees are not visited

tual vector x. These responsibilities and their gradients with respect to x are propagated from the top to the bottom. If a leaf is reached, its contribution is added to the global matrix, which has to be initialized with zeros before the procedure is called having the root, the vector x, the root's responsibility and gradient, and (a reference to) the matrix as parameters. If a neuron is reached, recursive subcalls are initiated for one or both of the descendants depending on the value of the decision function.

Average case runtime analysis can be performed in the same way as the one in Sect. 4. The difference is that here vector operations are executed at inner nodes and matrix operations are executed at the leafs that are visited during the evaluation. An upper bound for the number of leafs that have to be visited in the average case can be derived [18]:

$$B(N,P) \leq \frac{N+1}{2} \left(\frac{1+P}{2} \right)^{\log_2(N+1)-1} . \qquad (16)$$

$B(N,P)$ is much smaller than $A(N,P)$ and ranges between $(N+1)/2$ and 1 as P ranges between 1 and 0.

6   Evaluating Derivatives of Arbitrary Order

Many relationships between functions in nature and sciences can be described by differential or integral or integro-differential equations. A lot of tasks to be solved in diverse application areas of computing involve initial value problems [19]. To be able to support efficient solutions to all these tasks it is extremely desirable that higher-order derivatives and integrals of given functions can be evaluated very fast, especially if the function in question is represented by some larger model as a Neural Tree.

In the following it is assumed that T is a Neural Tree with input dimension $d = 1$ and output dimension $c \geq 1$ using a lazy activation function and that $v_T^{(n)}(x)$ — the n-th order derivative of the evaluation function at x — is properly defined. The recursive evaluation algorithm DIFF for $v_T^{(n)}$ can be derived directly from the alternative definition of v_T (Equations (5) and (6)) and is depicted in Fig. 3. Initially, when the procedure is

DIFF := **proc** (k: node, ϕ: polyFunction, x: real, n: integer): vector(c)

(1)   **if**   leaf (k)   **then return** ($v_k \times \phi^{(n)}(x)$)
(2)   **elif** $x \leq c_k - R_k$ **then return** (DIFF (k_ℓ, ϕ, x, n))
(3)   **elif** $x \geq c_k + R_k$ **then return** (DIFF (k_r, ϕ, x, n))
(4)   **else if** $x \leq c_k$ **then** $\varphi := y \rightarrow s((y - c_k)/R_k)$
(5)   **else** $\varphi := y \rightarrow 1 - s((c_k - y)/R_k)$ **fi**;
(6)   **return** (DIFF (k_ℓ, $\phi \times \hat{\varphi}$, x, n) + DIFF (k_r, $\phi \times \varphi$, x, n))
(7)   **fi**
   end;

Figure 3. Procedure DIFF evaluates the n-th order derivative of a Neural Tree function

called the arguments are the root of T, the constant function $\phi \equiv 1$, the value of x, and the order of the derivative n. All polynomial functions ϕ and φ within DIFF may be represented as final sequences of Taylor coefficients. The main idea of the algorithm is to recursively construct polynomial functions that are equal to the responsibilities ϕ_L at x for all leafs L that have $\phi_L(x) \neq 0$.

The contribution of a neuron k to the responsibilities of the leafs in the subtrees rooted at the sons of k is either 0 or 1 if x is not from the interval $(c_k - R_k, c_k + R_k)$. Only one of both subtrees has to be visited in this case, see lines (2) and (3). However, if x belongs to the mentioned interval then both subtrees add to the result (6) where k contributes to the responsibilities of the leafs in both subtrees with a polynomial factor that depends on the values of the variables associated with k, the value of x, the definition of the polynomial s, and the position of the subtree, see lines (4), (5), and (6). If k is a leaf then variable ϕ describes a polynomial $\phi(x) = \sum_{i=0}^{m} \phi_i x^i$ that is equal to the responsibility of k at x. Then k adds to the derivative a portion of $v_k \phi^{(n)}(x) = v_k \sum_{i=0}^{m-n} \phi_{i+n}(i+n)! x^i / i!$, see line (1).

Algorithm DIFF is very efficient, because it visits only those nodes k that have $\phi_k(x) \neq 0$. The expensive operations are products of polynomials at the neurons and differentiations of polynomials at the leafs. These operations may be necessary each time DIFF is called if the Neural Tree is adapted between evaluations. But the polynomials describing $\phi_L(x)$ and its derivatives within certain intervals can be stored at each leaf L to speed up the evaluation. Then they have to be recalculated by a modified memorizing DIFF procedure only if flags indicate that the parameters of an ancestor node of L have been changed. The number of intervals is discussed in Sect. 8. There the main idea of the algorithm will be reused for the integration algorithm.

7 Choosing Lazy Activation Functions

Lazy activation functions are very convenient for Neural Tree implementations, because they support efficient evaluations, as it was shown. When a lazy activation function is used, the tree function is piecewise polynomial. Here it is shown how to determine for arbitrary $n \geq 0$ a minimum degree polynomial s_n such that the corresponding σ_n is a lazy activation function that guarantees the existence of continuous derivatives of order up to n. It is easy to deduce from (4) and (1) that in this case the function $\sigma_n^{(i)}$ has to be continuous for all i from 0 to n, which requires fulfillment of the following equations:

$$s_n^{(i)}(-1) = 0 \qquad (\forall i \in \{0 \ldots n\}) \tag{17}$$

$$s_n^{(i)}(0) = \hat{s}_n^{(i)}(0) \qquad (\forall i \in \{0 \ldots n\}) \tag{18}$$

Equation (18) implies that in $s_n(x)$ the coefficient of x^0 must be $1/2$ and the coefficients of x^i have to be 0 for all even i from 2 to n. Other coefficients are not affected. Equation (17) gives another set of $n+1$ linear constraints, so that $n+2$ nonzero coefficients are needed altogether to find a solution for s_n, the degree of which is then $\lceil (3n+1)/2 \rceil$. Table 1 displays $s_n(x)$ for n from 0 to 5. It is easy to see that the corresponding functions $\sigma_n(x)$ become steeper at $x = 0$ as n grows larger.

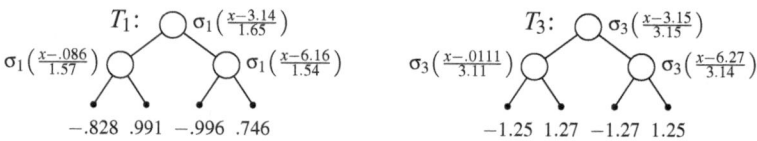

T_1: $\sigma_1\left(\frac{x-3.14}{1.65}\right)$

$\sigma_1\left(\frac{x-.086}{1.57}\right)$ $\sigma_1\left(\frac{x-6.16}{1.54}\right)$

$-.828$ $.991$ $-.996$ $.746$

T_3: $\sigma_3\left(\frac{x-3.15}{3.15}\right)$

$\sigma_3\left(\frac{x-.0111}{3.11}\right)$ $\sigma_3\left(\frac{x-6.27}{3.14}\right)$

-1.25 1.27 -1.27 1.25

Figure 4. Two Neural Trees representing approximations to the sine function and its derivatives in $[0, 2\pi]$.

The consequence of different abilities of two lazy activation functions can be observed at the trees in Fig. 4. Two trees T_1 and T_3 have the same structure but use different activation functions σ_1 and σ_3, respectively. Both trees were successfully trained

Table 1. Polynomials $s_n(x)$ defining the lazy activation functions $\sigma_n(x)$

n	$s_n(x)$
0	$1/2\,(x+1)^1$
1	$1/2\,(x+1)^2$
2	$1/2\,(1-x)\,(x+1)^3$
3	$1/4\,(2-3x)\,(x+1)^4$
4	$1/16\,(8+(15x-19)x)\,(x+1)^5$
5	$1/2\,(1+(3x-3)x)\,(x+1)^6$

to represent the sine function in the interval $[0,2\pi]$ with minimal error. The evaluation function of T_i has continuous derivatives of order up to i. The derivatives of T_3 are better in approximating the derivatives of the original function, but T_1 has the smaller evaluation costs.

8   Evaluating Integrals

Finding exact and efficient evaluation procedures for definite integrals is much more difficult than evaluating derivatives in the general case. Either combined symbolic and numeric [20] or purely numeric quadrature techniques [21] may be applied successfully. If, however, the function in question can be spatially decomposed into parts with known Taylor series expansions the evaluation of the integral is easy in principle. But in the worst-case all parts have to be inspected, which may be very time consuming. Integrating Neural Trees is much more efficient, if some additional information is stored in the tree.

The recursive algorithm INT for the evaluation of the definite integral $\int_a^b v_T(x)\,dx$ of a single input Neural Tree function is constructed along the same principles as DIFF and is partially displayed in Fig. 5. INT is initially called with the following arguments: the root of T, the constant function $\phi \equiv 1$, and the interval $[a,b]$. Again, all polynomial functions ϕ and φ within INT can be represented by final sequences of Taylor coefficients. The main idea of the algorithm is to construct the integral $I = \int_a^b \phi(x)v_k(x)\,dx$ piecewise. At each neuron k the given integration interval $[a,b]$ is divided into 4 (possibly empty) smaller intervals A, B, C, and D, that have different Taylor polynomial expansions of the responsibility functions at the son nodes, see lines (3) and (4). Up to 6 recursive procedure calls are then used to evaluate the contributions of these intervals in lines (5)–(10), which sum up to I. If k is a leaf, then variable ϕ contains the Taylor polynomial expansion $\phi(x) = \sum_{i=0}^m \phi_i x^i$ of $\phi_k(x)$ in $[a,b]$. The integral is then given by $v_k \int_a^b \phi(x)\,dx = v_k \sum_{i=1}^{m+1} \phi_{i-1}\,(b^i - a^i)/i$, see line (1).

The algorithm as described has a maximal branching factor of 6 and has to be modified to become more efficient[4]. For this purpose each boundary of an interval that is

---

[4] It is easy to see that the number of node evaluations using this first approach is not exponential but in the order of $N\log N$, if the tree is balanced.

```
INT := proc (k: node, φ: polyFunction, [a,b]: [real, real]): vector(c)
(1)     if leaf (k) then return (v_k × ∫_a^b φ(x) dx)
(2)     else        I_A := I_B := I_C := I_D := 0;
(3)     A := [a,b] ∩ [−∞, c_k − R_k];   B := [a,b] ∩ [c_k − R_k, c_k];
(4)     C := [a,b] ∩ [c_k, c_k + R_k];   D := [a,b] ∩ [c_k + R_k, ∞];
(5)     if A ≠ ∅ then I_A := INT (k_ℓ, φ, A) fi;
(6)     if B ≠ ∅ then  φ := y → s((y − c_k)/R_k);
(7)                    I_B := INT (k_ℓ, φ × φ̂, B) + INT (k_r, φ × φ, B) fi;
(8)     if C ≠ ∅ then  φ := y → 1 − s((c_k − y)/R_k);
(9)                    I_C := INT (k_ℓ, φ × φ̂, C) + INT (k_r, φ × φ, C) fi;
(10)    if D ≠ ∅ then I_D := INT (k_r, φ, D) fi;
(11)    return (I_A + I_B + I_C + I_D)
(12)    fi
    end;
```

Figure 5. Procedure INT evaluates the definite integral $\int_a^b \phi(x) v_k(x)\, dx$

given by c_k, $c_k - R_k$, or $c_k + R_k$ for some neuron k is marked and each interval with two marked boundaries is marked also. Each integral I_A, I_B, I_C, or I_D that belongs to a marked interval has to be evaluated only once and can be stored for later use together with the other data belonging to the neuron k where it emerges. Thereafter, the integral values of marked intervals can be simply read from the neurons.

For the evaluation of $\int_a^b v_T(x)\, dx$ only those nodes have to be visited by the memorizing version of the INT algorithm that are visited for the evaluation of both $v_T(a)$ and $v_T(b)$. That means that the average node visiting costs arc only twice the costs of the EVAL algorithm.

There are only two marked intervals at the root of the tree (B and C) and each node can have only 3 marked intervals more than its father node. Storing intervals and their integrals will therefore increase the storage complexity of the method by a logarithmic factor in the size of the tree if the tree is properly balanced.

As it was the case with the DIFF procedure, INT can be made to evaluate faster if the nontrivial (i.e. $\neq 0$) polynomials describing $\phi_L(x)$ and its indefinite integrals within certain intervals are stored at each leaf L. It is easy to see that a leaf in the i-th level of the tree can have no more than $2i + 1$ of such intervals. So again the mentioned modification increases the tree size only by a logarithmic factor but it avoids the expensive recalculations of responsibility functions.

9   Conclusions

Many more than 60 prominent or important varieties of trees have been developed or discovered for applications in computer science up to presence. Many interesting or promising ones are still awaiting their discovery.

We can regard the emergence of new varieties as a kind of evolutionary process with genetic operations such as mutation, crossover, and selection of course. The current contribution tried to point out some of the consequences of a single mutation that changed the inner nodes of decision trees to neurons with a slightly enhanced decision function. And what happened? More intelligence was introduced!

Neural Trees can represent approximations to arbitrary continuous functions. The lazy evaluation algorithms for the tree functions as well as for their derivatives and integrals are very efficient, since only a small part of the tree is effected by a particular input. Thus Neural Trees are extremely well suited for all applications where learning or adaptation is needed and where speed is a crucial factor. You can count on Neural Trees and — as far as selection is concerned — they surely will survive or only be superseded by some offspring.

References

1. Thomas Ottmann. Bäume in der informatik. Technical Report 13, Institut für Informatik, Universität Freiburg, June 1988.
2. Thomas Ottmann. Trees – a personal view. In Hermann A. Maurer, editor, *New Results and New Trends in Computer Science*, volume 555 of *Lecture Notes in Computer Science*, pages 243–255. Springer, 1991.
3. Alois P. Heinz. A tree-structured neural network for real-time adaptive control. In S. Amari, L. Xu, L.-W. Chan, I. King, and K.-S. Leung, editors, *Progress in Neural Information Processing, Proceedings of the International Conference on Neural Information Processing, Hong Kong*, volume 2, pages 926–931, Berlin, September 1996. Springer.
4. J. C. Burkill and H. Burkill. *A Second Course in Mathematical Analysis*. Cambridge University Press, Cambridge, England, 1970.
5. Alois P. Heinz. On a class of constructible neural networks. In Françoise Fogelman-Soulié and Patrick Gallinari, editors, *ICANN '95, Proceedings of the International Conference on Artificial Neural Networks, Paris, France*, volume 1, pages 563–568, Paris La Défense, France, October 1995. EC2 & Cie.
6. J. Kindermann and A. Linden. Inversion of neural networks by gradient descent. *Journal of Parallel Computing*, 14(3):277–286, 1992.
7. D. DeMers and K. Kreutz-Delgado. Solving inverse kinematics for redundant manipulators. In O. Omidvar and P. van Smagt, editors, *Neural Systems for Robotics*, pages 75–112. Academic Press, New York, 1997.
8. M. Brown and C. Harris. *Neurofuzzy Adaptive Modelling and Control*. Systems and Control Engineering. Prentice Hall International, London, UK, 1994.
9. J. E. Dennis and R. B. Schnabel. *Numerical Methods for Unconstrained Optimization and Nonlinear Equations*. Prentice-Hall, Englewood Cliffs, NJ, 1983.
10. R. E. Wengert. A simple automatic derivative evaluation program. *Comm. ACM*, 7(8):463–464, 1964.
11. A. E. Bryson and Y.-C. Ho. *Applied Optimal Control*. Blaisdell, New York, 1969. [Revised printing New York: Hemisphere, 1975].
12. W. Baur and V. Strassen. The complexity of partial derivatives. *Theoretical Computer Science*, 22:317–330, 1983.
13. D. E. Rumelhart, G. E. Hinton, and R. J. Williams. Learning representations by back-propagating errors. *Nature*, 323:533–536, 1986.
14. T. Yoshida. Derivation of a computational process for partial derivatives of functions using transformations of a graph. *Transactions of Information Processing Society of Japan*, 11(19):1112–1120, 1987.
15. A. Griewank and S. Reese. On the calculation of Jacobian matrices by the Markowitz rule. In *[17]*, pages 126–135. SIAM, 1991.

16. T. Yoshida. A node elimination rule for the calculation of Jacobian matrices. In *Proceedings of the Second International SIAM Workshop on Computational Differentiation, Santa Fe, NM*, Philadelphia, 1996. SIAM.

17. A. Griewank and G. F. Corliss, editors. *Automatic Differentiation of Algorithms: Theory, Implementation, and Application*. SIAM, Philadelphia, Penn., 1991.

18. Alois P. Heinz. Efficient top-down jacobian evaluation of tree-structured neural networks. In Lars Niklasson, Mikael Bodén, and Tom Ziemke, editors, *ICANN 98, Proceedings of the 8th International Conference on Artificial Neural Networks, Skövde, Sweden*, volume I, pages 87–92. Springer-Verlag, 1998.

19. Graeme Fairweather and Rick D. Saylor. The reformulation and numerical solution of certain nonclassical initial-boundary value problems. *SIAM Journal on Scientific and Statistical Computing*, 12(1):127–144, January 1991.

20. A. C. Norman and J. H. Davenport. Symbolic integration - the dust settles? In *Proc. EUROSAM 1979, Lecture Notes in Computer Science*, volume 72, pages 398–407. Springer-Verlag, 1979.

21. R. Cranley and T. N. L. Patterson. On the automatic numerical evaluation of definite integrals. *The Computer Journal*, 14(2):189–198, May 1971.

Java Applets for the Dynamic Visualization of Voronoi Diagrams*

Christian Icking[1], Rolf Klein[2], Peter Köllner[3], and Lihong Ma[1]

[1] FernUniversität Hagen, Praktische Informatik VI, 58084 Hagen, Germany
[2] Universität Bonn, Institut für Informatik I, 53117 Bonn, Germany
[3] Hiddenseer Str. 11, 10437 Berlin, Germany

Abstract. This paper is dedicated to Thomas Ottmann on the occasion of his 60th birthday. We discuss the design of several Java applets that visualize how the Voronoi diagram of n points continuously changes as individual points are moved across the plane, or as the underlying distance function is changed. Moreover, we report on some experiences made in using these applets in teaching and research. The applets can be found and tried out at
http://wwwpi6.fernuni-hagen.de/GeomLab/.

1 Introduction

Soon after personal computers became widely available, scientists and teachers started to think of the advantages this new technology could offer to research and teaching. In computer science, an early impulse to use computers in teaching originated from Hermann Maurer at Graz [25; 26]. In Germany, Thomas Ottmann became the leading pioneer in this field. He created a new approach to course developing called *authoring on the fly*, and put it to use in large projects many universities participated in [22].

It turned out that teaching can benefit greatly from computer support. On the one hand, complete computer-based courses allow students to carry on their studies independently of time and location; this is of particular importance to part-time or remote studying. On the other hand, both classical and computer-based courses can be enhanced by valuable tools that did not exist before.

Among such tools we find, at a basic level, *animation* by computer. Complicated, parameter-dependent behaviour can be illustrated by artificially created moving figures. One level up, *experiments* allow students to interactively manipulate certain structures, or apply certain methods, while being in complete control of both input and parameters. Even more advanced *virtual laboratories* enable students and researchers to create, and carry out, their own experiments.

In this paper we report on tools that enable experiments on *Voronoi diagrams*; for an example see Figure 1. The Voronoi diagram and its dual, the Delaunay triangulation, are among the most important structures in computational geometry; see e. g. Aurenhammer [3] or Aurenhammer and Klein [4] for surveys. The study of their structural properties belongs to the core of this area.

---

* This project was partially funded by the government of Nordrhein-Westfalen, under the heading of the "Virtual University".

R. Klein et al. (Eds.): Comp. Sci. in Perspective (Ottmann Festschrift), LNCS 2598, pp. 191–205, 2003.
© Springer-Verlag Berlin Heidelberg 2003

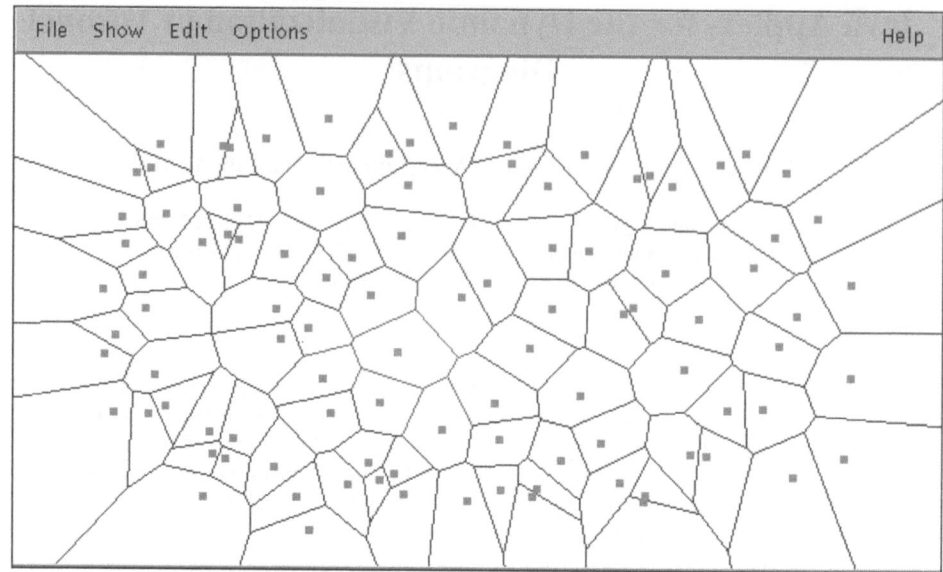

Figure 1. *VoroGlide* showing the Voronoi diagram of 100 points.

Given a set of point sites and a distance measure in the plane, the Voronoi diagram partitions the plane into regions, one to each site, containing all points of the plane that are closer to this site than to any other. Despite this rather simple definition, drawing a complex Voronoi diagram by pencil is an arduous task—particularly so for students who have not yet studied the algorithms—and computer based visualization is very desirable.

The paper is organized as follows. In 2 we discuss the intentions we had when this project[4] started at the FernUniversität Hagen in 1996. Next, in 3 we describe the algorithms implemented for dynamic and interactive Voronoi applets. Then, in 4, we report on some experiences we made once these tools were available in teaching and research.

2   Motivation

There exist a good number of tools that allow to insert point sites by mouse click and then display their Voronoi diagram; numerous references[1] can be found in the World

---

[1] For example:

    http://www.geom.umn.edu/software/cglist/
    http://www.ics.uci.edu/~eppstein/junkyard/software.html
    http://www.scs.carleton.ca/~csgs/resources/cg.html
    http://compgeom.cs.uiuc.edu/~jeffe/compgeom/demos.html
    http://wwwpi6.fernuni-hagen.de/GI/fachgruppe_ 0.1.2/software/

Wide Web. With some systems, like *XYZ GeoBench*[2] by Nievergelt et al. [28], *Mocha*[3] by Baker et al. [5], or *GeoLab*[4] by de Rezende and Jacometti [13], the user can even pick a point and move it across the scene. The Voronoi diagram of the altered point set is then computed and displayed, after some delay that depends on the size of the point set and the complexity of the animation.

We think that this dynamic feature is extremely helpful in teaching and understanding geometric structures and algorithms. The ideal system should respond *instantaneously* to the user's actions: While dragging a selected site we would like to see, at each point in time, the Voronoi diagram of the point sites that remain fixed and of the moving site in its current position.

Such a perfect dynamic animation, however, is hard to achieve. In practice, all we can hope for is to recompute the Voronoi diagram and update the display frequently enough so that the impression of a continuous motion results. Our experiments indicate that the *frame rate*, i. e. the number of screen updates per second, should not drop below 6 while the mouse is moved at normal speed. Between 10 and 18 frames per second result in a very satisfying animation. At a rate of 25, TV quality is reached. Note that extreme mouse speeds are not likely to create a problem: a slowly moving mouse can be served all the better, and if the user moves the mouse very fast he cannot expect to follow in detail the structural changes anyway.

The computer's graphical system keeps track of the moving mouse by generating a sequence of *drag events* that report the mouse's current coordinates. These drag events are generated at a rate that depends both on the system and on its current load. For technical reasons, at most 50 drag events per second are generated with a standard equipment, and a typical rate is less or equal to 33, which is largely sufficient.

As often as a new drag event has been created, we should quickly compute and display the current Voronoi diagram. The problem is that we can typically spend at most 30 milliseconds on this task before the next drag event arrives. If the computation is not finished by then, the next events get delayed or must be skipped. As a consequence, the frame rate drops.

Whereas, under a slight drop of the frame rate, the animation looks jerky, like a video running in slow motion, a severe drop destroys the impression of a continuous motion; only a sequence of seemingly unrelated pictures remains.

In a first and simple approach we used the randomized incremental algorithm supplied by the LEDA library [27] for constructing from scratch the whole Voronoi diagram after each drag event. As expected, this approach works quite well for point sets of moderate size (less than around $n = 20$ points), thanks to the efficient implementation and the speed of code compiled from C^{++}. But since the expected run time per event is in $\Theta(n \log n)$, larger point sets cannot be handled this way.

Furthermore, we wanted to create a tool that could easily be distributed over the Internet and that was independent of special compilers and certain versions of libraries. Therefore we have chosen Java as programming language; see Arold and Gosling [2] or Flanagan [17], for example. Java programs run virtually everywhere and can be eas-

---

[2] http://www.schorn.ch/geobench/XYZGeoBench.html

[3] http://loki.cs.brown.edu:8080/pages/

[4] http://www.cs.sunysb.edu/~algorith/implement/geolab/implement.shtml

ily accessed by the World Wide Web, the programmer need not give much thought to different run-time environments, and the code is instantaneously executable. On the other hand, Java programs run considerably slower than compiled code. This calls for a careful selection of algorithms and data structures.

3   Fast Algorithms for Interactive Voronoi Applets

Of the Voronoi applets we have implemented during the last years, two are described in this paper, namely the one for the usual Euclidean Voronoi diagram of point sites, called *VoroGlide*, and the variant for polygonal convex distance functions, called *Voro-Convex*. Other variants such as diagrams under additive weights or for sites that are line segments are also available from our web site at
http://wwwpi6.fernuni-hagen.de/GeomLab/.

When a new point site is inserted, we want to incrementally construct the new structure from the old one. To do this with the Delaunay triangulation is technically a lot easier than with the Voronoi diagram. For this reason, our algorithm computes and maintains the Delaunay triangulation as the primary structure. The Voronoi diagram, or those parts that must be updated, are directly derived from the Delaunay triangulation, as well as the convex hull, if needed.

In *VoroGlide* and *VoroConvex*, the user can insert, delete and move points with simple mouse click and drag operations. The *insert* operation uses a point location algorithm which is described in 3.1 and the well-known *edge flip* algorithm, see 3.2.

Deleting points from Delaunay triangulations or Voronoi diagrams is known to be a difficult task. With complicated algorithms, see e. g. Aggarwal et. al. [1], it can be performed in $O(d)$ time where d is the number of Voronoi neighbors of the point to removed. We are not using these techniques but prefer to simply recompute the triangulation resp. the diagram of the remaining points from scratch.

Special attention has been paid to *moving* points, i. e. handling the drag events from the mouse. Our strategy allows a drag event to be handled in time proportional to the number of Voronoi neighbors the moving point has at its current position, nearly independently of the number of points in the diagram.

In the following subsections we discuss these features of our algorithms that are crucial in achieving a good performance: a simple technique for localizing a point in a triangulation which works very well in practice, and an efficient strategy for maintaining the data structure and the display while a point is being dragged.

3.1   Simple and Efficient Point Location

Given a triangulation DT and a point p in the plane, which of the triangles contains the point?

One could think of using the very good data structures for dynamic point location in general subdivisions which have been developed in the last years [6; 7; 9; 10]. While dragging a point around, the structure would not even have to be changed, whereas for permanent insertions or deletions it would need some updates. For the point location tasks which occur very frequently during a drag operation, we would have logarithmic

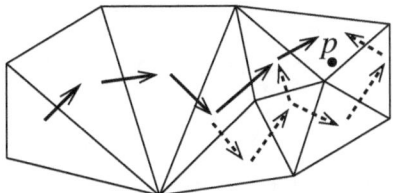

Figure 2. A walk for locating point p in a triangulation and some alternatives.

running time. But apart from the considerable effort for the implementation, we will see that such sophisticated data structures would not really be helpful in our case.

Instead, we are not using any auxiliary structure for point location, we just use the triangulation which is already there. We start at some triangle of DT. As long as the current triangle does not contain the point p, we switch to an adjacent triangle which is, in some sense, closer to p. This procedure is usually called a walk in a triangulation, and there is some ambiguity concerning which adjacent triangle is taken next in a walk, see Figure 2.

Devillers et al. [14] have looked at several different kind of walks. From practical experiments they conclude that all types of walks (so-called straight walks, stochastic walks, visibility walks, orthogonal walks etc., see also Devroye et al. [15]), are essentially equivalent, but visibility walks have the advantage of a very simple implementation and inherent robustness, at least for Delaunay triangulations.

In our setting the coordinates of two consecutive drag events are often very close together, in most cases the moving point is still in the same triangle as before or in an adjacent one. So if we start at the triangle where the moving point was located the last time, we can expect to locate point p in *constant time* in most cases.

The visibility walk is runs according to the following algorithm. It uses only simple tests for the position of p relative to the lines supported by the triangles' edges.

Algorithm VisibilityWalk
For each triangle we store the vertices a, b, and c in counter-clockwise order. We also store the triangles adjacent to the edges ab, bc, and ca.

1. Start with an arbitrary triangle T, preferably choose T as the result of the previous point location.

2. Let e be the edge ab of triangle T.

3. Check point p's position relative to edge e:

3a. If the point lies to the right side of the edge, we know that T does not contain p, so let T be the triangle adjacent to e and continue with step 2.

3b. Else, if not all three edges of T have been tested, let e be the next edge of T and perform step 3 again.

4. Otherwise, p does not lie on the right side of any of T's edges, so p is contained in T.

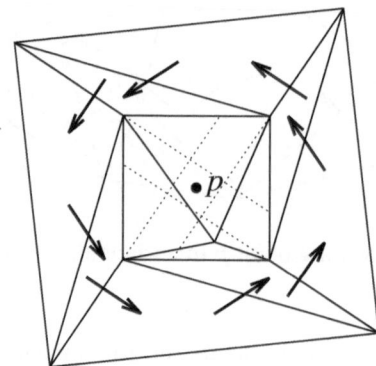

Figure 3. In this situation, VisibilityWalk can get into an infinite loop, but this does not happen with Delaunay triangulations.

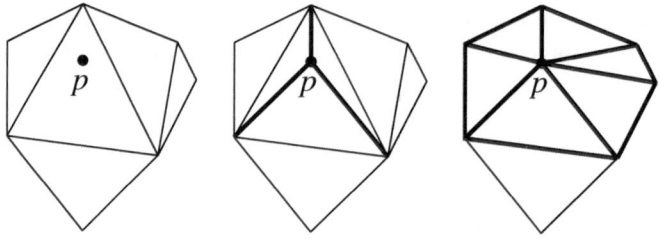

Figure 4. Three steps on inserting a new point into a Delaunay triangulation.

Remark that in practice we do not check all three edges of a triangle, because we have already tested the edge of the adjacent triangle we come from, except for the very first.

Surprisingly, VisibilityWalk may get into an endless loop in an arbitrary triangulation, see Figure 3. Fortunately, De Floriani et al. [12] have shown that this does not happen with Delaunay triangulations!

As was said before, the algorithm benefits from a good choice of the start triangle. But even for sets of several hundreds of points and a mouse moving very quickly, we have observed very short location times.

3.2   Temporary and Permanent Insert

Our algorithms are based on the standard procedure of inserting a new point site, p, into an existing Delaunay triangulation DT; see e. g. [4; 23] or use *VoroGlide* to provide an animation. Briefly, this procedure works as follows; cf.Figure 4.

First, the triangle T of DT containing p is determined; how this location is done has been explained in3.1. Then new edges are inserted between p and the vertices of T, and repeated edge flips are carried out for those neighboring triangles whose circumcircle contains p; in this way we have *permanently inserted* point p.

For a moving point, we proceed in a similar fashion. Picking and dragging a point and finally releasing it corresponds to one mouse-down event, many drag events, and one mouse-up event.

When a mouse-down event occurs, we look for a point at these coordinates and, if it is found, delete the point and compute the Delaunay triangulation DT of the remaining $n - 1$ points.[5]. For very large point sets, this recomputation may create some delay at the beginning of a drag operation. In the unlikely worst case it even requires time quadratic in n, and $O(n \log n)$ expected time, see [4], but we have to do it only once, at the beginning of each drag operation.

As long as the user keeps moving point p, we *keep* the triangulation DT. Each subsequent drag event is handled by *temporarily inserting* p, at its current position, into DT, as if it were a new point. At the end of a sequence of drag events, a mouse-up event occurs and the point is inserted permanently into DT.

Therefore, *temporary insert* means that we compute all necessary information to be able to display the updated structure, but we do not make any changes to the data structure DT of the other points, so there is no need for deleting the new point before treating the following drag event. In this way, we can generate the impression of a moving point by just temporarily inserting it into the same structure at many different locations.

To assess this algorithm from a theoretic point of view we would usually state that, once the moving point p has been located, a drag event is treated within time $O(d)$, where d is the degree of p in the current Delaunay triangulation.[6]

But with dynamic animation algorithms, the big-O notation is not precise enough. A closer analysis of the algorithm shows that, after locating the point, computing a drag event takes about $12 \cdot d$ steps. Here, a single step may consist of changing a pointer, computing the square of a distance, or constructing the circumcircle of three points, etc.

One could think of further speeding up the computation by using the following observation: if the graph structure of the Delaunay triangulation does not change over a sequence of consecutive drag events, it would not really be necessary to re-insert the moving point each time.

For example, if the point p moves along a line then the total number of structural changes is bounded by $2n$, independently of the number of drag events.[7] General bounds like this on the complexity of the Voronoi diagram of moving points can be found e. g. in Guibas et al. [18] or Huttenlocher et al. [19].

In the context of dynamic animation, however, we have no knowledge of the user's intentions, and there seems to be no easy way of predicting whether the actual drag event will cause a structural change or not. Therefore we handle all drag events as described above.

---

[5] If it is not found then a new point is inserted at these coordinates

[6] Or, equivalently, the number of Voronoi neighbors of p.

[7] A change can only occur when p becomes cocircular with three other points defining a Voronoi vertex.

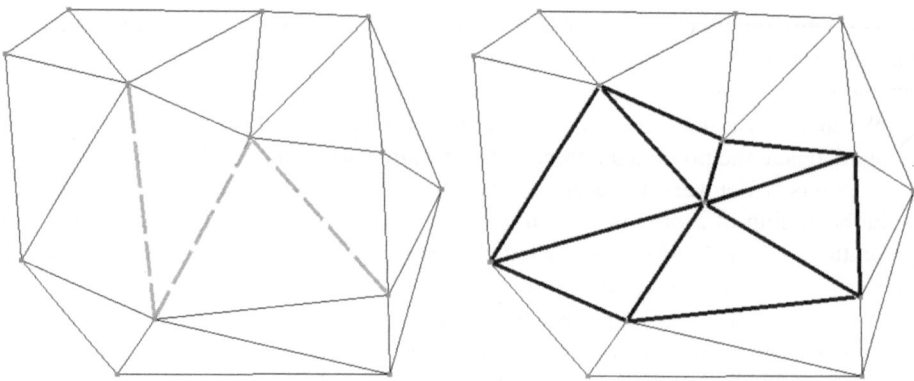

Figure 5. The right picture shows the star-shaped region with the edges that are redrawn for the moving point in the middle. The left picture shows the underlying structure without the moving point, three edges have to be erased before drawing the new ones.

3.3 Partial Redraw

We know that updates for a new point in the *Delaunay triangulation* affect only a star-shaped region composed of those triangles whose circumcircles contain the point, see Figure 5.

For maximum speed, we do not want to redraw the whole scene for every new position of the moving point. Instead, the display of the updates should be restricted to just the line segments within the star-shaped area, such that the update time is proportional to the degree of the new point, but independent of the number of points.

This is organized with the help of five layers of line segments which have to be drawn one over the other:

1. The lowest layer contains the Delaunay edges of the triangulation of all points except the moving one. This layer is displayed only once at the beginning of a drag operation (mouse-down) and once at the end (mouse-up). All draw and erase operations which occur during the motion are performed on the picture of this layer.
2. This layer contains the edges adjacent to the moving point at its previous position. These edges have to be erased, i. e. they will be drawn in background color (white).
3. The next layer stores the edges of layer 1 which have been erased for the previous picture. These edges must be redrawn in foreground color (black) for the new picture.
4. For the moving point at its new position some of the edges of layer 1 must be erased, they belong to this layer.
5. The uppermost layer contains the edges adjacent to the moving point at its new position. These edges are drawn in foreground color.

One *temporary insert* operation then consists of the following.

– Clear layers 2 and 3.
– Move all edges from layer 5 to layer 2 and all edges from layer 4 to layer 3.

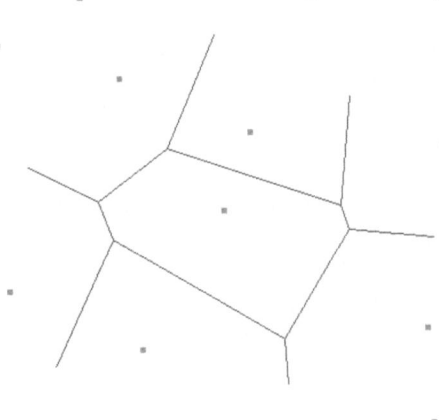

Figure 6. These edges of the Voronoi diagram must be redrawn while moving the point in the closed Voronoi region. The point set is the same as in the right part of Figure 5.

- Locate the moving point in the underlying triangulation.
- Determine the local changes. Edges that have to be added are put into layer 5, the edges that have to be erased go to layer 4. This is done by executing the edge flip algorithm, but without really changing the underlying structure.
- Draw the contents of layers 2 to 5 on the previous picture and display the new one on the screen.

As was mentioned in 3, we use the Delaunay triangulation as the primary structure and always derive the *Voronoi diagram* from this. This also holds for local changes. The edges which have to be changed in the Voronoi diagram are exactly those which are dual to the edges in the star-shaped region of the Delaunay triangulation, including the boundary of this region, see Figure 6.

A remarkable difference between the updates of the Delaunay triangulation and the Voronoi diagram is that new Delaunay edges are determined at the time when the edge flip occurs, but new Voronoi edges are only known after all edge flips are performed for a temporary insert. This is because Delaunay edges go from one of the points to another whereas Voronoi edges connect centers of circumcircles of triangles which may be affected by subsequent edge flips.

3.4 Buffered Display

Besides *computation cost*, each drag event handled causes *display cost*, too. After inserting the new point, p, into DT we need to display the new Delaunay triangulation. Since DT is still on the screen, it is sufficient to remove from the screen, by displaying them in background color, all edges of DT that have been flipped, and to display the new edges adjacent to p. After this picture has been visualized, the local changes have to be undone. Hence, the total number of edges to be displayed is about four times the degree, d, of p.

If the edges are drawn on the screen one by one, as they are output by the algorithm, the triangulation will sometimes flicker. A smoother impression results from using a buffered display, where only completed images appear on the screen.[8]

Using a buffer, it takes $2 \cdot d + B$ many steps to display the changes caused by a drag event. Here, a step consists of drawing a line segment, and B measures the time for displaying the buffer content.

VoroGlide can be run with or without buffering, which allows to estimate the time B. On our system, B seems to dominate both the computation and the display cost (indicating that faster hardware cannot always be replaced by better algorithms).

As far as the display cost is concerned, *VoroGlide* behaves optimally during drag operations: Only those edges of the Delaunay triangulation are redrawn that have changed.

3.5   Skipping Events

A well-known and useful technique for dynamic animation is used in *VoroGlide*, too. More drag events may be generated than we can digest. If we tried to treat them all, we would observe a growing difference between the mouse cursor and the location where the updates happen. To avoid this, we have to skip obsolete drag events. In other words, after the update for a drag event has been finished, we immediately go on with the *last* one of a sequence of drag events which may have occured in the meantime, and skip the others.

3.6   Special Requirements for Polygonal Convex Distance Functions

As a generalization, we also consider Voronoi diagrams under different metrics than the usual Euclidean distance. To define a *convex distance function* we take a compact, convex body and a point, called center, somewhere in its interior. All points on the boundary of the body, called unit circle, have distance 1 to the center, and the distance from a point p to point q is the factor by which the unit circle translated to p must be scaled such that q lies on its boundary. Such a distance function is not necessarily symmetric but fulfills the other properties of a metric.

There are a couple of interesting results about Voronoi diagrams under convex distance functions, e. g. see Chew and Drysdale [8], Drysdale [16], Skyum [29], Corbalan et al. [11], Jünger et al. [21], and Icking et al. [20]. To summarize, we can say that convex distance functions inherit some properties from the Euclidean metric but differ at others; the differences increase in higher dimensions.

To actually see many of such diagrams and to verify some conjectures we have realized a Java applet called *VoroConvex*, for a screen shot see Figure 7. It is capable to deal with arbitrary polygonal convex distance functions. The restriction to polygonal distances is reasonable because any convex set can be approximated with arbitrary ε-accuracy by a convex polygon and because each bisector has a polygonal shape itself and can be easily constructed.

---

[8] This technique is sometimes called *double-buffering*. In Java, it can be applied by using an Image object as the buffer which is displayed by the drawImage method.

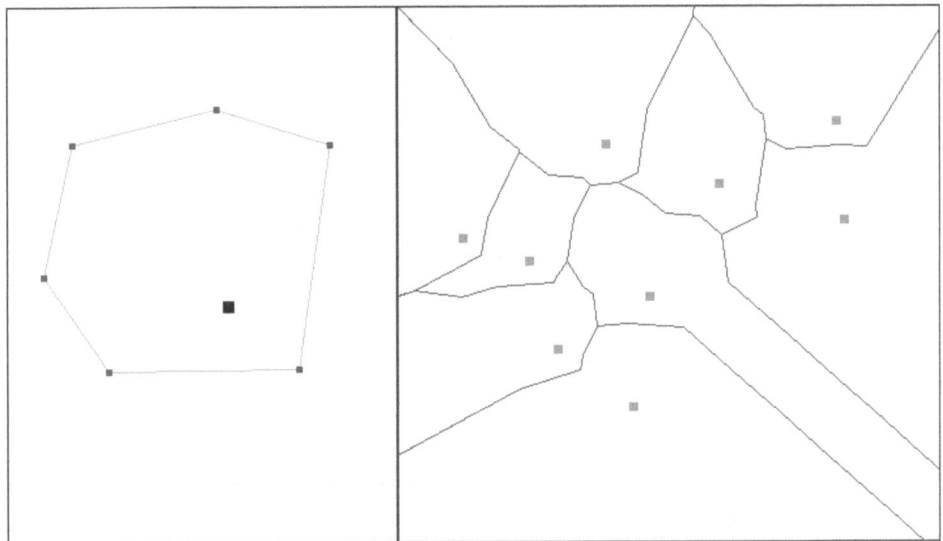

Figure 7. *VoroConvex* showing the unit circle on the left and a Voronoi diagram under this polygonal distance.

To compute this generalized Voronoi diagram we have taken a similar approach as described above for *VoroGlide*. We use the dual graph (which is not called Delaunay in this setting) as the basic structure; this is a triangulation but not necessarily of the convex hull of the point set, see Figure 8 for an example. The algorithm is again incremental and uses point location and edge flip techniques as before.

Extra difficulties result from the fact that bisectors are polygonal chains instead of lines and from the requirement that we want to be able to dynamically change the unit circle and its center in addition to the sites. For some special properties of these diagrams that also have to be taken care of see 4.

VoroConvex opens two windows: one is for the sites and the other is for the unit circle. The user can add as many points as he likes to that window, and the unit circle is always taken as the convex hull of them. The center point is marked in a special colour and can be moved just like the other points. Whenever the distance function is modified in this way, the whole Voronoi diagram is recomputed.

4   Experiences

VoroGlide and *VoroConvex* have turned out as valuable tools for computer-based experiments in teaching and research. In particular, *VoroGlide* has been downloaded thousands of times and has been been used very often, for example by students preparing examinations or theses, by university teachers and researchers not only from computer science and mathematics but also from geology, biology, engineering, meteorology, and even by school teachers. *VoroGlide* has surprised many users by its speed. Typically, it

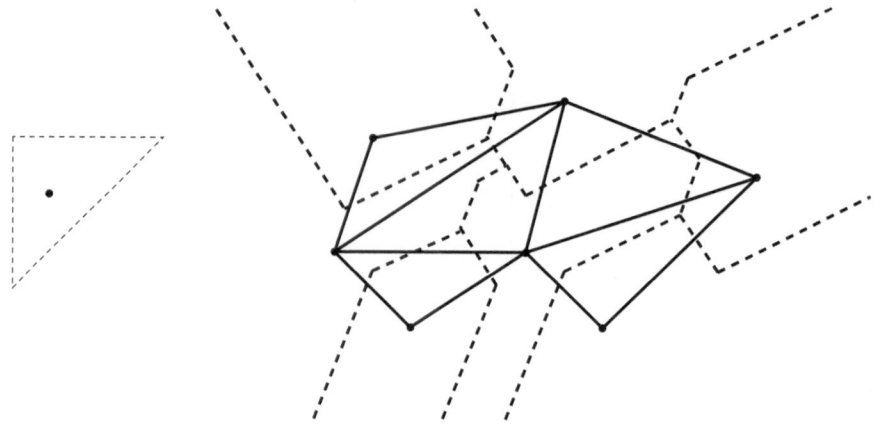

Figure 8. The Voronoi diagram and its dual under a triangular distance function whose unit circle is shown on the left.

delivers a rate of at least 20 frames per second on "average" point sets, nearly independently of the number of points in the diagram. One often achieves a frame rate equal to the rate of drag events. Of course, a machine with a faster processor will perform better, but even on low level computers with a processor speed of less than 100 MHz one gets very satisfying results.

But *VoroGlide* is capable of more. There are two somewhat hidden but useful features, especially for teaching. One is the step by step execution of the edge flip algorithm, to see this choose "Delaunay" in the "Show" menu and "Step Mode" in the "Options Dclaunay" menu. Typing the "s" key reveals a slider to accelerate or slow down the animation. The other feature is the record and replay function, for an example choose "Run Demo" from the "Edit" menu. Typing the "r" key reveals the user interface for recording. To prepare a lesson, the teacher can record any sequence of user actions, store this in one of the six slots provided and replay it at any later time.

In the following we want to illustrate some benefits of such computer-based experiments for teaching and research in computational geometry.

Many properties of Voronoi diagrams become obvious when one observes their dynamic behaviour under interactive changes. For example, taking a closer look at a Voronoi diagram of points under the standard Euclidean distance, one realizes that some point sites have a bounded Voronoi region, while others own an unbounded domain. At first glance, it may not be quite clear which property grants a site an unbounded Voronoi region.

We posed this question to graduate students, and encouraged them to use *VoroGlide* for experiments. Since it can optionally show the convex hull of the current point set, it did not take the students very long to come up with the following conjecture.

Theorem 1. *A point site's Voronoi region is unbounded if and only if the point lies on the convex hull of the set of all sites.*

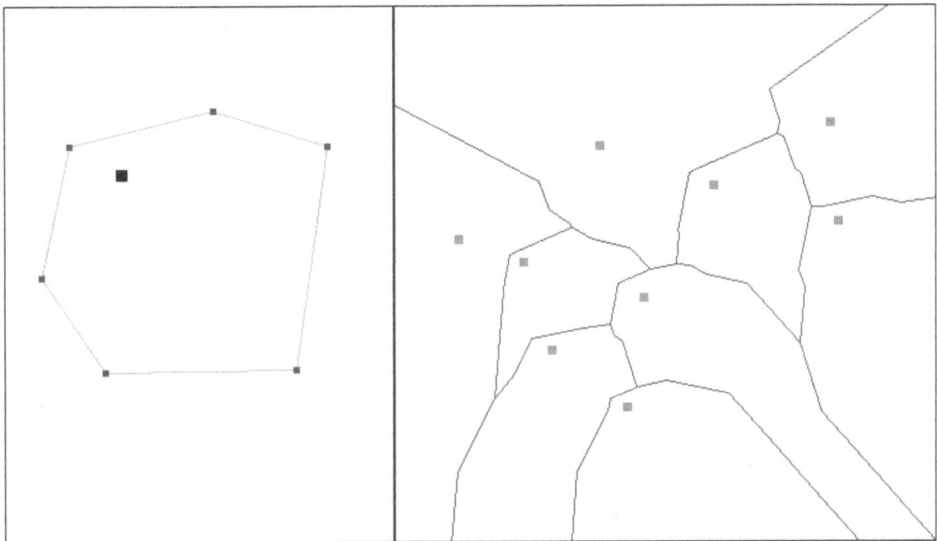

Figure 9. *VoroConvex* showing the Voronoi diagram of the same point set as in Figure 7, only the center of the unit circle is moved to a different place. One can check that the structures of the two diagrams are identical.

At the same time it became clear that observing this fact is quite different from being able to prove it; but encouraged by their conjecture, the students were now far better motivated to follow the proof than in a standard classroom lecture.

Our experimental tools have proven valuable to research, too. When working with *VoroConvex* we realized that moving the center of the convex polygon that defines the distance function does only change the shape of the Voronoi diagram of a fixed set of sites, but not its combinatorial structure. This led to the following theorem whose proof was published in [24], see also Figure 9.

Theorem 2. *For two unit circles which differ only in the position of the center, the two Voronoi diagrams of the same set of sites under the respective distances have the same combinatorial structure, which means that their dual graphs are identical.*

Other properties of Voronoi diagrams which can be observed with the help of our applets are, for example:

- The edges of the Delaunay triangulation connect sites which are adjacent in the Voronoi diagram.
- The Delaunay triangulation covers the convex hull of the sites.
- A Delaunay edge may not cross its dual Voronoi edge.
- The circumcircle of a Delaunay triangle does not contain other sites.
- The incremental method for inserting sites flips edges of the triangulation until the circumcircle property is fulfilled.

- For convex distances, there can be sites not on the convex hull which have an unbounded region, see Figures 7 through 9.
- The bisector of two sites is the locus of the intersection of two expanding circles which are translated to the sites.
- The bisector of two sites under a polygonal convex distance consists of line segments, one less than the unit circle has.
- For convex distances, the bisector of three sites can be empty, and the diagram can be disconnected, and this is *not* a phenomenon of degeneracy.

5   Concluding Remarks

The performance of *VoroGlide* leads us to believe that dynamic visualization with a sufficiently high frame rate is possible in Java. We are convinced that interactive tools like *VoroGlide* and *VoroConvex* are very valuable in both teaching and research, and that the considerable effort in implementing such applets is well invested.

References

1. A. Aggarwal, L. J. Guibas, J. Saxe, and P. W. Shor. A linear-time algorithm for computing the Voronoi diagram of a convex polygon. *Discrete Comput. Geom.*, 4(6):591–604, 1989.
2. K. Arnold and J. Gosling. *The Java Programming Language*. Addison-Wesley, Reading, MA, 1996.
3. F. Aurenhammer. Voronoi diagrams: A survey of a fundamental geometric data structure. *ACM Comput. Surv.*, 23(3):345–405, Sept. 1991.
4. F. Aurenhammer and R. Klein. Voronoi diagrams. In J.-R. Sack and J. Urrutia, editors, *Handbook of Computational Geometry*, pages 201–290. Elsevier Science Publishers B.V. North-Holland, Amsterdam, 2000.
5. J. E. Baker, I. F. Cruz, G. Liotta, and R. Tamassia. A new model for algorithm animation over the WWW. *ACM Computing Surveys*, 27(4):568–572, 1995.
6. H. Baumgarten, H. Jung, and K. Mehlhorn. Dynamic point location in general subdivisions. *J. Algorithms*, 17:342–380, 1994.
7. S. W. Cheng and R. Janardan. New results on dynamic planar point location. *SIAM J. Comput.*, 21:972–999, 1992.
8. L. P. Chew and R. L. Drysdale, III. Voronoi diagrams based on convex distance functions. In *Proc. 1st Annu. ACM Sympos. Comput. Geom.*, pages 235–244, 1985.
9. Y.-J. Chiang, F. P. Preparata, and R. Tamassia. A unified approach to dynamic point location, ray shooting, and shortest paths in planar maps. *SIAM J. Comput.*, 25:207–233, 1996.
10. Y.-J. Chiang and R. Tamassia. Dynamization of the trapezoid method for planar point location. In *Proc. 7th Annu. ACM Sympos. Comput. Geom.*, pages 61–70, 1991.
11. A. G. Corbalan, M. Mazon, T. Recio, and F. Santos. On the topological shape of planar Voronoi diagrams. In *Proc. 9th Annu. ACM Sympos. Comput. Geom.*, pages 109–115, 1993.
12. L. De Floriani, B. Falcidieno, G. Nagy, and C. Pienovi. On sorting triangles in a Delaunay tessellation. *Algorithmica*, 6:522–532, 1991.
13. P. J. de Rezende and W. R. Jacometti. Animation of geometric algorithms using GeoLab. In *Proc. 9th Annu. ACM Sympos. Comput. Geom.*, pages 401–402, 1993.
14. O. Devillers, S. Pion, and M. Teillaud. Walking in a triangulation. In *Proc. 17th Annu. ACM Sympos. Comput. Geom.*, pages 106–114, 2001.

15. L. Devroye, E. P. Mücke, and B. Zhu. A note on point location in Delaunay triangulations of random points. *Algorithmica*, 22:477–482, 1998.

16. R. L. Drysdale, III. A practical algorithm for computing the Delaunay triangulation for convex distance functions. In *Proc. 1st ACM-SIAM Sympos. Discrete Algorithms*, pages 159–168, 1990.

17. D. Flanagan. *Java in a Nutshell*. O'Reilly & Associates, Sebastopol, fourth edition, 2002.

18. L. J. Guibas, J. S. B. Mitchell, and T. Roos. Voronoi diagrams of moving points in the plane. In *Proc. 17th Internat. Workshop Graph-Theoret. Concepts Comput. Sci.*, volume 570 of *Lecture Notes Comput. Sci.*, pages 113–125. Springer-Verlag, 1992.

19. D. P. Huttenlocher, K. Kedem, and J. M. Kleinberg. On dynamic Voronoi diagrams and the minimum Hausdorff distance for point sets under Euclidean motion in the plane. In *Proc. 8th Annu. ACM Sympos. Comput. Geom.*, pages 110–119, 1992.

20. C. Icking, R. Klein, N.-M. Lê, and L. Ma. Convex distance functions in 3-space are different. *Fundam. Inform.*, 22:331–352, 1995.

21. M. Jünger, V. Kaibel, and S. Thienel. Computing Delaunay triangulations in Manhattan and maximum metric. Technical Report 174, Zentrum für Angewandte Informatik Köln, 1994.

22. P.-T. Kandzia and T. Ottmann. VIROR: The virtual university in the Upper Rhine Valley, a new challenge for four prestigious universities in Germany. In *Proc. Role of Universities in the Future Information Society*, 1999.

23. R. Klein. *Algorithmische Geometrie*. Addison-Wesley, Bonn, 1997.

24. L. Ma. *Bisectors and Voronoi Diagrams for Convex Distance Functions*. PhD thesis, Department of Computer Science, FernUniversität Hagen, Technical Report 267, 2000.

25. H. Maurer and D. Kaiser. AUTOOL: A new system for computer assisted instruction. IIG-Report 218, Inst. Informationsverarb., Tech. Univ. Graz, 1986.

26. H. Maurer and F. S. Makedon. COSTOC: Computer supported teaching of computer science. In *Proc. of IFIP Conference on Teleteaching*, pages 107–119. North-Holland, 1987.

27. K. Mehlhorn and S. Näher. *LEDA: A Platform for Combinatorial and Geometric Computing*. Cambridge University Press, Cambridge, UK, 2000.

28. J. Nievergelt, P. Schorn, M. de Lorenzi, C. Ammann, and A. Brüngger. XYZ: A project in experimental geometric computation. In *Proc. Computational Geometry: Methods, Algorithms and Applications*, volume 553 of *Lecture Notes Comput. Sci.*, pages 171–186. Springer-Verlag, 1991.

29. S. Skyum. A sweepline algorithm for generalized Delaunay triangulations. Technical Report DAIMI PB-373, CS Dept., Aarhus University, 1991.

Online Courses Step by Step

Paul-Th. Kandzia

Institut für Informatik, Albert-Ludwigs-Universität
D-79085 Freiburg, Germany
`kandzia@informatik.uni-freiburg.de`

Abstract. Virtual University is a buzzword. However, in spite of manifold activities e-learning, that is the usage of multimedia and network technology to benefit or profit the students, will not inevitably be adopted in the university. On the contrary, these technologies have to be strongly pushed and the characteristics and specific culture of traditional universities have to be taken into account. Various methods and techniques have proven to be best practice, which can succeed in a traditional university and open up a way for the alma mater to find its place in cyberspace, too.

1 Introduction

In the last years it has become common sense that also traditional universities have to open up their well-known campus-based teaching and learning scenarios towards enhancing them by net-based distance learning in the near future. This vision of *Virtual University* (however, denoting rather a virtual branch than a full-fledged university) focuses on the students' benefit by providing for more flexibility of teaching and learning in terms of time and space. The introduction of e-learning in a greater scale, which requires significantly less time spent in lecture rooms, may enable people to absolve university studies successfully, who otherwise would have been hindered by their personal circumstances like growing kids or professional workload.

Since the late nineties of the last century enormous activity has emerged in Germany and Switzerland, in its overwhelming majority organized like research projects initialised and financed by state funding. At present, this approach is running into serious problems: Providing regular students of a traditional German or Swiss university with convincing e-learning offerings which match the goals sketched above, can not be done by single university teachers, even when they are joined in greater units as it is usual in Virtual University projects. Strategic decisions and planning for the university as a whole as well as the allocation of central resources are a necessary prerequisite for succesful online courses. However, since incentives (i.e. money in the first place) have been given to university *teachers* only, appropriate administrative, organizational, and infrastructural measures still are found far too seldom.

Besides those political obstacles [1; 2] which are beyond the scope of this paper, there is a second hindrance to a sustaining and growing success of e-learning within universities: the means and methods used in e-learning projects do not fit well in the "production process" of a campus-based alma mater. In the following those problems and ways towards avoiding them are discussed.

R. Klein et al. (Eds.): Comp. Sci. in Perspective (Ottmann Festschrift), LNCS 2598, pp. 206–215, 2003.

2   E-learning and the University

When reasoning about e-learning most people bear in mind the idea of a website with an appealing design and many multimedia elements like audio, video, interactive animations, and integrated exercises which allow for (self-)assessment of the learner. Actually, this view focuses on only one facet of e-learning, the *web-based* or *computer-based training* (abbreviated WBT). Indeed, WBT has important advantages. Using web programming with HTML, Flash, or Java, advanced didactical concepts can be realized, changing deeply the traditional roles of teachers and learners. Of course, planning, designing, and implementing a WBT follows the logic of a software project, moreover, not a trivial one, since the different levels of content, didactics, design, and implementation very often require fairly complex interdisciplinary work.

However, a university is an institution never meant and not at all prepared for doing software development. Nevertheless, a majority of the Virtual University projects is exactly doing this: (ab)using scientific staff for software implementation, paid by project funds. This is not a convincing approach: First, routine work like Macromedia programming collides with the scientific motivation of the employee, usually a PHD student. Second, in the beginning usually the employee is not a highly skilled programmer, but later on, when he or she has gained a lot of experience, the university will not profit from it, since the project funding ends and the employee is discharged. Consequently, the production of WBTs in university projects neither is cost and time effective, nor does it yield optimal products, nor does it pay off as a kind of staff qualification.

But developing the WBT is only one half of the story. As one of the very characteristics of a university the subject matter taught is frequently and rapidly changing. For example, in Computer Science nearly never exactly the same lecture is held twice. When using WBTs this means updating software (or at least HTML code)! Who will do that work after the project staff has left the university? Quite to often projects start an implementation without solving issues like that (or even without taking care about them ...)

Anyway, WBTs for university teaching should not be rejected in general. In certain cases developing such a product may very well be reasonable, if thorough investigation and planning results in satisfying answers of the following questions:

– Is the subject matter suitable, particularly, is the content stable enough?
– Is a broad usage ensured, justifying the high effort of the development?
– Is sufficient support ensured, also in the long run?
– Is the development necessary, or could an existing product be used?
– Is the implementation work carried through effectively? Can it be bought from a
  professional partner rather than be done by scientific staff?

Besides a lot of questionable products also innovative and convincing examples of WBTs have emerged out of Virtual University projects, for instance in the area of *problem based learning* with a "virtual patient" in Medicine or animations of algorithms in Computer Science.

3   Virtual Elements for the University

As shown above WBT production probably may not be the best way for a campus university to build a substantial net-based distance learning program. In the following alternative methods will be discussed which much better fit in the structure of a university and can be integrated in the workflows of the campus. Moreover, they maintain the character of university teaching: highly up-to-date content, delivered with the personal flavor of the teacher together with coaching of the learner by the teacher or tutors.

As a further advantage, unlike a WBT development, those elements do not require a "do all or nothing" decision, but allow for a stepwise progression from some enhancement of classroom events to full online courses, with each step giving an added value in its own right. In sync with the goals of the institution, its resources, and the upcoming experience, a steadily growing "virtual branch" of the university is emerging.

3.1   Learning Management Systems

A *Learning Management System* (LMS) provides for the basic functions needed for composition and delivery of any web-based educational offer: administration of users and courses, composition and delivery of content, communication, assessment, and grading. As a central service an LMS must be installed and run by a central department within the university. Leaving this task to single small projects is extremely inefficient, especially if different projects run their courses on different platforms that are not interoperable (a situation found in many universities).

In most cases it is unreasonable for a university to develop its own LMS. There are many commercial products on the market, some of them open-source, and even when their drawbacks are taken into account, universities will find it difficult to develop a better solution and support it over many years [3].

Obviously an LMS is needed if an institution plans to offer full-fledged courses in online mode. However, it is also reasonable to establish an LMS very early as one of the first steps towards enhancing teaching and learning with multimedia, even when "proper" e-learning is not in the main focus at all. This way, all activities are running with a clear structure and a uniform interface from the very beginning. Later on, more advanced projects profit from and are planned according to the existing infrastructure. Otherwise, a very miserable situation will most probably arise: A lot of material already exists when finally the question of an appropriate LMS is tackled. Since the respective requirements are very different or even contradictory, it is impossible to find a really satisfying solution. Moreover, the long process of evaluating and establishing the LMS will delay the development of the projects waiting for it.

Last, a well designed interface with a clear visibility of all e-learning projects serves as a powerful means to attract students, the public, and new project activists within the institution.

3.2   E-learning University Style I: Lecture Recording

As shown above, universities are not too capable of developing WBT software, moreover, in many cases a WBT may not be the best approach to e-learning anyway. In the next sections methods are discussed which are more appropriate to a university teacher's "production process". Such scenarios transfering basic activities of the campus into cyberspace have proven to be best practice of many projects and increasingly are adopted by teachers outside of any project funding.

First, the "virtualisation" of a lecture is described: Electronic Lecture Recording. Its basic idea is to record a quite regular live lecture in order to obtain versatile digital material at negligible additional cost [4].

Lecture recording allows for a wide range of flexibility. Data formats depend on the expected user behaviour. If easy access over a browser is important, streaming formats like Real or Flash can be used. On the other hand, if the provision of functions such as effective "fast forward" and "fast backward" is necessary (since difficult content requires intensive searching, scrolling and jumping within the recording by the learner) a so-called *random access* format like AOF [5] is needed. As a further advantage, AOF-like data formats maintain the symbolic information of the presentation slides. In particular, the text objects presented are stored as text, hence enabling full text search within a recorded presentation.

The teacher's style (which may be restricted by the available technical equipment) covers a wide range of variety, too. As the basic case a laptop with a microphone is sufficient. Even if a teacher refuses to use a computer for his or her presentation, video and snapshots of a blackboard may be recorded (see figure 1). Of course, the usage of an electronic whiteboard (see figure 2) yields much better results. Provided a skilled teacher such graphical input devices allow for a highly interactive and lively style (featuring even hand-written annotations, see figure 3). A specifically designed portable speakert's desk with built-in pencil sensitive input device (e.g a *Wacom tablet*[1]), computer, and audio equipment has proven very useful, particularly facilitating lecture recording for newbies (see figure 4).

If appropriate, the presentation may contain computer animations and simulations.

In any case all data streams like audio, video, whiteboard activity, and animations are recorded automatically. They are stored on a server, and can be replayed in a synchronized way immediately after the live event. The recordings serve as an excellent basis for online courses. To that aim the recording is edited and exercises, self-assessments, or other supplementary material is added (see below).

When post-processing a recorded lecture one can adopt an arbitrary quality in a spectre between pure recordings without any editing and, for example, a Macromedia-WBT that comprises only short patches of thoroughly edited recordings. Anyway, for university level education, up-to-date content and interesting material for the student is much more important than appealing design. Hence, a teacher can start at a very low level which alleviates his or her principal decision for recording lectures, but nevertheless obtain reasonable results. As resources allow or if the course is repeated the material may be reworked and enhanced.

Besides its flexibility the main advantages of lecture recording are its cost effectiveness and its rapidity (which provides for very up-to-date material and swift updating when necessary). Also, due to the flexibility of lecture recording, the teacher's normal work is only slightly changed. The teacher is still preparing and holding lectures in

---

[1] see http://www.wacom.com

his or her individual style, not designing WBTs. Compared with a WBT the pedagogical potential of lecture recording may be restricted, but our experience is that students greatly appreciate lecture recordings if an interesting theme, a skilled teacher and appropriate support is made sure.

Nowadays, several lecture recording solutions are available: from open source, over cheap commercial solutions with restricted features (e.g. *Camtasia*[2] and *Real Presenter*[3]) to the state-of-the-art product *Lecturnity*[4] based on the AOF principle.

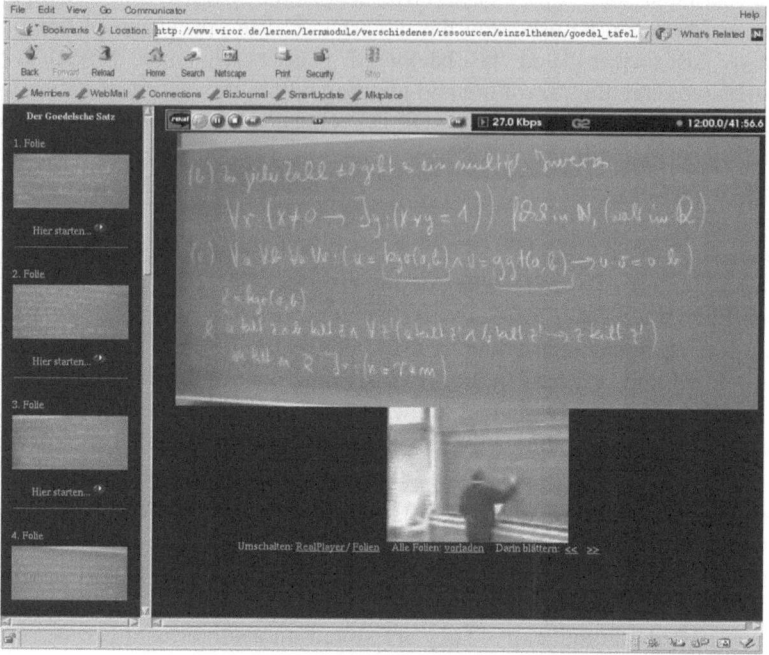

Figure 1. Recorded blackboard and chalk presentation

3.3   Course Production University Style II: Assignments and Tutoring

In studies of the natural sciences, Computer Science, or Mathematics during lectures *conceptualisation* of a subject takes place. As a pedagogical counterpart tutored assignments are necessary and foster *dialogue* and *construction*. Characteristically, depending on the lecture course, students are expected to solve assignments, either alone or in a group of two or three persons. After a fixed amount of time, often a week, their solution is presented to a tutor for assessment. Usually tutors are advanced students responsible

---

[2] see http://www.techsmith.com/products/camtasia/camtasia.asp

[3] see http://www.real.com

[4] see http://www.im-c.de

Figure 2. Using an electronic whiteboard

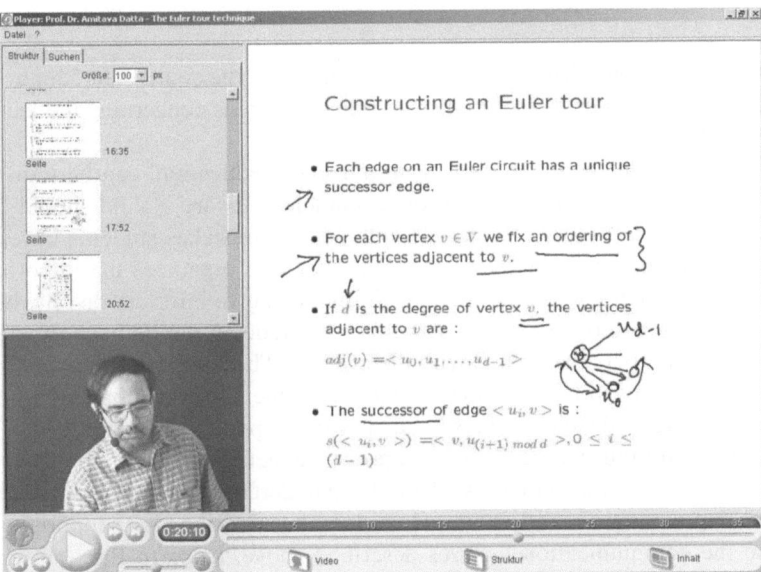

Figure 3. Recorded lecture with handwritten annotations

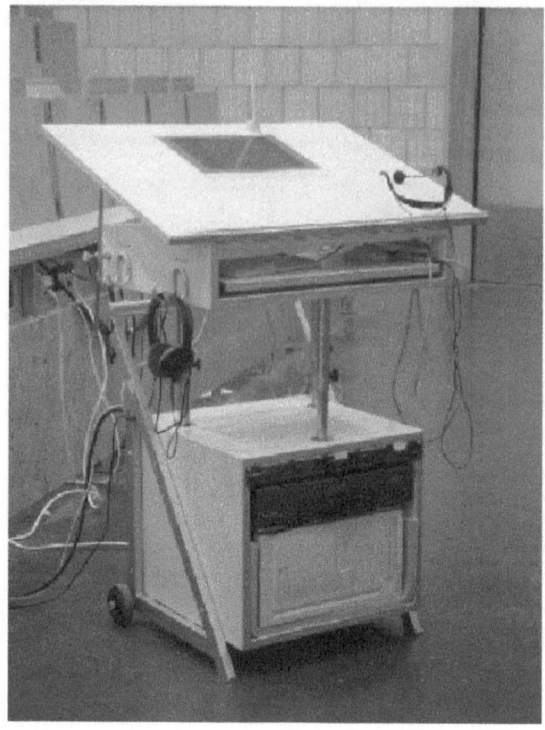

Figure 4. Speaker's desk for lecture recording

for ten to thirty students. The assessment is often part of the course accreditation. Afterwards the tutor discusses the solutions and other questions concerning the lecture with the student group.

This process, which is very typical in university education, can be transferred to web-based distance learning. In addition, a suitable software system should include a report pool where all tutors of a course note general mistakes students have made as well as good solutions to the exercises. This tool also organizes the important feedback between teacher and tutors. Statistical functions give information on whether the exercises have been too easy or too difficult. The students should have the possibility of comparing their personal marks with the average. The statistical functions are much easier to implement and to use in the virtual environment than in a paper-based setting.

Because exercise development is valuable work, a pool is needed in which all exercises are stored for reuse, together with appropriate metadata.

Surprisingly, although a lot of work has been undertaken on intelligent tutoring systems and adaptive exercises [6], few systems have been designed to support the specific tasks and the communication features described above. In particular, the features of widespread commercially available LMS like *Lotus Learning Space*[5] or *WebCT*[6] are

---

[5] see http://www.lotus.com

[6] see http://www.webct.com

not sufficient (however, WebCT is working on it for the next release, at least). The Fer-nUni Hagen (a German distance teaching university) was forced to optimize its paper and snail-mail based communication, thus it developed *WebAssign*[7] and uses the system with thousands of students. Over thirty universities also use that open-source software. WebAssign is specifically designed for creating exercises in programming [7].

Another popular system is also called *WebAssign* and was developed by the North Carolina State University. It has similar features as the FernUni Hagen' WebAssign package. One common drawback of both systems is that the integration into an overall LMS system (e.g. access to user data) has to be done by the adopting institution.

Recently, the German company imc has incorporated the functions described above in its LMS, called CLIX[8]. In a joint project the implemented tutoring workflow has been specified by the computer science department of the university of Freiburg [8].

Of course, delivering and correcting exercise solutions in a digital form increases the workload of both students and tutors. Though, experience shows that the "virtual" form is very much appreciated by the students, since the quality of the solutions increases. In particular, tutors are far less bothered with botched programs, because WebAssign automatically only accepts solutions after they have passed basic checks.

Tutored assignments over the web fit very comfortably in the everyday work of the alma mater. They are a necessary and logical completion of lecture recording with which they share similar advantages. Bringing both principles together enables a uni-versity to transfer a major part of its teaching and learning in the virtual world.

3.4   Net-Based Seminars

In the last sections the "virtual counterparts" of lectures and tutored assignments have been discussed. But how about more communication-oriented settings like a seminar? In principle, a net-based seminar can be realized either as a *videoconference* or using the usual *communication tools* of the web, as mail, newsgroups, shared working spaces, or chat. We will call the latter form a *web seminar*. Of course, both modes can be mixed. Furthermore, some face-to-face meetings, particularly a kickoff event may be useful.

It has turned out that for a seminar as videoconference, due to the more active partci-pation of the learners, video quality is less important than in the case of a lecture, where e.g. MPEG2-hardware and ATM-connections are used. Hence, standard conferencing software is sufficient.

In contrast to videoconferencing, web seminars mainly rely on textual communi-cation. Material in normal web-formats is shared easily. A web seminar is cheap and, from a technical viewpoint, is easy to use. It is fully flexible in terms of time and can be used from the student's home PC. The organisational effort required on the university side is quite limited. Web seminars are found in the philological, political and social sciences, or law (see e.g. [10] for the use of a MOO in law education) - all of which involve working with texts. Succesful web seminars show again that useful content is much more appreciated by university students than fancy technology. However, it would appear teachers of other disciplines consider the pedagogical potential is too restricted.

---

[7] see http://www.campussource.de

[8] see http://www.im-c.de, again

A very appealing didactical feature of web-based seminars is that professionals from outside the university can contact and work with students very easily.

4 The Virtual University Step by Step

We have discussed several elements of e-learning within the university. Using these elements a "virtual branch" of an alma mater can be constructed succesively.

As the first step a LMS is established for the institution as a whole. While more advanced steps towards e-learning depend on the respective characteristics of the different departments, the university management has to make sure that all educational activities – those in the classroom as well as those in cyberspace – are brought together and are made public over the LMS as a common portal. In the very near future students will regard as a matter of course that administrative services like enrolment, study programs, or exam results are found on the web. Hence, also a traditional university will have to intgrate such services into its web portal. Definitely, relying on an office opened from 9.00 to 12.00 will not be sufficient! Furthermore, some central support for teachers on technical, pedagogical, and organizational questions has to be provided.

Based on this central activities the departments and teachers tackle the next steps. Naturally, all digital material used in a lecture, the syllabus, etc. is delivered and managed using the LMS, which also benefits the teacher by alleviating reuse. Where it is reasonable, teachers record their classroom events and upload the recordings in the LMS as well. So students are quite well supported when revising a lecture (even if the live event has been missed) or when preparing themselves for an exam.

As a second step a teacher can design and organize net-based assignments and tutoring. Putting it all together, a course can be run completely online during its first or second repetition [9].

Independently from the elements of course production teachers can offer net-based seminars.

Anyway, departments and teachers have to decide to what extent they want to "go virtual" - maybe a full masters' program in distance learning mode, selected courses, simple enhancement of classroom-based work or merely the organizational services. In all cases basic issues have to be resolved before any implementation begins:

– Who are the users of the planned product?
– Is it a convincing offer for them?
– Are enough appropriate partners supporting the project?
– Are the partners able to finance the project both short and long term?

Clearing up these issues early on seems fundamental but at present still surprisingly few projects are able to give satisfactory answers.

In the future evidently the alma mater will not be swept away. Most of us will be glad about that. The approach favoured above will not cause a revolution in universities. Some will regret this. Some will regard the elements described as modest and low-level. However, their strength is based on the close relation to traditional university work, the successive proceeding, and the high grade of flexibility. Hence, the approach offers a realistic chance for universities to build a programme of web-based distance courses with a scope broad enough to give students noticeably more flexibility, without losing the advantages of a campus-based education.

References

1. Brake, C.: *Politikfeld Multimedia*. Waxmann, Münster, 2000.
2. Kandzia, P.-T.: *E-Learning an Hochschulen - Von Frustration und Innovation*. Proc. 7. Europäische Fachtagung der GMW (CAMPUS), Basel, 2002 (to appear).
3. Trahasch, S., Lienhard, J.: *Virtual Learning Environment in a University Infrastructure*. Proc. 3rd International Conference on New Learning Technologies (NLT/NETTIES), Fribourg, 2001.
4. Kandzia, P.-T., Maass, G.: *Course Production - Quick and Effective*. Proc. 3rd International Conference on New Learning Technologies (NLT/NETTIES), Fribourg, 2001.
5. Müller, R., Ottmann, T.: *Electronic Note-Taking Systems, Problems and their Use at Universities*. In: Adelsberger, H. H., Collis, B., Pawlowski, J. M. (eds.): *Handbook on Information Technologies for Education and Training*, Springer, Heidelberg, 2001.
6. Brusilovsky, P., Miller, P.: *Course Delivery Systems for the Virtual University*. In: Tschang, F. T., Della Senta, T. (eds.): *Access to Knowledge: New Information Technologies and the Emergence of the Virtual University*. Elsevier, Amsterdam, 2001.
7. Brunsmann, J. Homrighausen, A., Six, H.-W., Voss, J.: *Assignments in a Virtual University - The WebAssign System*. Proc. 19th World Conference on Open Learning and Distance Education (ICDE). Vienna, Austria, 1999.
8. Kandzia, P.-T., Trahasch, S.: *Tutored Assignments Going Online*, Proc. 4th International Conference on New Educational Environments (ICNEE), Lugano, 2002.
9. Lauer, T., Trahasch, S., Zupancic, B.: *Virtualizing University Courses: From Registration till Examination*. Proc. 4th International Conference on New Educational Environments (ICNEE), Lugano, May 2002.
10. Nett, B., Röhr, F.: *JurMOO - Cooperative Spaces in Computer and Law teaching*, Proc. Int. Conf. on Advances in Infrastructure for e-Business, e-Education, e-Science, and e-Medicine on the Internet (SSGRR), L'Aquila, Jan. 2002.

Spatial Data Management for Virtual Product Development

Hans-Peter Kriegel*, Martin Pfeifle*, Marco Pötke**,
Matthias Renz*, Thomas Seidl***

University of Munich, Institute for Computer Science
http://www.dbs.informatik.uni-muenchen.de
*{kriegel,pfeifle,renz}@dbs.informatik.uni-muenchen.de
**marco.poetke@sdm.de
***seidl@informatik.rwth-aachen.de

Abstract: In the automotive and aerospace industry, millions of technical documents are generated during the development of complex engineering products. Particularly, the universal application of Computer Aided Design (CAD) from the very first design to the final documentation created the need for transactional, concurrent, reliable, and secure data management. The huge underlying CAD databases, occupying terabytes of distributed secondary and tertiary storage, are typically stored and referenced in Engineering Data Management systems (EDM) and organized by means of hierarchical product structures. Although most CAD files represent spatial objects or contain spatially related data, existing EDM systems do not efficiently support the evaluation of spatial predicates. In this paper, we introduce spatial database technology into the file-based world of CAD. As we integrate 3D spatial data management into standard object-relational database systems, the required support for data independence, transactions, recovery, and interoperability can be achieved. Geometric primitives, transformations, and operations on three-dimensional engineering data will be presented which are vital contributions to spatial data management for CAD databases. Furthermore, we will present an effective and efficient approach to spatially index CAD data by using the concepts of object-relational database systems and the techniques of relational access methods. The presented techniques are assembled to a complete system architecture for the Database Integration of Virtual Engineering (DIVE). By using relational storage structures, the DIVE system provides three-dimensional spatial data management within a commercial database system. The spatial data management and the query processor is fully embedded into the Oracle8i server and has been evaluated in an industrial environment. Spatial queries on large databases are performed at interactive response times.

1 Introduction

In mechanical engineering, three-dimensional Computer Aided Design (CAD) is employed throughout the entire development process. From the early design phases to the final production of cars, airplanes, ships, or space stations, thousands to millions of CAD files and many more associated documents including technical illustrations and

R. Klein et al. (Eds.): Comp. Sci. in Perspective (Ottmann Festschrift), LNCS 2598, pp. 216–230, 2003.

business documents are generated. Most of this data comprises spatial product components or spatially related content. Recently, new CAD applications have emerged to support virtual engineering on this data, i.e. the evaluation of product characteristics without building even a single physical prototype. Typical applications include the digital mock-up (DMU) [BKP98] or haptic rendering of product configurations [MPT99].

Engineering Data Management (EDM) systems organize the huge underlying CAD databases by means of hierarchical product structures. Thus, structural queries as "retrieve all documents that refer to the current version of the braking system" are efficiently supported. If we look at CAD databases from a spatial point of view, each instance of a part occupies a specific region in the three-dimensional product space (cf. Fig. 1). Together, all parts of a given product version and variant thereby represent a virtual prototype of the constructed geometry. Virtual engineering requires access to this product space by spatial predicates in order to "find all parts intersecting a specific query volume" or to "find all parts in the immediate spatial neighborhood of the disk brake". Unfortunately, the inclusion of the respective spatial predicates is not supported efficiently by common, structure-related EDM systems.

This paper is organized as follows: In the next section, we shortly review the common file-based organization of spatial CAD data and describe three important industrial applications which benefit from a relational index. In section 3 we discuss the transformation and approximation of high-resolution CAD data originating from heterogeneous sources. In section 4 we shortly review the RI-tree, which is an efficient index structure for intervals. Finally, in the last section we present an architecture for the Database Integration of Virtual Engineering (DIVE) for existing Engineering Data Management systems (EDM).

2 Spatial Engineering

In the file-based world of CAD applications, the huge amount of spatial data is typically organized by hierarchical product structures which efficiently support selections with respect to variants, versions, and functions of components. In this section, we propose the integration of spatial data management into existing EDM systems. Existing EDM systems are based on fully-fledged object-relational database servers. They organize and synchronize concurrent access to the CAD data of an engineering product. The distributed CAD files are linked to global product structures which allow a hierarchical

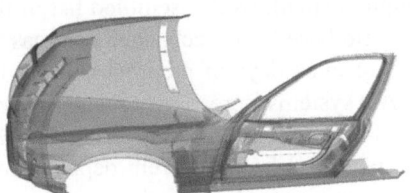

Fig.1: Virtual prototype of a car.

view on the many possible product configurations emerging from the various versions and variants created for each component [Pö98]. Although the EDM system maintains a consistent knowledge about the storage location of each native CAD file, only a few spatial properties, including the position and bounding box of the respective part in the product space, are immediately accessible. But many applications of virtual engineering require a more fine-grained spatial selection. Thus, the EDM system has to be extended by a high-resolution representation of the product geometry which is organized by spatial indexes in order to achieve interactive response times.

2.1 Integrated Spatial Data Management

In order to supplement an existing EDM system with spatial representation and selection of CAD data, one of the most important components required from the underlying object-relational database system is a three-dimensional spatial index structure. By introducing an efficient spatial access path besides the existing support for structural evaluations, we can achieve an integrated database management for both, spatial and non-spatial engineering data. Relational access methods on spatial databases can be seamlessly integrated into the object-relational data model while preserving the functionality and performance of the built-in transaction semantics, concurrency control, recovery services, and security. The following section discusses realistic applications to illustrate the practical impact of the proposed concepts.

2.2 Industrial Applications

We present three industrial applications of virtual engineering which immediately benefit from a relational index on spatial CAD data. We have analyzed and evaluated them in cooperation with partners in the automotive and aerospace industry, including the Volkswagen AG, Wolfsburg, the German Aerospace Center DLR e.V., Oberpfaffenhofen, and the Boeing Company, Seattle.

Digital Mock-Up of Prototypes. In the car industry, late engineering changes caused by problems with fit, appearance or shape of parts already account for 20-50 percent of the total die cost [CF91]. Therefore, tools for the digital mock-up (DMU) of engineering products have been developed to enable a fast and early detection of colliding parts, purely based on the available digital information. Unfortunately, these systems typically operate in main-memory and are not capable of handling more than a few hundred parts. They require as input a small, well-assembled list of the CAD files to be examined. With the traditional file-based approach, each user has to select these files manually. This can take hours or even days of preprocessing time, since the parts may be generated on different CAD systems, spread over many file servers and are managed by a variety of users [BKP98]. In a concurrent engineering process, several cross-functional project teams may be recruited from different departments, including engineering, production, and quality assurance to develop their own parts as a contribution to the

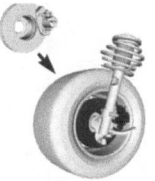

a) Box volume query   **b)** Collision query

Fig.2: Spatial queries on CAD data.

whole product. However, the team working on section 12B of an airplane may not want to mark the location and the format of each single CAD file of the adjacent sections 12A and 12C. In order to do a quick check of fit or appearance, they are only interested in the *colliding* parts. Moreover, the internet is gaining in importance for industrial file exchange. Engineers, working in the United States, may want to upload their latest component design to the CAD database of their European customer in order to perform interference checks. Thus, they need a fast and comfortable DMU interface to the EDM system. Fig. 2 depicts two typical spatial queries on a three-dimensional product space, retrieving the parts intersecting a given box volume (*box volume query*), and detecting the parts colliding with the geometry of a query part (*collision query*). A spatial filter for DMU-related queries on huge CAD databases is easily implemented by a spatial access method which determines a tight superset of the parts qualifying for the query condition. Then, the computationally intensive query refinement on the resulting candidates, including the accurate evaluation of intersection regions (cf. Fig. 2a), can be delegated to an appropriate main memory-based CAD tool.

Haptic Rendering. The modern transition from the physical to the digital mock-up has exacerbated the well-known problem of simulating real-world engineering and maintenance tasks. Therefore, many approaches have been developed to emulate the physical constraints of natural surfaces, including the computation of force feedback, to capture the contact with virtual objects and to prevent parts and tools from interpenetrating [GLM96] [LSW99] [MPT99]. Fig. 3a [Re00] depicts a common haptic device to transfer the computed force feedback onto a data glove. The simulated environment, along with the force vectors, is visualized in Fig. 3b. By using this combination of haptic algorithms and hardware, a realistic force loop between the acting individual and the virtual scene can be achieved. Naturally, a real-time computation of haptic rendering

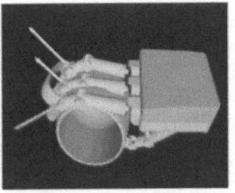

a) Haptic device   **b)** Virtual environment

Fig.3: Sample scenario for haptic rendering.

Fig.4: Virtual environment of the International Space Station.

requires the affected spatial objects to reside in main-memory. In order to perform haptic simulations on a large scale environment comprising millions of parts, a careful selection and efficient prefetching of the spatially surrounding parts is indispensable. Fig. 4 illustrates the complexity of usual virtual environments by the example of the International Space Station (ISS). In order to simulate and evaluate maintenance tasks, e.g. performed by autonomous robots, an index-based prefetching of persistent spatial objects can be coupled with real-time haptic rendering [Re02].

Spatial Document Management. During the development, documentation, and maintenance of complex engineering products, many other files besides the geometric surfaces and solids of product components are generated and updated. Most of this data can also be referenced by spatial keys in the three-dimensional product space (cf. Fig. 5), including kinematic envelopes which represent moving parts in any possible situation or spatial clearance constraints to reserve unoccupied regions, e.g. the minimal volume of passenger cabins or free space for air circulation around hot parts. Furthermore, technical illustrations, evaluation reports or even plain business data like cost accounting or sales reports for specific product components can be spatially referenced. Structurally referencing such documents can become very laborious. For example, the meeting minutes concerning the design of a specific detail of a product may affect many different components. Spatial referencing provides a solution by simply attaching the meeting minutes to a spatial key created for the region of interest. A very intuitive query could be: "retrieve all meeting minutes of the previous month concerning the spatial region between parts A and B". Such queries can be efficiently supported by spatial indexes.

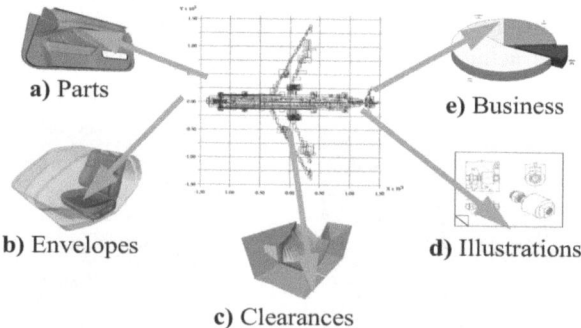

a) Parts
b) Envelopes
c) Clearances
d) Illustrations
e) Business

Fig.5: Spatial referencing of engineering documents.

3 Operations on Spatial Engineering Data

We consider an engineering product as a collection of individual, three-dimensional parts, while each part potentially represents a complex and intricate geometric shape. The original surfaces and solids are designed at a very high precision. In order to cope with the demands of accurate geometric modeling, highly specialized CAD applications are employed, using different data primitives and native encodings for spatial data. For our spatial database, we defined a set of universal representations which can be derived from any native geometric surface and solid. The supported geometric data models include triangle meshes for visualization and interference detection, voxel sets and interval sequences as conservative approximations for spatial keys, and dynamic point shells to enable real-time haptic rendering.

3.1 Triangle Meshes

Accurate representations of CAD surfaces are typically implemented by parametric bicubic surfaces, including Hermite, Bézier, and B-spline patches. For many operations, such as graphical display or the efficient computation of surface intersections, these parametric representations are too complex [MH99]. As a solution, approximative polygon (e.g. triangle) meshes can be derived from the accurate surface representation. These triangle meshes allow for an efficient and interactive display of complex objects, for instance by means of VRML encoded files, and serve as an ideal input for the computation of spatial interference.

For the digital mock-up (DMU), collision queries are a very important database primitive. In the following, we assume a multi-step query processor which retrieves a candidate part S possibly colliding with a query part Q. In order to refine such collision queries, a fine-grained spatial interference detection between Q and S can be implemented on their triangle meshes. We distinguish three actions for interference detection [MH99]: *collision detection*, *collision determination*, and *collision response*:

Collision detection: This basic interference check simply detects if the query part Q and a stored part S collide. Thus, collision detection can be regarded as a geometric intersection join of the triangle sets for S and Q which already terminates after the first intersecting triangle pair has been found.

Collision determination: The actual intersection regions between a query part and a stored part are computed. In contrast to the collision detection, all intersecting triangle pairs and their intersection segments have to be reported by the intersection join.

Collision response: Determines the actions to be taken in consequence of a positive collision detection or determination. In our case of a spatial database for virtual engineering, a textual or visual feedback on the interfering parts and, if computed, the intersection lines seems to be appropriate.

a) Triangulated surface   **b)** Voxelized surface

Fig.6: Scan conversion on a triangulated surface.

3.2 Interval Sequences

The Relational Interval Tree (RI-tree), as presented in section 4, significantly outperforms competing techniques with respect to usability and performance. In order to employ the RI-tree as a spatial engine for our CAD database, we propose a conversion pipeline to transform the geometry of each single CAD part to an interval sequence by means of voxelization. A basic algorithm for the *3D scan-conversion* of polygons into a voxel-based occupancy map has been proposed by Kaufmann [Ka87]. Similar to the well-known 2D scan-conversion technique, the runtime complexity to voxelize a 3D polygon is $O(n)$, where n is the number of generated voxels. If we apply this conversion to the given triangle mesh of a CAD object (cf. Fig. 6a), a conservative approximation of the part surface is produced (cf. Fig. 6b). In the following, we assume a uniform three-dimensional voxel grid covering the global product space.

If a triangle mesh is derived from an originally solid object, each triangle can be supplemented with a normal vector to discriminate the interior from the exterior space. Thus, not only surfaces, but also solids could potentially be modeled by triangle meshes. Unfortunately, triangle meshes generated by most faceters contain geometric and topological inconsistencies, including overlapping triangles and tiny gaps on the surface. Thus, a robust reconstruction of the original interior becomes very laborious. Therefore, we follow the common approach to voxelize the triangle mesh of a solid object first (cf. Fig. 7a), which yields a consistent representation of the object surface. Next, we apply a 3D flood-fill algorithm [Fo00] to compute the exterior voxels of the object (cf. Fig. 7b), and thus, determine the outermost boundary voxels of the solid. We restrict the flood-fill to the bounding box of the object, enlarged by one voxel in each direction. The initial fill seed is placed at the boundary of this enlarged bounding box.

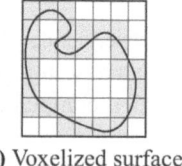   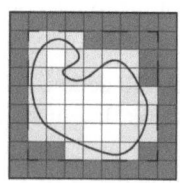   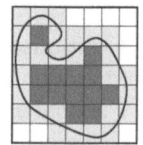

a) Voxelized surface   **b)** Filled exterior   **c)** Inverted result

Fig.7: Filling a closed voxelized surface.

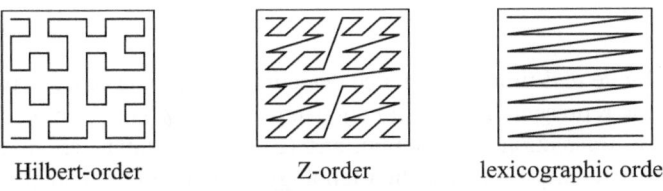

Hilbert-order   Z-order   lexicographic order

Fig.8: Examples of space-filling curves in the two-dimensional case

In the final step, we simply declare all voxels as interior which are neither boundary nor exterior voxels (cf. Fig. 7c). In consequence, we obtain a volumetric reconstruction of the original solid, marking any voxels behind the outermost surface as interior. The above algorithm has a runtime complexity of $O(b)$, where b is the number of voxels in the enlarged bounding box.

The derived voxel set of an arbitrary surface or solid represents a consistent input for computing interval sequences. The voxels correspond to cells of a grid, covering the complete data space. By means of space filling curves, each cell of the grid can be encoded by a single integer number, and thus an extended object is represented by a set of integers. Most of these space filling curves achieve good spatial clustering properties. Therefore, cells in close spatial proximity are encoded by similar integers or, putting it another way, contiguous integers encode cells in close spatial neighborhood. Examples for space filling curves include Hilbert-, Z-, and the lexicographic-order, depicted in Fig. 8. The Hilbert-order generates the minimum number of intervals per object [Ja90] [FR89] but unfortunately, it is the most complex linear order. Taking redundancy and complexity into consideration, the Z-order seems to be the best solution. Therefore, it will be used throughout the rest of this paper.

Voxels can be grouped together to *Object Interval Sequences*, such that an extended object can be represented by some continuous ranges of numbers. Thereby, the storage of spatial CAD objects as well as box volume queries, collision queries, and clearance queries can be efficiently supported by the RI-tree. Fig. 9 summarizes the complete transformation process from triangle meshes over voxel sets to interval sequences. The major advantage of this conversion pipeline lies in the fact that it can be universally applied for any CAD system comprising a triangle faceter. The restriction of the possible input to triangle meshes naturally yields a suboptimal processing cost, but significantly reduces the interface complexity of our spatial CAD database.

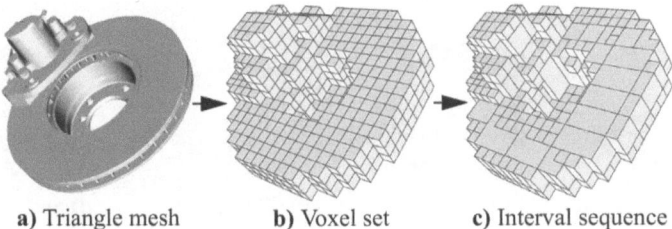

a) Triangle mesh   b) Voxel set   c) Interval sequence

Fig.9: Conversion pipeline from triangulated surfaces to interval sequences.

3.3 Point Shells

In order to achieve real-time interference detection for moving objects, two properties of the geometric representation are of major importance: (*1*) efficient geometric transformations, e.g. translation and rotation, and (*2*) efficient intersection tests. Triangle meshes naturally qualify for (*1*), as their topology is invariant to geometric transformations of triangle vertices. If consecutive triangle intersection joins rely on hierarchical indexes as OBBTrees or k-DOPTrees [GLM96][Kl98], criterion (*2*) is principally fulfilled as well. In the case of haptic rendering, a constant refresh rate of at least 1,000 Hz has to be guaranteed to create a realistic force feedback [MPT99]. The performance of triangle-based intersections is still too slow and unstable for this purpose [Pö01]. For voxel sets, on the other hand, intersections can be computed very efficiently by using bitmap operations, thus fulfilling (*2*) even for the high requirements of haptic rendering. But a voxelized representation of a moving object has to be recomputed for each single position and orientation, and therefore fails for (*1*). As a solution, McNeely, Puterbaugh and Troy [MPT99] have proposed the *Voxmap PointShell* technique (*VPS*), combining the high performance of voxel intersections with an efficient representation for dynamic objects. Thus, both criteria (*1*) and (*2*) are fulfilled for real-time haptic rendering of objects moving in a static environment.

As the basic idea of VPS, point shells representing the moving objects are checked for interference with a voxelized environment. We compute the point shell for a moving object in four steps: first, a triangle mesh for the object surface is derived (cf. Fig. 10a). Next, we voxelize the resulting mesh (Fig. 10b) and get a first approximation of the point shell by the center points of all boundary voxels (Fig. 10c). As an extension to the original algorithm, we further increase the accuracy of the point shell in a final step to generate a smoother surface representation as proposed in [Re01]: within each boundary voxel we interpolate the closest surface point to the voxel center (cf. Fig. 10d). As a result, we obtain a set of accurate surface points which are uniformly distributed over the surface of the moving object. In addition, the respective normal vector $n(p)$ is computed for each surface point p, pointing inwards. The set of all resulting surface points along with the normal vectors comprises the *point shell*. Its accuracy is determined by the resolution of the underlying voxel grid and the triangle mesh. In the example of Fig. 10, the sphere represents a fingertip of the hand depicted in Fig. 3b. Due to the haptic device of Fig. 3a, it is physically exposed to the computed force feedback.

 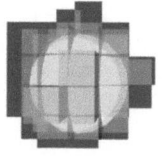 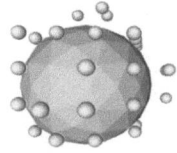 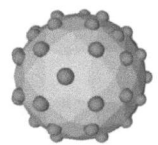

a) Triangulation   **b)** Voxelization   **c)** Center points   **d)** Surface points

Fig.10: Computation of point shells.

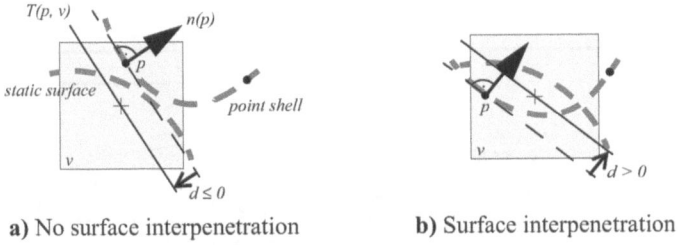

a) No surface interpenetration   **b)** Surface interpenetration

Fig.11: Computation of force feedback.

A practical setup for VPS comprises a voxelized representation of the static environment and a point shell for the moving object. Therefore, the interference detection is reduced to point-voxel intersections which can be computed by a plain three-dimensional address computation for each point. If the resolution of the voxel grid and the dynamic point shell are chosen properly, point-voxel intersections are guaranteed to occur in the case of a surface interpenetration. Fig. 11 [MPT99] depicts a local interference of a surface point p of the dynamic point shell with a voxel v of the static environment. In this case, the depth of interpenetration d is calculated as the distance from p to the tangent plane $T(p, v)$. This tangent plane is dynamically constructed to pass through the center of v and to have $n(p)$ as normal vector. If p has not penetrated below the tangent plane, i.e. in the direction to the interior of the static object, we obtain $d \leq 0$ and produce no local force feedback (cf. Fig. 11a). According to Hooke's law, the contact force is proportional to $d > 0$ (cf. Fig. 11b), and the force direction is determined by $n(p)$. Both determine the force vector of this local interpenetration. The average over all local force vector in the point shell is transferred to the haptic device to exert the resulting force feedback. In the haptic exploration scene depicted in Fig. 12 the average force vectors for each fingertip are visualized as arrows pointing towards the effective contact force.

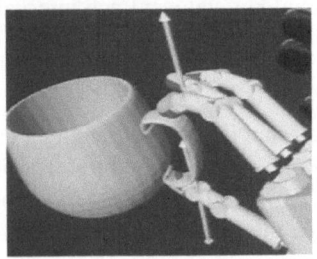 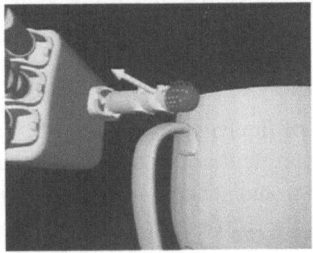

Fig.12: Display of the contact forces in a haptic exploration scene.

4 The Relational Interval Tree as Efficient Spatial Access Method

The Relational Interval Tree (RI-tree) [KPS00] is an application of extensible indexing for interval data. Based on the relational model, intervals can be stored, updated and

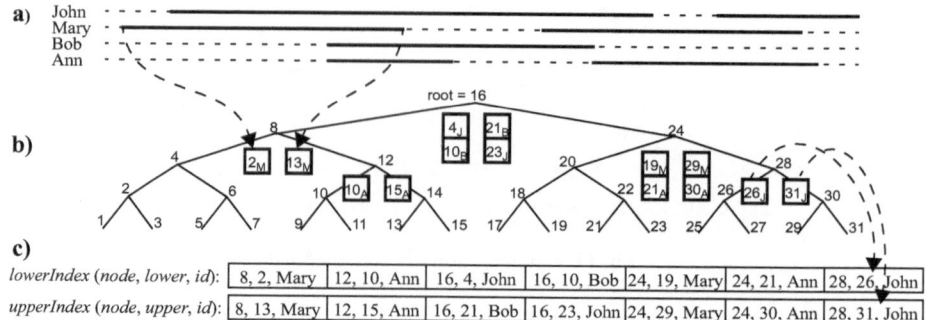

Fig.13: a) Four sample interval sequences. **b)** The virtual backbone positions the intervals. **c)** Resulting relational indexes.

queried with an optimal complexity. In this section, we briefly review the basic concepts of the RI-tree.

The RI-tree strictly follows the paradigm of relational storage structures since its implementation is restricted to (procedural) SQL and does not assume any lower level interfaces. In particular, the built-in index structures of a DBMS are used as they are, and no intrusive augmentations or modifications of the database kernel are required.

The conceptual structure of the RI-tree is based on a virtual binary tree of height h which acts as a backbone over the range $[0 \ldots 2^h - 1]$ of potential interval bounds. Traversals are performed purely arithmetically by starting at the root value 2^h and proceeding in positive or negative steps of decreasing length 2^{h-i}, thus reaching any desired value of the data space in $O(h)$ time. This backbone structure is not materialized, and only the root value 2^h is stored persistently in a metadata tuple. For the relational storage of intervals, the nodes of the tree are used as artificial key values: Each interval is assigned to a fork node, i.e. the first intersected node when descending the tree from the root node down to the interval location.

An instance of the RI-tree consists of two relational indexes which in an extensible indexing environment are at best managed as index-organized tables. The indexes then obey the relational schema *lowerIndex* (*node, lower, id*) and *upperIndex* (*node, upper, id*) and store the artificial fork node value *node*, the bounds *lower* and *upper* and the *id* of each interval. Any interval is represented by exactly one entry in each of the two indexes and, thus, $O(n/b)$ disk blocks of size b suffice to store n intervals. For inserting or deleting intervals, the *node* values are determined arithmetically, and updating the indexes requires $O(\log_b n)$ I/O operations per interval. We store an interval sequence by simply labelling each associated interval with the sequence identifier. Fig. 13 illustrates the relational interval tree by an example.

To minimize barrier crossings between the procedural runtime environment and the declarative SQL layer, an interval intersection query (*lower, upper*) is processed in two steps. In the procedural query preparation step, range queries are collected in two transient tables, *leftNodes* and *rightNodes*, which are obtained in the following way: By a purely arithmetic traversal of the virtual backbone from the root node down to *lower* and to *upper*, respectively, at most $2 \cdot h$ different nodes are visited. Nodes left of *lower* are collected in *leftNodes,* since they may contain intervals who overlap *lower*. Analogously, nodes right of *upper* are collected in *rightNodes* since their intervals may contain the value of *upper*. As a third class of affected nodes, the intervals registered at nodes between *lower* and *upper* are guaranteed to overlap the query and, therefore, are reported without any further comparison by a so-called *inner query*. The query preprocessing procedure is purely main memory-based and, thus, requires no I/O operations.

In the second step, the declarative query processing, the transient tables are joined with the relational indexes *upperIndex* and *lowerIndex* by a single, three-fold SQL statement (Fig. 14). The upper bound of each interval registered at nodes in *leftNodes* is checked against *lower*, and the lower bounds of the intervals from *rightNodes* are checked against *upper*. We call the corresponding queries *left queries* and *right queries*. The *inner query* corresponds to a simple range scan over the nodes in (*lower, upper*). The SQL query requires $O(h \cdot \log_b n + r/b)$ I/Os to report r results from an RI-tree of height h. The height h of the backbone tree depends on the expansion and resolution of the data space, but is independent of the number n of intervals. Furthermore, output from the relational indexes is fully blocked for each join partner.

```
SELECT id FROM upperIndex i, :leftNodes left
    WHERE i.node = left.node AND i.upper >= :lower
UNION ALL
SELECT id FROM lowerIndex i, :rightNodes right
    WHERE i.node = right.node AND i.lower <= :upper
UNION ALL
SELECT id FROM lowerIndex i   /* or upperIndex i */
    WHERE i.node BETWEEN :lower AND :upper;
```

Fig.14: SQL statement for a single query interval with bind variables *leftNodes, rightNodes, lower, upper.*

The naive approach disregards the important fact that the intervals of an interval sequence represent the same object. As a major disadvantage, many overlapping queries are generated. This redundancy causes an unnecessary high main memory footprint for the transient query tables, an overhead of query time, and lots of duplicates in the result set which have to be eliminated. The basic idea in [KPS01] is to avoid the generation of redundant queries, rather than to discard the respective queries after their generation. This optimized RI-tree leads to an average speed-up factor of 2.9 compared to the naive RI-tree, and outperforms competing methods by factors of up to 4.6 (Linear Quadtree) and 58.3 (Relational R-tree) for query response time [KPS01].

5 DIVE: Database Integration for Virtual Engineering

Finally, we present an architecture for the Database Integration of Virtual Engineering (DIVE) into existing Engineering Data Management (EDM) systems. A prototype of this architecture has been evaluated in cooperation with the Volkswagen AG, Wolfsburg [Kr01a][Kr01b].

5.1 Spatial Data Management

The geometry of a part occupies a specific region in the product space. By using this region as a spatial key, related documents such as native CAD files, VRML scenes, production plans or meeting minutes may be spatially referenced. The key challenges in developing a robust and dynamic database layer for virtual engineering have been (1) to store spatial CAD data in a conventional relational database system and (2) to enable the efficient processing of the required geometric query predicates. The above presented Relational Interval Tree (*RI-tree*) is a light-weight access method that efficiently manages extended data on top of any relational database system while fully supporting the built-in transaction semantics and recovery services. Among different approaches the RI-tree seems very promising as spatial engine for the DIVE system. Therefore, we propose the conversion pipeline, presented in section 3, to transform the geometry of each single CAD part into the required interval sequence and store this spatial key in the RI-tree. The redundancy and accuracy of each interval sequence can be controlled individually by size-bound or error-bound approximation.

The DIVE server maps geometric query predicates to region queries on the indexed data space. Our multi-step query processor performs a highly efficient and selective filter step based on the stored interval sequences. The non-spatial remainder of the query, e.g. structural exclusions, is processed by the EDM system. The current DIVE release contains filters for the following spatial queries:

Volume query: Determine all spatial objects intersecting a given rectilinear box volume.

Collision query: Find all spatial objects that intersect an arbitrary query region, e.g. a volume or a surface of a query part.

Clearance query: Given an arbitrary query region, find all spatial objects within a specified Euclidean distance.

To demonstrate the full potential of our approach, we have integrated an optional refinement step for the digital mock-up to compute intersections on high-accurate triangulated surfaces. A part resulting from the previous spatial query may be used as query object for the next geometric search. Thus, the user is enabled to spatially browse a huge persistent database at interactive response times. The DIVE server also supports the ranking of query results according to the intersection volume of the query region and the found spatial keys. For digital mock-up, this ranking can be refined by computing

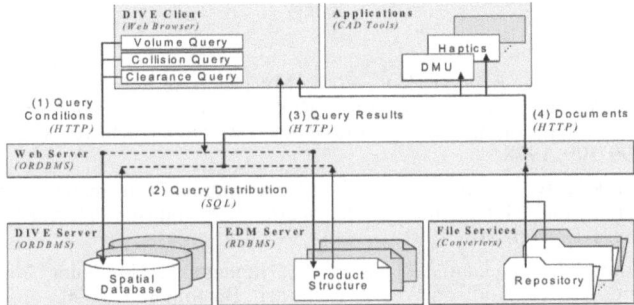

Fig.15: Processing a query on the DIVE system.

the shape and length of the intersection segments on the part surfaces. Thus, the attention of the user is immediately guided to the most relevant problems in the current product design.

5.2 System Architecture

Fig. 15 presents the three-tier client/server architecture of the DIVE system. The client application runs on a conventional web browser and enables the user to specify spatial and non-spatial query conditions (1). The query evaluation is distributed to the DIVE and EDM servers (2). The DIVE server can be implemented on top of any relational database system, whereas extensible object-relational database systems facilitate the seamless embedding of complex spatial datatypes and operators. We integrated the DIVE server into Oracle8i by using PL/SQL and Java Stored Procedures. Therefore, the queries are simply submitted in the standard SQL syntax via Oracle's Net8 protocol. After completion of the spatial and structural filter steps and the optional query refinement, the query result is returned to the client as a table of document URLs (3). Finally, the browser may be used to display the contents of the corresponding documents or, alternatively, their content may be downloaded to a specialized application (4).

5.3 Industrial Evaluation

We have evaluated the DIVE server in an industrial environment on real product data. An installation on an Athlon/750 machine with IDE hard drives and a buffer pool of 800 KB performed average volume and collision queries in 0.7 seconds response time on a database containing 11.200 spatial keys (2 GB of compressed VRML data). Due to the logarithmic scaleup of our query processor, interactive response times can still be achieved for much larger databases.

References

[BKP98]   Berchtold S., Kriegel H.-P., Pötke M.: Database Support for Concurrent Digital Mock-Up. Proc. IFIP Int. Conf. PROLAMAT, Globalization of Manufacturing in the Digital Communications Era of the 21st Century, Kluwer Academic Publishers, 499-509, 1998.

[CF91]   Clark K. B., Fujimoto T.: Product Development Performance – Strategy, Organization, and Management in the World Auto Industry. Harvard Business Scholl Press, Boston, MA, 1991.

[Fo00]   Foley J. D., van Dam A., Feiner S. K., Hughes J. F.: Computer Graphics: Principles and Practice. Addison Wesley Longman, Boston, MA, 2000.

[FR89]   Faloutsos C., Roseman S.: Fractals for Secondary Key Retrieval. Proc. ACM Symposium on Principles of Database Systems (PODS), 247-252, 1989.

[GLM96]   Gottschalk S., Lin M. C., Manocha D.: OBBTtree: A Hierarchical Structure for Rapid Interference Detection. Proc. ACM SIGGRAPH Int. Conf. on Computer Graphics and Interactive Techniques, 171-180, 1996.

[Ja90]   Jagadish H. V.: Linear Clustering of Objects with Multiple Attributes. Proc. ACM SIGMOD Int. Conf. on Management of Data, 332-342, 1990.

[Ka87]   Kaufman A.: An Algorithm for 3D Scan-Conversion of Polygons. Proc. Eurographics, 197-208, 1987.

[Kl98]   Klosowski J. T., Held M., Mitchell J. S. B., Sowizral H., Zikan K.: Efficient Collision Detection Using Bounding Volume Hierarchies of k-DOPs. IEEE Transactions on Visualization and Computer Graphics, 4(1), 21-36, 1998.

[KPS00]   Kriegel H.-P., Pötke M., Seidl T.: Managing Intervals Efficiently in Object-Relational Databases. Proc. 26th Int. Conf. on Very Large Databases (VLDB), 407-418, 2000.

[KPS01]   Kriegel H.-P., Pötke M., Seidl T.: *Interval Sequences: An Object-Relational Approach to Manage Spatial Data.* Proc. 7th Int. Symposium on Spatial and Temporal Databases (SSTD), LNCS, 2001.

[Kr01a]   Kriegel H.-P., Müller A., Pötke M., Seidl T.: DIVE: Database Integration for Virtual Engineering (Demo). Demo Proc. 17th Int. Conf. on Data Engineering (ICDE), 15-16, 2001.

[Kr01b]   Kriegel H.-P., Müller A., Pötke M., Seidl T.: Spatial Data Management for Computer Aided Design (Demo). Proc. ACM SIGMOD Int. Conf. on Management of Data, 2001.

[LSW99]   Lennerz C., Schömer E., Warken T.: A Framework for Collision Detection and Response. Proc. 11th European Simulation Symposium (ESS), 309-314, 1999.

[MH99]   Möller T., Haines E.: Real-Time Rendering. A K Peters, Natick, MA, 1999.

[MPT99]   McNeely W. A., Puterbaugh K. D., Troy J. J.: Six Degree-of-Freedom Haptic Rendering Using Voxel Sampling. Proc. ACM SIGGRAPH Int. Conf. on Computer Graphics and Interactive Techniques, 401-408, 1999.

[Pö01]   M. Pötke: Spatial Indexing for Object-Relational Databases. Doctoral thesis, University of Munich, 2001.

[Pö98]   Pötke M.: Database Support for the Digital Mockup in Mechanical Engineering (in german). Diploma Thesis, University of Munich, 1998.

[Re00]   Renz M.: Dynamic Collision Detection in Virtual Environments (in german). Advanced Term Project, University of Munich, 2000.

[Re02]   Renz M.: Database Prefetching for Large Scale Haptic Simulations (in german). Diploma Thesis, University of Munich, to appear, 2002.

[Re01]   M. Renz, C. Preusche, M. Pötke, H.-P. Kriegel, G. Hirzinger: Stable Haptic Interaction with Virtual Environments Using an Adapted Voxmap-PointShellTM Algorithm. Proc. Int. Conf. Eurohaptics 2001, Birmingham, UK, 2001.

On the Containment of Conjunctive Queries

Georg Lausen and Fang Wei

University of Freiburg, Institute for Computer Science, Germany
{lausen,fwei}@informatik.uni-freiburg.de

Abstract. Testing containment of conjunctive queries is the question whether for any such queries Q_1, Q_2, for any database instance D, the set of answers of one query is contained in the set of answers of the other. In this paper, we first introduce into the general approach for testing for the case, when both queries do not contain negation. Based on these techniques we discuss the question for the case of queries with negation. We propose a new method and compare with the currently known approach. Instead of always testing the exponential number of possible canonical databases, our algorithm will terminate once a certain canonical database is constructed. However, in the worst case, still an exponential number of canonical databases has to be checked.

1 Introduction

Testing containment of conjunctive queries is the question whether for any such queries (CQ) Q_1, Q_2, for any database instance D, the set of answers of one query is contained in the set of answers of the other.

Recently, in the database community, there is a renewed interest in containment checking of conjunctive queries. The main motivation lies in its tight relation to the problem of *answering queries using views* [9; 1], which arises as one of the central problems in data integration and data warehousing (see [15] for a survey). Furthermore, query containment has also been used for checking integrity constraints [6], and for deciding query independence of updates [10].

Based on the NP-completeness result proposed by Chandra and Merlin [2], many researchers have been working on extensions of the containment question. Containment of CQs with inequalities is discussed in [8; 16]. Containment of unions of CQs is treated in [12], containment of CQs with negated subgoals in [10; 15], containment over complex objects in [11], and over semi-structured data with regular expressions in [5].

The containment problem for conjunctive queries with safe negated subgoals has drawn considerably less attention in the past. In [10] uniform containment is discussed, which is a sufficient, however not necessary condition for containment. In [15] it is argued that the complexity of the containment test is Π_2^p-complete. The algorithm proposed always needs to check an exponential number of so called *canonical databases*. In this paper, we propose a new method for testing the containment problem for conjunctive queries with safe negated subgoals. Instead of always testing the exponential number of possible canonical databases, our algorithm will terminate once a certain

R. Klein et al. (Eds.): Comp. Sci. in Perspective (Ottmann Festschrift), LNCS 2598, pp. 231–244, 2003.

canonical database is constructed. However, in the worst case, still an exponential number of canonical databases has to be checked.

The rest of the paper is organized as follows. In Section 2 we introduce into the containment problem by discussing the case of no negation. In Section 3 we review the known approach for the general case with negation. Then in Section 4 we introduce our new approach which improves the currently known techniques. Finally, Section 5 concludes the paper.

2   Conjunctive Queries

Any expression of the form $p(\bar{X})$, where p is a predicate whose vector of arguments $\bar{X}$ is built out of variables and constants, is called an *atom*. An atom is called *ground*, whenever it does not contain any variable. A *conjunctive query* (*CQ*) Q is an expression built out of atoms in the following way:

$$h(\bar{X}) :\text{-} p_1(\bar{Y}_1), ..., p_n(\bar{Y}_n)$$

where $h(\bar{X})$ is the *head*, and $p_1(\bar{Y}_1), ..., p_n(\bar{Y}_n)$ are the *subgoals* forming the *body* of the query. We assume that the variables appearing in the head also appear in the body. The query presented in Example 1 is a CQ.

Example 1. In this example we are interested in the types T of parts P, such that there exists a sale of P from supplier S to customer C, where the customer and the supplier have the same address A. This query can be posed in a conjunctive form:

$$Types(T) :\text{-} Sales(P,S,C), Part(P,T), Cust(C,A), Supp(S,A)$$

$\square$

A *database* D is given by a finite set of ground atoms, which, as usual in the context of databases, are called *tuples*. We assume, that for each predicate p with arity n, appearing in the body of a query Q posed on D, the set of corresponding tuples in D are of the form $p(a_1, ..., a_n)$, i.e. are of the same arity. For each p, the set of corresponding tuples is also called p's *relation*.

A CQ Q is *applied* to a database D (written as $Q(D)$) by considering all possible substitutions of values for the variables in the body of Q. If a substitution makes all the subgoals true, then the same substitution, applied to the head, defines an element of the *answer set* $Q(D)$. Thus, if $h(\bar{X})$ is the head of Q, then $Q(D)$ contains a finite set of corresponding tuples, where each tuple is of h's arity. In Example 2 we see a database of four relations and the corresponding answer set which is defined by the query stated in Example 1.

Example 2. Consider the query Q introduced in Example 1:

$$Q: \qquad Types(T) :\text{-} Sales(P,S,C), Part(P,T), Cust(C,A), Supp(S,A)$$

Let a database D be given by the following corresponding set of relations:

Sales	Part	Cust	Supp
$p_1\ s_1\ c_1$	$p_1\ t_1$	$c_1\ a_1$	$s_1\ a_1$
$p_2\ s_1\ c_2$	$p_2\ t_2$	$c_2\ a_1$	

Types
Then $Q(D) = \quad t_1$
$\quad t_2$

□

A conjunctive query Q_1 is *contained* in another conjunctive query Q_2, denoted as $Q_1 \sqsubseteq Q_2$, if for all databases D, $Q_1(D) \subseteq Q_2(D)$. Two CQs are *equivalent*, if and only if each query is contained in the other.

Let us consider conjunctive queries Q_1, Q_2 as follows:

$$Q_1 : \quad h(\bar{X}) :- r_1(\bar{X}_1), \ldots, r_n(\bar{X}_n)$$
$$Q_2 : \quad h(\bar{U}) :- s_1(\bar{U}_1), \ldots, s_m(\bar{U}_m)$$

Further, let a substitution θ be a mapping from variables to variables and constants, which is extended to be the identity for constants and predicates. θ is called a *containment mapping* from Q_2 to Q_1, if θ turns Q_2 into Q_1 in the following way:

- $\theta(h(\bar{U})) = h(\bar{X})$,
- For each i, $i = 1, \ldots, m$, there exists a j, $j \in \{1, \ldots, n\}$, such that $\theta(s_i(\bar{U}_i)) = r_j(\bar{X}_j)$.

This definition is examplified in Example 3.

Example 3. Let us return to

$$Q : \quad Types(T) :- Sales(P,S,C), Part(P,T), Cust(C,A), Supp(S,A)$$

and let us also consider

$$Q' : \quad Types(T) :- Sales(P,S,C), Part(P,T), Cust(C,A), Supp(S,A),$$
$$\qquad\qquad\qquad Sales(P',S',C'), Part(P',T)$$

Then θ defined as:

X	P	P'	S	S'	C	C'	T	A
$\theta(X)$	P	P	S	S	C	C	T	A

is a containment mapping from Q_2 to Q_1.

□

Theorem 1 (Containment of CQs [2]). *Consider two CQs Q_1 and Q_2:*

$$Q_1 : h(\bar{X}) :- p_1(\bar{X}_1), \ldots, p_n(\bar{X}_n).$$
$$Q_2 : h(\bar{U}) :- q_1(\bar{U}_1), \ldots, q_l(\bar{U}_l).$$

Then $Q_1 \sqsubseteq Q_2$ if and only if there exists a containment mapping ρ from Q_2 to Q_1.   □

The containment problem for CQs has been shown to be NP-complete [2].

The condition of the theorem is sufficient, because we can argue as follows. Let θ be a containment mapping, D be a database and $\mu \in Q_1(D)$. Then there exists a substitution τ, such that $\tau(p_j(\bar{X}_j)) \in D$, $j \in \{1,\ldots,n\}$ and also $\mu = \tau(h(\bar{X}))$. Consider then the substitution $\tau' = \theta \circ \tau$ and in addition $\tau'(q_i(\bar{U}_i))$. It then holds $\tau'(q_i(\bar{U}_i)) \in D$, $i \in \{1,\ldots,m\}$ and also $\mu = \tau'(h(\bar{U}))$, i.e., $\mu \in Q_2(D)$.

To proof the other direction we first introduce the notion of a *canonical database*.

Let Q be a conjunctive query of the form $h(\bar{X}):\text{-}p_1(\bar{X}_1),\ldots,p_n(\bar{X}_n)$. The *canonical database* $D(Q)$ is constructed as follows

- Let τ be a substitution, which maps each variable X to a distinct constant a_X.
- Then $D(Q)$ is the set of tuples, which can be derived by applying τ on Q's subgoals:

$$D(Q) = \{\tau(p_i(\bar{X}_i)) \mid 1 \leq i \leq n\}$$

We call τ a *canonical substitution*.

Example 4. Here we show for the canonical databases for given queries Q, Q'.

$Q:$　　　$Types(T)$:- $Sales(P,S,C),Part(P,T),Cust(C,A),Supp(S,A)$
$Q':$　　　$Types(T)$:- $Sales(P,S,C),Part(P,T),Cust(C,A),Supp(S,A),$
　　　　　　　　$Sales(P',S',C'),Part(P',T)$

$D(Q):$

Sales	Part	Cust	Supp
$a_P\ a_S\ a_C$	$a_P\ a_T$	$a_C\ a_A$	$a_S\ a_A$

$D(Q'):$

Sales	Part	Cust	Supp
$a_P\ a_S\ a_C$	$a_P\ a_T$	$a_C\ a_A$	$a_S\ a_A$
$a_{P'}\ a_{S'}\ a_{C'}$	$a_{P'}\ a_T$		

$\square$

Having introduced the notion of a canonical database, it is now easy to proof the necesssary part of the theorem. Let $Q_1 \sqsubseteq Q_2$, $D(Q_1)$ a canonical database of Q_1 and τ a corresponding canonical substitution. We then have $\tau(h(\bar{X})) \in Q_1(D(Q_1))$. Because $Q_1 \sqsubseteq Q_2$, in addition, there holds $\tau(h(\bar{X})) \in Q_2(D(Q_1))$. Therefore there exists a substitution ρ, such that $\rho(h(\bar{U})) = \tau(h(\bar{X}))$ and for each $i, 1 \leq i \leq l$, there is a j, $j \in \{1,\ldots,n\}$, such that $\rho(q_i(\bar{U}_i)) = \tau(p_j(\bar{X}_j))$. But this means, that $\rho \circ \tau^{-1}$ is a containment mapping. $\square$

3　Conjunctive Queries with Negation

A *conjunctive query with negation* ($CQ^{\neg}$) extends a *conjunctive query CQ* by allowing negated subgoals in the body. It has the following form:

$$h(\bar{X}) :\text{-} p_1(\bar{X}_1),\ldots,p_n(\bar{X}_n),\neg s_1(\bar{Y}_1),\ldots,\neg s_m(\bar{Y}_m).$$

where $h, p_1, \ldots, p_n, s_1, \ldots, s_m$ are predicates whose arguments are variables or constants, $h(\bar{X})$ is the *head*, $p_1(\bar{X}_1), \ldots, p_n(\bar{X}_n)$ are the positive subgoals, and $s_1(\bar{Y}_1), \ldots, s_m(\bar{Y}_m)$ are the negated subgoals. We assume that, firstly, the variables occurring in the head also occur in the body; secondly, all the variables occurring in the negated subgoals also occur in positive ones, which is also called the *safeness* condition for $CQ^\neg$. The examples in this paper are *safe* $CQ^\neg$s if not mentioned otherwise.

A $CQ^\neg$ is applied to a set of finite database relations by considering all possible substitutions of values for the variables in the body. If a substitution makes all the positive subgoals true and all the negated subgoals false (i.e. they do not exist in the database), then the same substitution, applied to the head, composes one answer of the conjunctive query. As before, the set of all answers to a query Q with respect to a certain database D is denoted by $Q(D)$.

There exists a straightforward, however time-consuming test for the containment problem of $CQ^\neg$s. This test is attributed to Levy and Sagiv ([14]) and is described as follows:

> The problem is to check whether for $CQ^\neg$s Q_1, Q_2, there holds $Q_1 \sqsubseteq Q_2$.
> - Consider all databases D for the predicates in the body of Q_1, that contain
>   - No more tuples for a predicate p's relation than Q_1 has positive subgoals with the same predicate symbol p.
>   - Only tuples made from the symbols $1, 2, \ldots, n$, where n is the number of variables in Q_1.
> - Test for all such D, whether $Q_1(D) \subseteq Q_2(D)$; $Q_1 \sqsubseteq Q_2$, if all these containments hold.

The number of databases to be considered obviously is finite. Moreover, if Q_1 is not contained in Q_2, then there must exist at least one database, which can be used as counterexample for the containment. If there exists a counterexample, then the Levy-Sagiv-Test will find a counterexample, as well.

To see this assume D' is a counterexample, i.e. $Q_1(D') \not\subseteq Q_2(D')$. Then there exists a substitution θ, which makes all subgoals in the body of Q_1 true with respect to D', however $\theta(h(\bar{X})) \notin Q_2(D')$, where $h(\bar{X})$ is the head of Q_1. Let D'' be a database, which only contains those tuples, which θ derives when applied on the body of Q_1. Obviously, D'' is also a counterexample. The number of constants in D'' is limited by the number of variables in Q_1. Therefore, we can consistently rename the constants in D'' by symbols $1, 2, \ldots, n$, where n is the number of variables in Q_1. Thus, the Levy-Sagiv-Test will find a counterexample as well.

Although the Levy-Sagiv-Test works correctly, in case the containment between Q_1 and Q_2 does hold, the complete set of possible counterexamples has to be checked. In [15] Ullman proposes a refinement of this test, which, however, still requires to check all possible counterexamples.

Example 5. The following discussion explains Ullman's approach: Consider the following queries Q_1 and Q_2:

$$Q_1 : q(X,Z) :\text{-} a(X,Y), a(Y,Z), \neg a(X,Z).$$
$$Q_2 : q(A,C) :\text{-} a(A,B), a(B,C), a(B,D), \neg a(A,D).$$

$\square$

Table 1. The five canonical databases and their answers to Q_1 and Q_2

Partition	Canonical Databases	Answers
$\{X\}\{Y\}\{Z\}$	$\{a(0,1),a(1,2)\}$	$Q_1 : q(0,2); Q_2 : q(0,2)$
$\{X,Y\}\{Z\}$	$\{a(0,0),a(0,1)\}$	$Q_1 : \texttt{false}$
$\{X\}\{Y,Z\}$	$\{a(0,1),a(1,1)\}$	$Q_1 : \texttt{false}$
$\{X,Z\}\{Y\}$	$\{a(0,1),a(1,0)\}$	$Q_1 : q(0,0); Q_2 : q(0,0)$
$\{X,Y,Z\}$	$\{a(0,0)\}$	$Q_1 : \texttt{false}$

In order to show that $Q_1 \sqsubseteq Q_2$, all five partitions of $\{X,Y,Z\}$ in Table 1 are considered: all variables in one set of a certain partition are replaced by the same constant. From each partition, a canonical database, built out of the positive subgoals, is generated according to the predicates in the body of Q_1. At first, Q_1 has to be applied to the canonical database D from each partition, and if the answer set is not empty, then the same answer set has to be obtained from $Q_2(D)$. Next, for each canonical database D which results in a nonempty answer, we have to extend it with "other tuples that are formed from the same symbols as those in D". In fact, for each specific predicate, let k be the number of arguments of the predicate, n be the number of symbols in the canonical database, and r be the number of subgoals of Q_1 (both positive and negative), there will be $2^{(n^k-r)}$ sets of tuples which have to be checked. Taking the partition $\{X\}\{Y\}\{Z\}$, we need to consider 6 other tuples: $\{a(0,0),a(1,0),a(1,1),a(2,0),a(2,1),a(2,2)\}$. At the end, one has to check 2^6 canonical databases, and if for each database D', $Q_2(D')$ yields the same answer as Q_1, it can then be concluded that $Q_1 \sqsubseteq Q_2$, which is true in this example.

The following queries Q_1 and Q_2 demonstrate the case when containment does not hold.

Example 6.

$$Q_1 : q(X,Z) :\text{-} a(X,Y),a(Y,Z),\neg a(X,Z).$$
$$Q_2 : q(A,C) :\text{-} a(A,B),a(B,C),\neg b(C,C).$$

□

The application of Q_2 to the canonical databases shown in Table 1 yields the same answer as Q_1. Similar to the above example, extra tuples have to be added into the canonical database. Taking the partition $\{X\}\{Y\}\{Z\}$, we have 15 other tuples (9 tuples with $b(0,0),\ldots,b(2,2)$ and 6 tuples as in Example 5), such that 2^{15} canonical databases have to be verified. Since the database $D = \{a(0,1),a(1,2),b(2,2)\}$ is a counterexample, i.e., $Q_2(D)$ does not generate the same answer as $Q_1(D)$, the test terminates with the result $Q_1 \not\sqsubseteq Q_2$.

In the rest of the paper we shall introduce an improved containment test, which will only need to test all counterexamples in the worst case.

4   Query Containment for CQ^-s

Unlike CQs with only positive subgoals, which are always satisfiable, CQ^-s might be unsatisfiable.

Proposition 1. *A CQ^- is unsatisfiable if and only if there exist $p_i(\bar{X}_i)(1 \leq i \leq n)$ and $s_j(\bar{Y}_j)(1 \leq j \leq m)$ such that $p_i = s_j$ and $\bar{X}_i = \bar{Y}_j$.*   □

From now on, we only refer to satisfiable CQ^-s, if not otherwise mentioned.

The containment of CQ^-s is defined in the same manner as for positive ones: a CQ^- Q_1 is contained in another one Q_2, denoted as $Q_1 \sqsubseteq Q_2$, if for all databases D, $Q_1(D) \subseteq Q_2(D)$. Two CQ^-s are equivalent if and only if they are contained in each other.

4.1   Some Necessary Conditions

Definition 1 (Super-Positive SP Q^+). *Given a CQ^- Q as follows:*

$$Q : h(\bar{X}) \mathrel{:-} p_1(\bar{X}_1), \dots, p_n(\bar{X}_n), \neg s_1(\bar{Y}_1), \dots, \neg s_m(\bar{Y}_m).$$

The SP of Q, denoted as Q^+ is: $h(\bar{X}) \mathrel{:-} p_1(\bar{X}_1), \dots, p_n(\bar{X}_n).$ □

Lemma 1. *Given a CQ^- Q with negated subgoals and its SP Q^+, $Q \sqsubseteq Q^+$ holds.*   □

Proposition 2. *Let Q_1 and Q_2 be two CQ^-s, let Q_1^+ and Q_2^+ be their SP respectively. $Q_1 \sqsubseteq Q_2$ only if $Q_1^+ \sqsubseteq Q_2^+$.*

Proof. Assume $Q_1 \sqsubseteq Q_2$ and a tuple $t \in Q_1^+(D)$ where D is any canonical database (i.e. each variable is assigned to a unique constant) of Q_1^+. We show that $t \in Q_2^+(D)$: Let ρ be the substitution from variables of Q_1 to distinct constants in D. Let $s_i(\bar{Y}_i)(1 \leq i \leq m)$ be any negated subgoal in Q_1. Since Q_1 is satisfiable, therefore we obtain that $\rho(s_i(\bar{Y}_i)) \notin D$. Consequently, $t \in Q_1(D)$ and $t \in Q_2(D)$ are obtained. Following Lemma 1 it is obvious that $t \in Q_2^+(D)$.

Proposition 2 provides a necessary but not sufficient condition for query containment of CQ^-s. Next we give a theorem, stating a condition for $Q_1 \not\sqsubseteq Q_2$.

Theorem 2. *Let Q_1 and Q_2 be two CQ^-s. Assume $Q_1^+ \sqsubseteq Q_2^+$, and let $\rho_1, \dots, \rho_r$ be all containment mappings from Q_2^+ to Q_1^+, such that $Q_1^+ \sqsubseteq Q_2^+$. Q_1 and Q_2 are given as follows:*

$$Q_1 : h(\bar{X}) \mathrel{:-} p_1(\bar{X}_1), \dots, p_n(\bar{X}_n), \neg s_1(\bar{Y}_1), \dots, \neg s_m(\bar{Y}_m).$$
$$Q_2 : h(\bar{U}) \mathrel{:-} q_1(\bar{U}_1), \dots, q_l(\bar{U}_l), \neg a_1(\bar{W}_1), \dots, \neg a_k(\bar{W}_k).$$

If for each $\rho_i(1 \leq i \leq r)$, there exists at least one $j(1 \leq j \leq k)$, such that $\rho_i(a_j(\bar{W}_j)) \in \{p_1(\bar{X}_1), \dots, p_n(\bar{X}_n)\}$, then $Q_1 \not\sqsubseteq Q_2$.

Proof. A canonical database D could be constructed as follows: freeze the positive subgoals of Q_1 and replace each variable in Q_1 with a distinct constant. We call this substitution σ. Let t be any tuple such that $t \in Q_1(D)$, we have to show that $t \notin Q_2(D)$: that is, for each substitution θ which makes $t \in Q_2^+(D)$ true, there is at least one negated subgoal $a_j(\bar{W}_j)$, where $1 \leq j \leq k$, such that $\theta(a_j(\bar{W}_j)) \in D$.

Since $\rho_1, \ldots, \rho_r$ are *all* the containment mappings from Q_2^+ to Q_1^+, it is true that $\theta \in \{\rho_1 \circ \sigma, \ldots, \rho_r \circ \sigma\}$. [1] Assume $\theta = \rho_i \circ \sigma$ ($1 \leq i \leq r$). Since for each ρ_i, there exists a $j(1 \leq j \leq k)$, such that $\rho_i(a_j(\bar{W}_j)) \in \{p_1(\bar{X}_1), \ldots, p_n(\bar{X}_n)\}$, thus we have $\rho_i \circ \sigma(a_j(\bar{W}_j)) \in \{\sigma(p_1(\bar{X}_1)), \ldots, \sigma(p_n(\bar{X}_n))\}$ As a result, $\theta(a_j(\bar{W}_j)) \in D$ is obtained.

4.2   Containment of $CQ^\neg$s

The following theorem states a necessary and sufficient condition for the containment checking of $CQ^\neg$s, which is one of the main contributions of this paper.

Theorem 3. *Let Q_1 and Q_2 be two $CQ^\neg$s as follows:*

$$Q_1 : h(\bar{X}) :\text{-} p_1(\bar{X}_1), \ldots, p_n(\bar{X}_n), \neg s_1(\bar{Y}_1), \ldots, \neg s_m(\bar{Y}_m).$$
$$Q_2 : h(\bar{U}) :\text{-} q_1(\bar{U}_1), \ldots, q_l(\bar{U}_l), \neg a_1(\bar{W}_1), \ldots, \neg a_k(\bar{W}_k).$$

Then $Q_1 \sqsubseteq Q_2$ if and only if

1. *there is a containment mapping ρ from Q_2^+ to Q_1^+ such that $Q_1^+ \sqsubseteq Q_2^+$, and*
2. *for each $j(1 \leq j \leq k)$, $Q' \sqsubseteq Q_2$ holds, where Q' is as follows:*

$$Q' : h(\bar{X}) :\text{-} p_1(\bar{X}_1), \ldots, p_n(\bar{X}_n), \neg s_1(\bar{Y}_1), \ldots, \neg s_m(\bar{Y}_m), \rho(a_j(\bar{W}_j)).$$

Proof.

- $\Leftarrow$: Let D be any database and t the tuple such that $t \in Q_1(D)$, we have to prove that $t \in Q_2(D)$.
  Since $t \in Q_1(D)$, we have immediately $t \in Q_1^+(D)$ and $t \in Q_2^+(D)$. Let σ be the substitution from the variables in Q_1^+ to the constants in D such that $t \in Q_1^+(D)$. Let $\theta = \rho \circ \sigma$. It is apparent that $\{\theta(q_1(\bar{U}_1)), \ldots, \theta(q_l(\bar{U}_l))\} \subseteq D$.
  If for each $j(1 \leq j \leq k)$, $\theta(a_j(\bar{W}_j)) \notin D$, then the result is straightforward. Otherwise, if there is any $j(1 \leq j \leq k)$, such that $\theta(a_j(\bar{W}_j)) \in D$, then we have $\rho \circ \sigma(a_j(\bar{W}_j)) \in D$. it can be concluded that $t \in Q'(D)$ where Q' is

$$Q' : h(\bar{X}) :\text{-} p_1(\bar{X}_1), \ldots, p_n(\bar{X}_n), \neg s_1(\bar{Y}_1), \ldots, \neg s_m(\bar{Y}_m), \rho(a_j(\bar{W}_j)).$$

  From the assumption that $Q' \sqsubseteq Q_2$ in the above theorem, $t \in Q_2(D)$ then can be obtained.
- $\Rightarrow$: The proof is via deriving a contradiction.
  1. If $Q_1^+ \not\sqsubseteq Q_2^+$, then from Proposition 2, $Q_1 \not\sqsubseteq Q_2$ can be obtained immediately.
  2. Otherwise, if for each containment mapping ρ from Q_2^+ to Q_1^+, such that $Q_1^+ \sqsubseteq Q_2^+$, there is at least one Q' as given in the above theorem, such that $Q' \not\sqsubseteq Q_2$, then there exists at least one database D, such that $t \in Q'(D)$, but $t \notin Q_2(D)$. Since Q' has only one more positive subgoal than Q_1, it is obvious that $t \in Q_1(D)$. This leads to the result that $Q_1 \not\sqsubseteq Q_2$.

---

[1] $\rho \circ \sigma$ denotes the composition of substitutions ρ and σ.

Theorem 3 involves a recursive containment test. In each round, the containment $Q' \sqsubseteq Q_2$ (the definition of Q' see the above theorem) has to be verified. This might lead to one of the two results: (1) Q' is unsatisfiable – the test terminates with the result $Q_1 \sqsubseteq Q_2$. The reason is simple: if Q' is unsatisfiable, then Q' is contained in any other query; (2) $Q' \not\sqsubseteq Q_2$. This can be verified according to Theorem 2. If this test succeeds, the result of $Q_1 \not\sqsubseteq Q_2$ can be obtained. Otherwise, a new $CQ^\neg$ is generated with one more positive subgoal and a new round has to be executed. The following example illustrates the algorithm. Note that we intentionally omit the variables in the head in order to *generate* more containment mappings from Q_2^+ to Q_1^+. It is not difficult to understand that the less containment mappings are there from Q_2^+ to Q_1^+, the simpler the test will be.

Example 7. Given the queries Q_1 and Q_2:

$$Q_1 : h :\text{-} \ a(X,Y), a(Y,Z), \neg a(X,Z).$$
$$Q_2 : h :\text{-} \ a(A,B), a(C,D), \neg a(B,C)$$

There are four containment mappings from Q_2^+ to Q_1^+, one of which is $\rho_1 = \{A{\to}Y, B{\to}Z, C{\to}X, D{\to}Y\}$. Now a new conjunctive query is generated as follows:

$$Q' : h :\text{-} \ a(X,Y), a(Y,Z), a(Z,X), \neg a(X,Z).$$

Note that the subgoal $a(Z,X)$ is generated from $\rho_1(a(B,C))$. One of the containment mappings from Q_2 to Q' is $\rho_2 = \{A{\to}Z, B{\to}X, C{\to}Z, D{\to}X\}$. Since the newly generated subgoal $\rho_2(a(B,C))$ is $a(X,Z)$, this leads to a successful unsatisfiability test of Q''.

$$Q'' : h :\text{-} \ a(X,Y), a(Y,Z), a(Z,X), a(X,Z), \neg a(X,Z).$$

it can then be concluded that $Q' \sqsubseteq Q_2$. In the sequel we have $Q_1 \sqsubseteq Q_2$.   □

The detailed algorithm is given in Appendix. The idea behind the algorithm can be informally stated as follows: we start with the positive subgoals of Q_1 as root. Let r be the number of all containment mappings from Q_2^+ to Q_1^+, such that $Q_1^+ \sqsubseteq Q_2^+$. r branches are generated from the root, with sets of mapped negated subgoals as nodes (cf. Figure 1(a)). Next, each node might be marked as *Contained*, if it is identical to one of the negated subgoals of Q_1, or as *Terminal*, if it is identical to one of the positive subgoals of Q_1. If there exists one branch such that each node is marked as *Contained*, then the program terminates with the result $Q_1 \sqsubseteq Q_2$. Otherwise, if at least one node of each branch is marked as *Terminal*, then the program terminates too, however with the result $Q_1 \not\sqsubseteq Q_2$. If none of these conditions is met, the program continues with expansion of the non-terminal nodes, that is, the nodes mark with *Terminal* will not be expanded any more.

It can be shown that the algorithm terminates. The reasons are: (1) the expansion does not generate new variables; (2) the number of variables in Q_1 is finite.

The next example shows how the algorithm terminates when the complement problem $Q_1 \not\sqsubseteq Q_2$ is solved.

Example 8. Given the queries Q_1 and Q_2:

$$Q_1 : h :\text{-} \ a(X,Y), a(Y,Z), \neg a(X,Z).$$
$$Q_2 : h :\text{-} \ a(A,B), a(C,D), \neg a(A,D), \neg a(B,C)$$

In Figure 1, it is shown that there are four containment mappings from Q_2^+ to Q_1^+: $\rho_1, \ldots, \rho_4$. Each mapping contains two sub-trees since there are two negated subgoals in Q_2. The branches ρ_2 and ρ_3 are marked as *Terminal* because there is at least one *Terminal* node from each of the above branches (note that we denote a node *Terminal* with a shadowed box around it, and *Contained* with a normal box). The node $a(X,Z)$ from branch ρ_1 is marked as *Contained*, because it is the same as the negated subgoal in Q_1. (Note that in Figure 1 the node $a(Y,Y)$ of branch ρ_1 is marked as *Terminal*, but it is the result of the next round. Up to now, it has not been marked. The same holds for the nodes of branch ρ_4.)

Next the non-terminal node $a(Y,Y)$ is expanded. Five new containment mappings are generated as $\rho_5, \ldots, \rho_9$. Since all the branches are *Terminal*, and $a(Y,Y)$ is also a sub-node of ρ_4, it can be concluded that the expanded query

$$Q' \ h :\text{-} \ a(X,Y), a(Y,Z), a(Y,Y), \neg a(X,Z).$$

is not contained in Q_2. Because all the containment mappings from Q_2^+ to Q'^+ have been verified. As a result, $Q_1 \not\sqsubseteq Q_2$ is obtained.   □

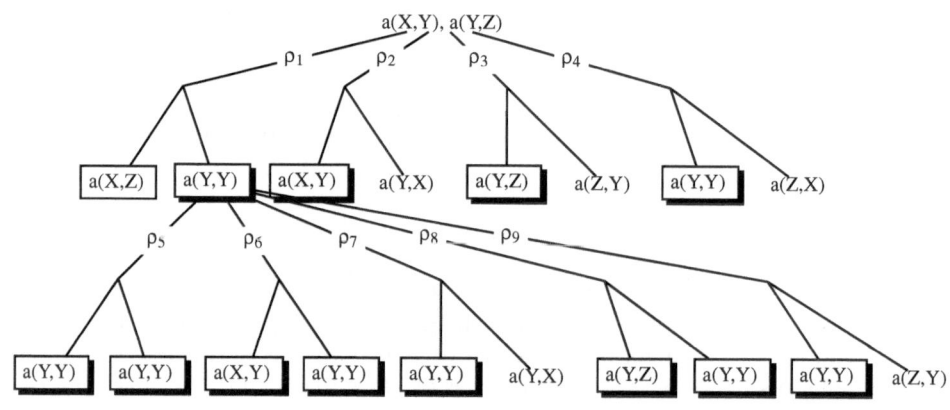

(a) The generated tree

ρ_1	$\{A \to X, B \to Y, C \to Y, D \to Z\}$	ρ_2	$\{A \to X, B \to Y, C \to X, D \to Y\}$
ρ_3	$\{A \to Y, B \to Z, C \to Y, D \to Z\}$	ρ_4	$\{A \to Y, B \to Z, C \to X, D \to Y\}$
ρ_5	$\{A \to Y, B \to Y, C \to Y, D \to Y\}$	ρ_6	$\{A \to X, B \to Y, C \to Y, D \to Y\}$
ρ_7	$\{A \to Y, B \to Y, C \to X, D \to Y\}$	ρ_8	$\{A \to Y, B \to Y, C \to Y, D \to Z\}$
ρ_9	$\{A \to Y, B \to Z, C \to Y, D \to Y\}$		

(b) The containment mappings

Figure 1. The graphic illustration of Example 8.

Comparison of the algorithms. We notice several interesting similarities and differences between our algorithm and the one in [15]: The partitioning of variables is similar to the step of checking of $Q_1^+ \sqsubseteq Q_2^+$. It can be proven that if the containment checking $Q_1^+ \sqsubseteq Q_2^+$ is successful, there exists at least one partition of variables, namely the partition with a distinct constant for each variable, such that when applied to the canonical database built from this partition, Q_2 yields the same answer set as Q_1.

The next step of the algorithm in [15] is to check an exponential number of canonical databases, as described in the Introduction. In contrast, our algorithm continues with the containment mappings of their positive counterparts and executes the specified test, which takes linear time in the size of Q_1. If the test is not successful, the query is extended with one more positive subgoal (but without new variables) and the next containment test continues. It is important to emphasize that in the worst case, the expanded tree generates all the nodes composed of variant combinations of the variables in Q_1, which coincides with the method in [15].

However, in the following examples we show that our algorithm may terminate earlier.

Example 9. Consider the following queries Q_1 and Q_2:

$$Q_1 : \mathsf{q}(X,Z) \text{ :- } \mathsf{a}(X,Y), \mathsf{a}(Y,Z), \neg \mathsf{a}(X,Z).$$
$$Q_2 : \mathsf{q}(A,C) \text{ :- } \mathsf{a}(A,B), \mathsf{a}(B,C), \mathsf{a}(B,D), \neg \mathsf{a}(A,D).$$

There is a containment mapping ρ from Q_2^+ to Q_1^+: $\{A{\rightarrow}X, B{\rightarrow}Y, C{\rightarrow}Z, D{\rightarrow}Z\}$. Next we construct the query Q':

$$Q' : \mathsf{q}(X,Z) \text{ :- } \mathsf{a}(X,Y), \mathsf{a}(Y,Z), \neg \mathsf{a}(X,Z), \mathsf{a}(X,Z).$$

It is apparent that Q' is unsatisfiable, thus we have proven that $Q_1 \sqsubseteq Q_2$. □

Example 10. Given the queries Q_1 and Q_2 as in Example 6. There is a containment mapping ρ_1 (cf. Figure 2). A new node $\rho_1(\mathsf{b}(C,C)) = \mathsf{b}(Z,Z)$ is then generated. The new query Q' is the following:

$$Q' : \mathsf{q}(X,Z) \text{ :- } \mathsf{a}(X,Y), \mathsf{a}(Y,Z), \neg \mathsf{a}(X,Z), \mathsf{b}(Z,Z).$$

Since ρ_1 is the *only* containment mapping from Q_2 to Q', we mark the node $\mathsf{b}(Z,Z)$ as *Terminal*. Following Theorem 2, $Q' \not\sqsubseteq Q_2$ can be obtained, which is followed by the result $Q_1 \not\sqsubseteq Q_2$. □

5 Conclusion

We have discussed the query containment problem for conjunctive queries with safe negated subgoals. We have introduced a new method for testing the containment of $CQ^\neg$s. In comparison to the previously known method in [15], our method avoids to always test an exponential number of canonical databases, however has the same worst case complexity. Therefore, it can be expected, that our method will behave more efficient in practical applications. Finally, it should be mentioned that the algorithm in [15] can deal with unsafe negations, while ours cannot. However, for databases, safety of queries is a generally accepted precondition.

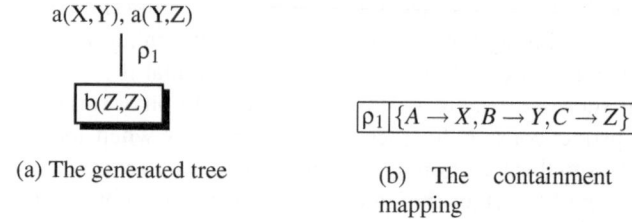

(a) The generated tree

(b) The containment mapping

Figure 2. The graphic illustration of Example 10.

References

1. S. Abiteboul and O. M. Duschka. Complexity of answering queries using materialized views. In *ACM Symp. on Principles of Database Systems (PODS)*, 1998.
2. A. K. Chandra and P. M. Merlin. Optimal implementations of conjunctive queries in relational data bases. In *ACM Symp. on Theory of Computing (STOC)*, pages 77–90. 1977.
3. O. M. Duschka and A. Y. Levy. Recursive plans for information gathering. In *International Joint Conference on Artificial Intelligence (IJCAI)*, pages 778–784, 1997.
4. S. Flesca and S. Greco. Rewriting queries using views. In *TKDE*, 13(6), 2001.
5. D. Florescu, A. Levy, and D. Suciu. Query containment for conjunctive queries with regular expressions. In *ACM Symp. on Principles of Database Systems (PODS)*. 1998.
6. A. Gupta, Y. Sagiv, J. D. Ullman, and J. Widom. Constraint checking with partial information. In *ACM Symp. on Principles of Database Systems (PODS)*. 1994.
7. A. Halevy. Answering queries using views: A survey. In *VLDB Journal*, 10:4, pp. 270–294, 2001.
8. A. Klug. On conjunctive queries containing inequalities. In *Journal of the ACM* 35:1, pp. 146–160, 1988.
9. A. Y. Levy, A. O. Mendelzon, Y. Sagiv, and D. Srivastava. Answering queries using views. In *ACM Symp. on Principles of Database Systems (PODS)*. 1995.
10. A. Levy and Y. Sagiv. Queries independent of updates. In *International Conference on Very Large Data Bases (VLDB)*, 1993.
11. A. Levy and D. Suciu. Deciding containment for queries with complex objects. In *ACM Symp. on Principles of Database Systems (PODS)*. ACM Press, 1997.
12. Y. Sagiv and M. Yannakakis. Equivalence among relational expressions with the union and difference operations. *Journal of the ACM, (4)*, 27(4):633–655, 1980.
13. J. Ullman. *Principles of Database and KnowledgeBase Systems, Volume II*. 1989.
14. J. Ullman. *Principles of Database Systems – Lecture Notes*. http://www.db-stanford.edu/ ullman/cs345-notes.html..
15. J. D. Ullman. Information integration using logical views. In *International Conference on Database Theory (ICDT)*, 1997.
16. X. Zhang and Z. Meral Özsoyoglu. On efficient reasoning with implication constraints. In *DOOD*, pages 236–252, 1993.

Appendix

A1: The Containment Checking Algorithm

Algorithm Containment Checking(Q_1, Q_2)
Inputs: Q_1 and Q_2 are CQ^-s with the form as follows:

$$Q_1 : h(\bar{X}) \text{ :- } p_1(\bar{X}_1), \ldots, p_n(\bar{X}_n), \neg s_1(\bar{Y}_1), \ldots, \neg s_m(\bar{Y}_m).$$
$$Q_2 : h(\bar{U}) \text{ :- } q_1(\bar{U}_1), \ldots, q_l(\bar{U}_l), \neg a_1(\bar{W}_1), \ldots, \neg a_k(\bar{W}_k).$$

Begin
1. Set $\{p_1(\bar{X}_1), \ldots, p_n(\bar{X}_n)\}$ as the root of a tree: c_0.
Let $\rho_1, \ldots, \rho_r$ be all the containment mappings from Q_2^+ to Q_1^+.
2. Generate r nodes $c_1, \ldots, c_r$ as children of root, and each node $c_i (1 \leq i \leq r)$ has a subtree with k children, in which each child $c_{i,j}$ with the form $\rho_i(a_j(\bar{W}_j))(1 \leq j \leq k)$.
3. Marking the nodes:
For i=1 to r **do**
  For j=1 to k **do**
   If $c_{i,j} \in \{p_1(\bar{X}_1), \ldots, p_n(\bar{X}_n)\}$ **Then** mark c_i as *Terminal*;
   If $c_{i,j} \in \{s_1(\bar{Y}_1), \ldots, s_m(ymbar)\}$ **Then** mark $c_{i,j}$ as *Contained*;
  EndFor
EndFor
4. Execute the containment checking:
If all nodes $c_1, \ldots, c_r$ are terminal nodes, **Then return** Not-contained;
For i=1 to r **do**
 If each $c_{i,j}(1 \leq j \leq k)$ is marked as *Contained*, **Then return** Contained;
EndFor
5. Continue expanding non-terminal nodes:
For i=1 to r **do**
 If c_i is not *Terminal* **Then**
  For j = 1 to k **do**
   If $c_{i,j}$ is not *Contained* **Then**
    let Q_1' be: h :- $p_1(\bar{X}_1), \ldots, p_n(\bar{X}_n), c_{i,j}, \neg s_1(\bar{Y}_1), \ldots, \neg s_m(\bar{Y}_m)$.;
    let $\rho_{i,j,1}, \ldots, \rho_{i,j,b_{i,j}}$ be the new containment mappings from Q_2 to Q_1';
    Generate $b_{i,j}$ nodes with children in the same way as in **Step 2**;
    Mark the nodes in the same way as **Step 3**;
   EndIf
  EndFor
 EndIf
EndFor
6. Execute the containment checking:
For i=1 to r **do**
 If each $c_{i,j}(1 \leq j \leq k)$ either is marked as *Contained*
 or has a child whose children are all marked *Contained*
 Then return Contained;

EndIf
from each non-terminal node c_i, choose one child $c_{i,j} (1 \le j \le k)$;
add all these $c_{i,j}$ to the body of Q_1, and let the new query be Q_1'';
If each branch w.r.t. Q_1'' is marked as *Terminal*, **Then return** Not-contained;
EndFor
Let r be all expanded nodes; **Go to** Step 5;
End

Important Aspect of Knowledge Management

Hermann Maurer

Institute for Information Processing and Computer Supported New Media
Graz University of Technology
Inffeldgasse 16c
A-8010 Graz, Austria
hmaurer@iicm.edu
http://www.iicm.edu/maurer

This paper is dedicated to my colleague and friend Thomas Ottmann on the occasion of his 60-th birthday. Happy birthday, Thomas, and lots of further great ideas and success stories!

Abstract. In this paper it is explained what Knowledge Management (KM) is and why it will play an important role in the future. This implies that KM is indeed more than just the sophisticated use of information systems or distributed databases for complex tasks. With the introduction of the Maurer-Tochtermann model for KM and the description of the basic elements of new techniques, it can be demonstrated quite readily that KM is indeed something fundamentally new. At the end of this paper a system developed in Europe called Hyperwave will be mentioned briefly as perhaps being the leading example of this new type of software, i.e. software for KM systems.

1 Introduction

For quite some time, the term "knowledge management" (KM) has been increasingly used in connection with the efficient exchange of structured information within large organizations, and for describing the use of heterogeneous sources of knowledge. Since information systems, databases, and networked systems, such as many internet or intranet applications, have similar objectives, the question arises whether KM can really be considered something new, or whether it is just "old wine in new bottles." In this paper, it will be argued that KM is in fact leading toward different objectives, and that important new concepts and technologies have already been developed in this area. The term KM is vague, since the underlying term "knowledge" itself is vague: even though we all know (!) what the word "knowledge" means, it is nonetheless difficult to describe it. The same problem occurs when trying to describe such words as data, information and knowledge.

If you ask "When was Archduke Johann of Austria made the imperial vice regent of the German empire?", the answer is "1848," a simple piece of data. And yet, if you put "Archduke Johann of Austria" together with "imperial vice regent of the German empire," then suddenly we know (!) something that we did not know before. Thus, given that new knowledge is created by proper linking and structuring of data, information and existing knowledge, it could be argued that the definition for knowledge is "structured information." Even though this definition is not totally satisfactory, (since we know(!)

R. Klein et al. (Eds.): Comp. Sci. in Perspective (Ottmann Festschrift), LNCS 2598, pp. 245–254, 2003.

that "knowledge" is more than that), it will be used as the working hypotheses for this paper. Since we already are having problems with trying to define "knowledge," how can we expect that defining terms such as "knowledge management" or "knowledge management systems" will be any easier?

With this in mind, it certainly comes as no surprise that there is currently a whole slew of different definitions for KM, e.g. [1] - [4], [8], [19].

In order to avoid further difficulties with definitions that we also encounter when trying to define terms such as "intelligence," "life," "consciousness," etc., it makes sense to steer away from precise definitions and proceed in a more pragmatic fashion. An approach that goes in this direction will be introduced next. It will be shown beyond a doubt that KM is in fact a new field of know(!)ledge, one in which information technology plays an essential part.

For the sake of completeness, it must be pointed out that the separation between KM and other disciplines is quite subtle: KM is viewed differently by the different groups. For example, book [3] treats many areas of IT as branches of KM that this author would have more likely handled separately. There are also positions that go to the other extreme, in which the definition of KM moves away from information technology, towards that of organizational science [4]. The author of this paper does not by any means ignore the organizational aspects that play a role in any good KM system, but rather concentrates on those parts having to do with information technology and computer science.

2   Pragmatic Starting Points for Knowledge Management

The saying "if only our employees knew what our employees know, then we would be the best organization in the world" articulates very well the most essential aspect of KM: a group of people always has more knowledge than any of the individuals, and the same knowledge will even be perceived in different ways by different people.

The challenge presented here is to come up with a way to capture the knowledge (or parts of it) that resides in peopleŠs heads; to somehow archive it in electronic form; and to make it easily accessible to other people. The advantages for the organization that manages to do this are considerable. It would eliminate duplication of work, cooperation between employees would be made easier, knowledge would not be lost when employees leave, the training of new employees would be made substantially easier, etc. Of course, the capturing of this knowledge must occur without being a significant burden to the employees and the system has to be able to deliver the knowledge quickly, when it is required, even before the employees are aware that they are in need of additional knowledge.

It will be shown that there are indeed methods that exist which are able to implement the above-mentioned features to some extent. KM of this type will be referred to as "KM for Organizations" in the following discussions. It should be noted that even very simple solutions can help quite a lot. For example, it may not be necessary for every employee to know what every other employee knows, but employees that are interested in a certain topic should at least know whether there are any other employees that could help them with that topic. However, knowledge management of this sort is less ambitious. This type of KM is already being used in some organizations as so-called "Knowledge Domains" [5] or "Yellow Pages".

On the other hand, a much more innovative use of "KM for Organizations" could be envisioned, which would be important not only for organizations, but for all of society as well. One in which a modern version of the internet and omni-present computers could be used to make large parts of the collective knowledge of mankind readily available to everyone, almost as an expansion of the human mind. Reflections on how and why something like this might not be all that unrealistic in the future are found in [6].

There is another more pragmatic approach to KM, that also finds its basis in a saying, but which comes from a different direction. In organizations that work with large amounts of data that come from heterogeneous sources there is the saying, "If only we were able to find a way to automatically link new data with existing data such that the data that belongs together would be recognized and classified as such, and that related information would also be linked together, then the problem of archiving information would finally be solved." This problem will be referred to as "KM for Archives," for short.

The following two examples will help to illustrate this situation. In the Journal of Universal Computer Science, J.UCS, (see [13] and [14]), the papers are not just classified by numerous attributes and are therefore able to be located using several different criterion, they are also designed to have "links into the future" added to them. The term "Link into the future" can best be explained using an example: say there is a contribution A, written in 1995 that is used in a new paper B, written in 2002: then B usually will quote A, this being a "link into the past". It is, however, possible in digital libraries [15] that paper A also contains a reference to B, i.e. a "link into the future", a feature clearly imposible in printed publications. Such "links into the future" are quite easy to create in digital libraries when literature references have a fixed format. However, generating these links is also possible beyond the scope of any one digital library using various means. The most obvious is the use of so-called "citation indexes." Another way is the use of a system that is able to recognize "similarities" or "logical connections" between documents. These methods will be described in more detail in section 3.

A further example is the electronic version of the Brockhaus Encyclopedia [16]. For each entry, there is a "knowledge web" that graphically displays other entries that have a logical connection with the current one. In the Brockhaus model, the recognition of these connections occurs through the previously mentioned "similarity concept," as well as through corresponding metadata [17]: every entry is classified under one or more categories. In this manner, references between contributions such as Bush (the politician) and Bush (the vegetation) can be eliminated. In the year 2003, a Brockhaus model for an electronic knowledge web containing the complete ontology [12] of the German language is being created for the first time!

It will be shown in the following that "KM for Archives" is included under the definition of "KM for Organizations," as one of its important branches. This is why the following discussion can be focused on the latter.

3   A Model for Knowledge Management in Organizations

In addition to the previously mentioned methods that described the main elements of a knowledge management system, the communication model by Maurer-Tochtermann in Figure 1 clearly illustrates in which way KM goes beyond (distributed) information systems and databases.

Figure 1 shows a group of people exchanging knowledge with one another, whereby a large portion of this knowledge is exchanged over a networked computer system (this being asynchronous, i.e. time-delayed, as well). Each of the arrows 1 through 7 has the following particular meaning attached to it:

Arrow 1 shows that people exchange knowledge directly with one another, e.g. during coffee breaks, on business trips, over the telephone, etc. It is this type of exchange that the organizational side of knowledge management is mainly concerned with and will not be handled further in this paper.

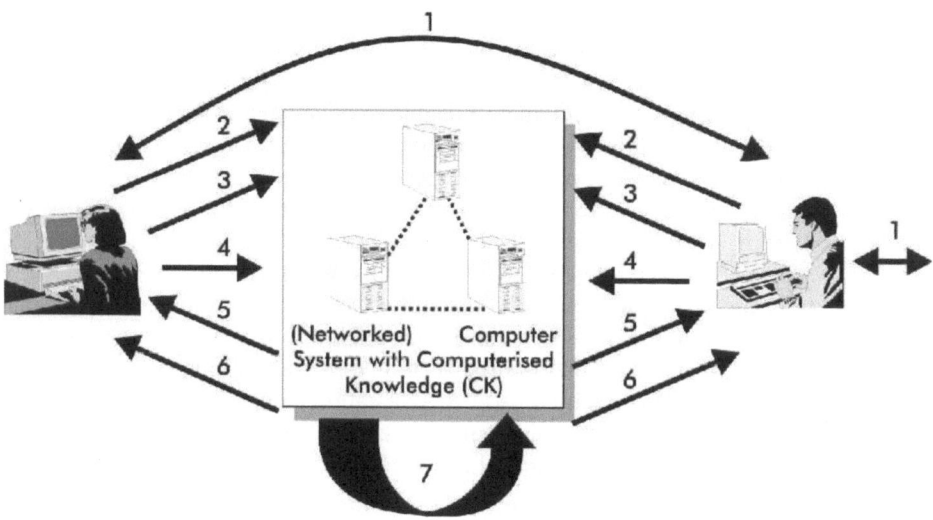

Figure 1.

Arrows 2 through 4 symbolize the different ways that information is put into the KM system, while arrows 5 and 6 indicate that there are at least two ways for knowledge to be delivered from the KM system to the users. More precisely, arrow 2 symbolizes the explicit input of information in the system and arrow 3 symbolizes the implicit input. It should be noted that the gathering of this information and knowledge occurs as a by-product of other activities that are carried out, i.e. new knowledge is created within the KM system without additional work being required. Arrow 4 signifies that a KM system is also able to "systematically" generate information by observing users. Arrow 5 symbolizes the traditional querying of information in an information system or database. Users enter explicit queries and receive the corresponding answers from the KM system, which can be anything from fragments of information to large coherent documents, such as manuals, books or teaching modules. Perhaps even "more interesting" is arrow 6, which symbolizes that a KM system can act on its own to deliver knowledge to the user, even without an explicit request. Finally, arrow 7 symbolizes that a KM system is able to generate new knowledge out of existing knowledge.

Figure 1 illustrates quite clearly the differences between classical information systems (databases) and KM systems. Taking away the arrows 3, 4, 6 and 7 away, then what is left is the exact model of a classical information system. Thus, the question of whether KM systems are really different from classical information systems depends on whether the behavior symbolized by arrows 3, 4, 6 and 7 is actually feasible. The answer to this question will be answered affirmatively in the following sections.

KM for Organizations, depicted in Figure 1, can also stand for KM for Archives, if arrows 1 and 3 are ignored. This confirms the previous statement that attention can be focused on KM for Organizations since KM for Archives is included within it by definition.

4   A Few Knowledge Management Techniques

In this section, it will be shown how the actions depicted by the arrows 3, 4, 6 and 7 in Figure 1 can be realized in a modern KM system.

Arrow 3 in Figure 1 symbolizes the implicit input of information. In other words, the increase of the amount of knowledge in KM systems occurs as a by-product of other non-related actions carried out by people. The KM system could be designed to react on even the simplest of actions, such as the announcement of an event that is printed and sent out by e-mail to a distribution list. The KM system would record the event in the event calendar, and automatically move it to the Events of the past year folder after the date of the event had passed. At the yearŠs end, it would also add it to the "Events since XXXX" folder. All information that is put into a web server (from telephone lists to the structure and activities of the company and its departments) would also make its way into the KM system. There is hardly any organization today that can stay competitive without being ISO 9000 certified. This brings its own benefits with it. The extensive amount information that has to be constantly maintained presents a powerful and natural source for the KM system. The information collected on current projects with their goals, team members, tools, milestones, problems, protocols, documentation, etc. carries with it extensive knowledge about what is going on within the organization. Even completed projects suddenly prove to be very useful, since much can be learnt about current and future trends from positive and negative experiences in earlier projects. Workflow procedures in an organization are also components that are quite often kept in electronic form and therefore are logical candidates to be automatically integrated into the KM system. The same goes for year-end reports, product lists with descriptions and prices, manuals and other internal reports, and so on, including everything up to the integration of existing information systems, for which so-called knowledge portals [20] are needed. E-mail (with the appropriate mechanisms for authorization) should be kept centrally in the KM system as well, rather than managed by the individual users. The reason why bringing this information together is so important is clear: only when an adequate store of information is available can the information in a KM system be linked together automatically, resulting in structured information, i.e. knowledge.

Arrow 4, which symbolizes the creation of systemic knowledge from the input information, is currently the weakest part of KM systems. The basic idea is that general rules can be derived from entries that come from certain sources (e.g., from specific employees or databases), which would then be able to be used in similar situations. Rules like these are often closely connected with the activities symbolized by arrow 5 (explicit data search). For example, a good KM system would recognize that some searches are carried out over and over again by one or more users. It would correspondingly design shortcuts and generate automatic bookmarks; or it would realize that almost all persons who search for X, for example, also search for Y. Thus, the next time a user searches for X, information about Y will be offered automatically as well, in the sense of arrow 6.

The power of the arrows 6 and 7 can perhaps be better illustrated when a few techniques are explained that are of key importance to a good KM system.

One of the most important concepts in a KM system is the idea propagated by Maurer [9] of the active document. In an exaggerated sense, the following could be envisioned: users can ask any question (typically by typing it in, but perhaps in the future by simply asking it aloud) about a document that they have called up and the KM system will provide the answer!

Although it is clearly impossible for a KM system to be able to answer every question, there are two methods that come very close to delivering the intended result. First, certain questions can be converted into database queries and answered in this way. Secondly, questions asked for the first time can be answered by experts (perhaps even with a delayed response). Afterwards, the use of a trick prevents that the same question has to be answered a second time by an expert: the KM system saves questions and the corresponding answers. If another question is asked about the same document, it will be checked whether this question is equivalent to a previous question. In this case, the question can be answered by the KM system without having to call on an expert. This behavior becomes most practical when there are many persons accessing the same document (say on an intranet or internet server), and where answers stay the same over an extended period. Of course, there is the (huge) problem of determining whether question X is equivalent to question Y.

This problem can be tackled in a number of ways. Question X and Y can be checked for similarities through a comparison of the words (with the support of thesauruses, semantic networks or ontologies if necessary). If it is suspected that question X is equivalent to question Y, then the KM system will ask whether question Y is an acceptable substitute for question X, in which case the question can be answered by the KM system. If question X does not appear to be equivalent to any of the previously asked questions, then and only then will the question be forwarded to an expert. A refinement of this method also works sometimes using approaches from artificial intelligence, such as "case-based reasoning." However such methods are, up to now, still of lesser practical importance.

Of course, it would be more elegant if it could firmly be proven whether two texts X and Y are equivalent. However, this is not possible with the current methods or even those of the foreseeable future, except under very restricted conditions, e.g. if only certain syntactical forms are allowed and the choice of subject is limited (see [9]).

However, another quite simple approach to solving the problem of active documents exists, which can be termed "localized FAQs." More precisely, when users have questions, they mark the corresponding section of the document and type in their question. The question will then be answered by an expert, either during given "consultation hours," or otherwise at a later time. An icon will be inserted before the marked section of the document, which will signal to later readers that "a question was asked about this section and was answered by an expert." Subsequent users can check the Question and Answer dialog before asking further questions about the same section. In large-scale experiments it has been shown that there are seldom more than a few questions asked about the same section of a document (and if more do occur, the author of the document is well-advised to revise it since it is obviously not clear enough). This list of questions quickly brings the document to a "stable" condition. In one company of 150,000 employees that required everyone to read extensive materials, it turned out that after the first 600 employees had read the information only .03% of those remaining had new questions! That means that for the first 600 employees, experts needed to be available (in this case, around the clock), but for the remaining 149,400 employees (i.e. for 99.6%) only 45 new questions arose, and these were able to be answered without causing many problems with a time-delayed response.

The above discussion makes it obvious that a good KM system has to include the concept of "active documents."

As was just discussed concerning the comparison of two questions, X and Y, a problem that is often encontered is to figure out whether X and Y are "equivalent" or "similar." There are a number of ways to do this, the most common of which being improvements on the approach of using the frequency of the most important words as a measure for similarity. This approach is able to determine with high probability whether two texts are similar. These methods not only allow for the automatic classification of texts (in the sense of arrow 7 in figure 1), but also for active notification when there is a suspicion that two documents are similar (in the sense of arrow 6 in figure 1). A few concrete examples will help illustrate the practical application of similarity tests such as these.

Example 1: Consider an organization that has branches world-wide. When a new project is started at a location A, a description of it is documented in accordance with ISO 9000. The KM system takes this document, translates it from the local language into English and then compares the resulting (poor) translation with the English descriptions of all the current projects in the organization. If a similarity is found with a project at a location B, then both A and B will be notified that a duplication of work might be happening. The KM system may be wrong in some cases, i.e. the similarities in A and B are already known or irrelevant, but even if just a few cases are found in which duplication can be avoided, it is still very significant financially.

Example 2: In an organization that has a large internal research department, an employee adds a new publication A to the digital library. Soon afterwards, the employee receives an e-mail notification because an analysis of similarities between documents shows that there are two similar contributions, B and C, in the library as well. At the same time, the authors of B and C are notified that a similar paper was recently added to the digital library. Just as in example 1, duplication can be avoided and cooperation is increased. Ideally, publication A would not only be compared to documents in the organization's internal digital library, but also against those in other digital libraries accessible over the internet (for example, as in J.UCS [13], [14]).

Example 3: Through the regular review of non-private e-mails, it is determined that two groups, A and B, seem to be working on similar problems. Both groups are notified accordingly.

Example 4: In a discussion forum, a topic is brought up that had already been thoroughly covered a few months ago. The KM system recognizes this and curtails an unnecessary discussion with a link to the previous entries.

All together, it is quite clear that the testing for similarity between documents is one of the most powerful of the set of tools that a KM system needs to have. Computed links can be graphically displayed quite well through a knowledge web, such as in Brockhaus Multimedial [16]. An example of this is Figure 2, which displays an automatically computed knowledge web for the word space probe ("Raumsonde" in German).

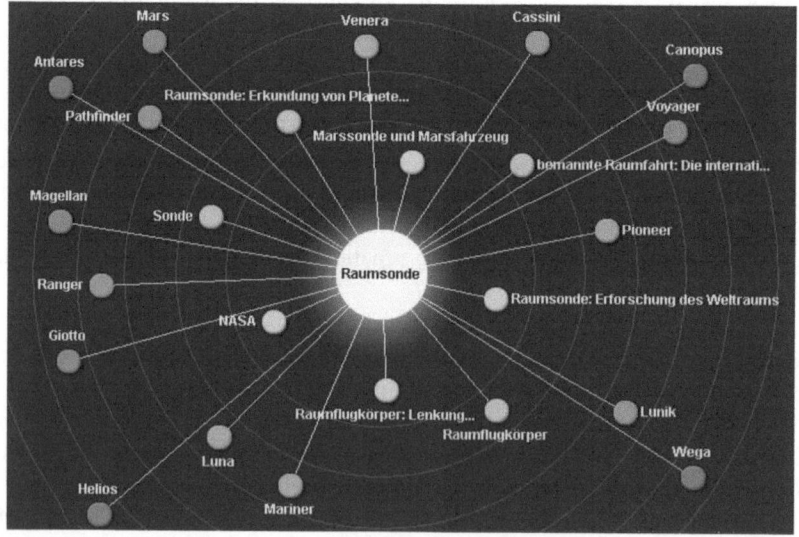

Figure 2.

A more complex variation on recognizing similarities is the recognition of logical connections or links. However, the methods for doing this are not yet generic enough for general use; they must instead be adapted to the particular requirements individually. For this reason, one example that follows will have to suffice to explain this variation.

Let us assume we have a very large store of data, for instance, all publicly accessible WWW servers and databases, including all reports available in electronic form from newspapers and news agencies. The task would be to determine which people are closely connected to a specified person X.

If a KM system finds the text, "X visited Nassau on October 15, 2002," then the information "October 15, 2002" will be entered in the "X database." In an analysis of an entry found elsewhere, "Maurer was on North Eluthera from the 10th to the 20th of October, 2002," the following occurs: (a) the word "Maurer" will be recognized as a proper name by syntax- analysis techniques, and (b) the entry will be compared with every entry in the X database. Since the KM system is able to identify certain geographical relationships, it knows (!) that North Eluthera belongs to the Bahamas and that Nassau is the capital of the Bahamas. Since October 15, 2000 falls between the 10th and 20th of October 2000, it is conceivable that Maurer ran into person X. In the "links database," it would be saved as (X, Maurer, 1), meaning that there is one circumstance in which Maurer and person X could have met. If the KM system later finds that "X met with person Z" and that "Maurer went to school with person Z," then the set (X, Maurer, 1) would be replaced by (X, Maurer, 2). Should the counter reach a critical size, e.g. it reaches (X, Maurer, 100), the system alarm is sounded. This now means that it is highly probable that person X and Maurer are connected in some way. The "sounding of the alarm" is another typical sort of action that is symbolized by the arrow 6 in Figure 1, while the process of calculating such connection is one of the actions symbolized by arrow 7.

It should be clear that the creation of the relationships described in this example assumes a rather complex system. However, a look at the past shows that it performs quite well in e.g. tracing person, with terrorism a currently very hot topic.

There are much less complex applications that also have to do with logical connections or links, for instance, in the area of E-commerce: similarities between customers A and B concerning e.g. their tastes in literature can be found out through their purchases. Now, once this has been established, if customer A buys a book written by author X, whose books A has not previously read, and soon afterwards buys another book by author X, then it can be assumed that customer A likes author X. Hence author X can be recommended (most often successfully) to customer B.

Research in the area of the recognition of logical connections is still very much going on, but it is already clear that this will be an additional tool that is of great importance to KM systems.

5 Summary

In this paper, it was shown that knowledge management systems (KM systems) differ in a number of decisive aspects from classical information systems. The methods described have already been realized in a few existing systems. Hyperwave[7], [18] is probably the most fully developed of these, since it offers active documents, automatic classification, recognition of similarities and many other essential components of a KM system.

References

1. Woods, E., Sheina, M.: Knowledge Management - Applications, Markets and Technologies. Ovum Report (1998)
2. Studer, R., Abecker, A., Decker, S.: Informatik-Methoden für das Wissensmanagement Angewandte Informatik und Formale Beschreibungsverfahren, Teubner-Texte zur Informatik, vol. 29 (1999).
3. Karagiannis, D., Telesko, R.: Wissensmanagement - Konzepte der Künstlichen Intelligenz und des Softcomputing, Lehrbücher Wirtschaftsinformatik, Oldenbourg Verlag (2001).
4. Davenport, T., Prusak, L.: Working Knowledge: How Organizations Manage What They Know. Harvard Business School Press, Boston, (1998).
5. Helic, D., Maurer, H., Scerbakov, N.: Knowledge Domains: A Global Structuring Mechanism for Learning Resources in WBT Systems. Proceedings of WEBNET 2001, November 2001, AACE, Charlottesville, USA (2001) 509-514
6. Maurer, H.: Die (Informatik) Welt in 100 Jahren. Informatik Spektrum, Springer Verlag (2001) 65-70
7. Maurer, H. (ed.): HyperWave: The Next Generation Web Solution. Addison-Wesley Longman, London (1996)
8. Sivan, Y.: The PIE of Knowledge Infrastructure: To Manage Knowledge We Need Key Building Blocks. WebNet Journal 1,1 (1999) 15-17
9. Heinrich, E., Maurer, H.: Active Documents: Concept, Implementation and Applications. Journal of Universal Computer Science 6, 12 (2000) 1197- 1202.
10. Tochtermann, K., Maurer,H.: Umweltinformatik und Wissensmanagement. Tagungsband des 14. Internationalen GI-Symposiums Informatik für den Umweltschutz, Bonn, Metropolis Verlag (2000) 462-475
11. Meersman, R., Tari, Z., Stevens, S. (eds.): Database Semantics. Kluwer Academic Publishers, USA, (1999).
12. Uschold, M., Gruninger, M.: Ontologies: principles, methods and applications. The Knowledge Engineering Review 11,2 (1996) 93-136
13. www.jucs.org
14. Krottmaier, H., Maurer, H.: Transclusions in the 21st Century. Journal of Universal Computer Science 7,12 (2001) 1125-1136
15. Endres, A., Fellner, D.: Digitale Bibliotheken. dpunkt Verlag Heidelberg, Germany (2000)
16. Der Brockhaus Multimedial 2002 Premium. DVD, Brockhaus Verlag, Mannheim (2001)
17. Weibel, S.: "The state of the Dublin core metadata initiativ" D-Lib Magazine 5, 2 (1999) 300
18. www.Hyperwave.de
19. Petkoff, B.: Wissensmanagement. Addison Wesley (1998)
20. Hyperwave: Hyperwave Information Portal White paper (1999)
   http://www.hyperwave.de/publish/downloads/Portalwhitepaper.pdf

The Reliable Algorithmic Software Challenge RASC
Dedicated to Thomas Ottmann on the Occasion of His 60th Birthday

Kurt Mehlhorn

Max-Planck-Institut für Informatik, 66123 Saarbrücken, Germany
www.mpi-sb.mpg.de/~mehlhorn

1   Introduction

When I was asked to contribute to a volume dedicated to Thomas Ottmann's sixtieth birthday, I immediately agreed. I have known Thomas for more than 25 years, I like him, and I admire his work and his abilities as a cyclist. Of course, when it came to start writing, I started to have second thoughts. What should I write about? I could have taken one of my recent papers. But that seemed inappropriate; none of them is single authored. It had to be more personal.

A sixtieth birthday is an occasion to look back, but it is also an occasion to look forward, and this is what I plan to do. I describe what I consider a major challenge in algorithmics, and then outline some venues of attack.

2   The Challenge

Algorithms are the heart of computer science. They make systems work. The theory of algorithms, i.e., their design and their analysis, is a highly developed part of theoretical computer science [OW96].

In comparison, algorithmic software is in its infancy. For many fundamental algorithmic tasks no reliable implementations are available due to a lack of understanding of the principles underlying reliable algorithmic software, see Section 3 for some examples. *The challenge is*

- *to work out the principles underlying reliable algorithmic software and*
- *to create a comprehensive collection of reliable algorithmic software components.*

3   State of the Art

I give examples of basic algorithmic problems for which no truly reliable software is available.

Linear Programming is arguably one of the most useful algorithmic paradigms. It allows one to formulate optimization problems over real-valued variables that have to obey a set of linear inequalities.

$$\text{maximize} \quad c^T x \quad \text{subject to} \quad Ax \leq b$$

R. Klein et al. (Eds.): Comp. Sci. in Perspective (Ottmann Festschrift), LNCS 2598, pp. 255–263, 2003.

where x is a vector of n real variables, A is an $m \times n$ matrix with $m > n$, b is an m-vector, and c is an n-vector. A large number of problems can be formulated as linear programs, e.g., shortest paths, network flow, matchings, convex hull, ..., and hence a linear programming solver is an extremely useful algorithmic tool. *There is no linear programming solver that is guaranteed to solve large-scale linear programs to optimality. Every existing solver may return suboptimal or infeasible solutions.*[1] For example, the current version of CPLEX does not find the optimal solution for problems etamacro and scsd6 in the Netlib library, a popular collection of benchmark problems; see Table 1.

Problem				CPLEX solution		Exact Verification		
Name	C	R	NZ	RelObjErr	TC	V	Res	TV
degen3	1504	1818	26230	6.91e-16	8.08	0	opt	8.79
etamacro	401	688	2489	1.50e-16	0.13	10	feas	1.11
fffff800	525	854	6235	0.00e+00	0.09	0	opt	4.41
pilot.we	737	2789	9218	2.93e-11	3.8	0	opt	1654.64
scsd6	148	1350	5666	0.00e+00	0.1	13	feas	0.52
scsd8	398	2750	11334	7.54e-16	0.48	0	opt	1.52

Table 1. Behavior of CPLEX for problems in the Netlib library. The first four columns give the name of the instance, and the number of constraints, variables, and non-zeroes in the constraint matrix. The columns labeled RelObjErr, T, and Res give information about the solution computed by CPLEX: the column labeled TC shows the time (in seconds) for solving the LP, and the column labeled Res shows whether the basis (= a symbolic representation of a solution) computed by CPLEX is optimal or not. An entry "feas" indicates that the computed basis is feasible but *not* optimal. RelObjErr is the relative error in the objective value at the basis returned by CPLEX. For example, for problem pilot.we CPLEX found the optimal basis and returned an objective value with relative error 2.93e-11. The meaning of the remaining columns is explained later in the text.

Computer aided design systems manipulate 3-dimensional solids under boolean operations (and other operations); see Figure 1 for an example. *No existing system is guaranteed to compute the correct result, not even for solids bounded by plane faces*; see Table 2.

There are systems that solve large-scale problems on graphs and networks. These systems are trustworthy because their authors are. *They come with no formal guarantee.*

The situation is even worse for parallel and distributed algorithms.

Many algorithmic libraries exist, e.g., Maple and Mathematica for symbolic computation, STL for data structures, LEDA for data structures, graph and network algorithms, and computational geometry, CGAL for computational geometry, ACIS for computer

---

[1] There are solvers that solve small problems to optimality.

Figure 1. The figure on the left shows a red and a blue solid and the figure on the right shows their union.

System	n	α	time	output
ACIS	1000	1.0e-4	5 min	correct
ACIS	1000	1.0e-6	30 sec	incorrect answer
Rhino3D	200	1.0e-2	15sec	correct
Rhino3D	400	1.0e-2	–	CRASH

Table 2. Runs for the example of Figure 1. We used cylinders whose basis is a regular n-gon and that were rotated by α degrees relative each other. ACIS and Rhino3D are popular commercial CAD-engines.

aided design, LAPACK for linear algebra problems, MATLAB for numerical computation and visualization, CPLEX and Xpress for optimization, and ILOG solvers for constraint solving. None of the implementations in any of these systems comes with a formal guarantee of correctness. Moreover, for problems such as linear programming and boolean operations on solids the existing implementations are known to be incorrect.

Given that algorithms (and frequently in the form of implementations from these libraries) form the core of any software system, we are facing a major challenge: Develop a theory for reliable algorithmic software and construct reliable algorithmic components based on it.

4   Approaches

I discuss some approaches for addressing the challenge. I hope to demonstrate that viable roads of attack exist. I do not claim and, in fact, do not believe that the approaches outlined are sufficient for solving the challenge.

4.1   Program Verification

Formal program verification is the obvious approach. However, there are several obstacles to applying it. I mention just two: (1) the non-trivial mathematics underlying

Figure 2. The figure on the left shows a red and a blue polygon with curved boundaries and the figure on the right shows their union. The computation takes about 1 minute for input polygons with 1000 vertices, see[BEH$^+$02] for details.

the algorithms must be formalized, and (2) verification must be applicable to languages in which algorithmicists want to formulate their algorithms. In particular, we would need a formal semantics for languages like C++.In view of these obstacles, the direct applicability of program verification is doubtful, but see Section 4.5 below.

4.2 The Exact Computation Paradigm

Algorithms are frequently designed for ideal machines that are assumed to be able to calculate with real numbers in the sense of mathematics. However, real machines offer only crude approximations, namely fixed precision integers and floating point numbers. Arbitrary-precision integers and floating point systems go beyond, but are still approximate.

The *exact computation paradigm* goes a step further. It aims to exploit the fact that computations with algebraic numbers can be performed exactly. There are symbolic [Yap99; VG99] and numerical methods [BFM$^+$01] to discover, for example, that

$$3\sqrt{\sqrt[3]{5} - \sqrt[3]{4}} - \sqrt[3]{2} - \sqrt[3]{20} + \sqrt[3]{25} = 0$$

In principle, the computations required in computational geometry, computer aided design, and for a large class of optimization problems stay within the algebraic numbers and hence can be solved by the exact computation paradigm. The caveat is that the cost of exact algebraic computation is enormous, so enormous, in fact, that the approach was considered to be doomed to failure until recently.

Significant progress has been made in the last years. The geometry in LEDA [LED] and CGAL [CGA] follows the exact computation paradigm, but these systems deal only with linear objects. CORE [KLPY99] and LEDA [BFM$^+$01] offer (reasonably) efficient computation with algebraic numbers that can be represented as expressions involving radicals, and ESOLID [KCF$^+$02] takes a step towards exact boundary evaluation of curved solids. Exact boolean operations on 2-dimensional curved objects of low degree are now feasible; see Figure 2. However, the problem is much harder in 3D.

One of the earliest paper on the exact computation paradigm was co-authored by Thomas Ottmann; see [OTU87].

The recent progress makes it plausible (this might be wishful thinking on my side) to build a CAD-system that is exact and efficient at the same time. *I consider the construction of such a system the litmus test for the exact computation paradigm.*

4.3 Program Result Checking

Instead of ensuring that a program computes the correct answer for *any* input (= verification), one may also try to verify that it computed the correct answer for a *given* input. The latter approach is called program result checking [BK89; WB97; BW96].

In its pure form, a checker for a program computing a function f takes an instance x and an output y, and returns true if $y = f(x)$. Of course, this is usually a simpler algorithmic task than computing $f(x)$.

We give an example: The multiplication problem is to compute the product $y = x_1 \cdot x_2$ of two given numbers x_1 and x_2. A (probabilistic) checker for the multiplication problem gets x_1, x_2, and y, chooses a random prime p and verifies that $(x_1 \bmod p) \cdot (x_1 \bmod p) \bmod p = y \bmod p$. If the equality holds, it returns true, otherwise it returns false. By repeating the test with independently chosen primes, the error probability can be made arbitrarily small. Some readers were taught this check (with $p = 9$) by their math teachers.

4.4 Certifying Algorithms

A looser version of program result checking is provided by certifying algorithms [SM90; MNS+96; KMMS02]. Such algorithms return additional output (frequently called a *witness*) that simplifies the verification of the result. More precisely, a certifying program for a function f returns on input x a value y, the alleged value $f(x)$, and additional information I that makes it easy to check that $y = f(x)$. By "easy to verify" we mean two things. Firstly, there must be a simple program C (a checking program) that given x, y, and I checks[2] whether indeed $y = f(x)$. The program C should be so simple that its correctness is "obvious". Secondly, the running time of C on inputs x, y, and I should be no larger than the time required to compute $f(x)$ from scratch. This guarantees that the checking program C can be used without severe penalty in running time.

We give some examples.

Consider a program that takes an $m \times n$ matrix A and an m vector b and is supposed to determine whether the linear system $A \cdot x = b$ has a solution. As stated, the program is supposed to return a boolean value indicating whether the system is solvable or not. This program is not certifying. In order to make it certifying, we extend the interface. On input A and b the program returns either

– "the system is solvable" and a vector x such that $A \cdot x = b$ or
– "the system is unsolvable" and a vector c such that $c^T A = 0$ and $c^T \cdot b \neq 0$. Observe that such a vector c certifies that the system is unsolvable. Assume otherwise, say $Ax_0 = b$ for some x_0. Multiplying this equation with c from the left, gives $c^T Ax_0 = c^T b$ and hence $0 \neq 0$. Gaussian elimination is easily modified to compute the vector c in the case of an unsolvable system.

---

[2] Of course, if $y \neq f(x)$ then there should be no I such that the triple (x, y, I) passes checking.

The certifying program is easy to check. If it answers "the system is solvable", we check that $A \cdot x = b$ and if it answers "the system is unsolvable" we check that $c^T A = 0$ and $c^T \cdot b \neq 0$. Thus the check amounts to a matrix-vector and a vector-vector product which are fast and also easy to program.

The second example id primal-dual algorithms for problems that can be formulated as linear programs. Consider the problem of finding a maximum cardinality matching M in a bipartite graph $G = (V, E)$. A matching M is a set of edges no two of which share an endpoint. A node cover C is a set of nodes such that every edge of G is incident to some node in C. Since edges in a matching do not share endpoints, $|M| \leq |C|$ for any matching M and any node cover C. It can be shown [CCPS98] that for every maximum-cardinality matching M there is a node cover C of the same cardinality. A certifying algorithm for the matching problem in bipartite graphs returns a matching M and a node cover C with $|M| = |C|$.

The third example is planarity testing. The task is to decide whether a graph G is planar. A witness of planarity is a planar embedding and a witness of non-planarity is a Kuratowski subgraph, see [MN99, Chapter 8] for details. It was known since the early 70s that planarity of a graph can be tested in linear time [LEC67; HT74]. Linear time algorithms to compute witnesses for planarity [CNAO85] and non-planarity [MN99, Section 8.7] were found much later.

For any algorithmic problem, we may ask the question whether a certifying algorithm exists for it. Short witnesses that can be checked in polynomial time can only exist for problems in NP∩co-NP. This holds for decision problems and optimization problems.

4.5 Verification of Checkers

We postulated in the preceding section that the task of verifying a triple (x, y, I) should be so simple that the correctness of the program implementing the checker is "obvious". In fact, formal verification of checkers is probably a feasible task for program verification. Observe, that the verification problem is simplified in many ways: (1) the mathematics required for verifying the checker is usually much simpler than that underlying the algorithm for finding solutions and witnesses, (2) the checkers are simple programs, and (3) algorithmicists may be willing to code the checkers in languages that ease verification, e.g., in a functional language.

For a correct program, verification of the checker is as good as verification of the program itself.

4.6 Cooperation of Verification and Checking

Consider the following example[3]: a sorting routine working on a set S

(a) must not change S and
(b) must produce a sorted output.

---

[3] The author does not recall where he learned about this example.

The first property is hard to check (provably as hard as sorting), but usually trivial to prove, e.g., if the sorting algorithm uses a *swap*-subroutine to exchange items. The second property is easy to check by a linear scan over the output, but hard to prove (if the sorting algorithm is complex). A combination of verification and checking provides a simple solution for both parts.

4.7   A Posteriori Analysis

Despite the approaches outlined in the preceding sections, there will be many situations where we have to be content with inexact algorithms. It is likely always to be true that exact and verified methods are significantly less efficient than inexact methods.

In this realm, the question of *a posteriori* analysis arises. Given an instance of the problem and a solution, we may want to analyze the quality of the solution. As an example, consider the problem of finding the roots of a univariate polynomial $f(x)$ of degree n. Given approximate solutions $x_1, \ldots, x_n$, compute the quantities [Smi70]

$$\sigma_i = \frac{f(x_i)}{\prod_{j \neq i}(x_i - x_j)} \quad \text{for } 1 \leq i \leq n .$$

Let Γ_i be the disk in the complex plane centered at x_i with radius $n|\sigma_i|$. Then the union of the disks contains all roots of f. Moreover, a connected component consisting of k disks contains exactly k roots of f. The disks Γ_i give a posteriori analysis of the quality of root estimates and the σ_i are easily computed with controlled error using multi-precision floating point arithmetic.

Are there analogous examples in the realm of combinatorial computing. Clearly, in the area of approximation algorithms, one frequently computes *a priori* bounds. The nice feature of the example above is that the approximate solution plays essential part in estimating its quality.

4.8   Test and Repair

Sometimes we can even do better than just a posteriori analysis. We might be able to take the approximate solution returned by an inexact algorithm as the starting point for an exact algorithm.

Consider the linear programming problem

$$\text{maximize} \quad c^T x \quad \text{subject to} \quad Ax = b, \, x \geq 0$$

where x is a vector of n real variables, A is an $m \times n$ matrix, $m < n$, b is an m-vector, and c is an n-vector. For simplicity, we assume A to have rank m. It is well known that one can restrict consideration to basic solutions. A basic solution is defined by an $m \times m$ non-singular sub-matrix B of A and is equal to (x_B, x_N) where the x_B are the variables corresponding to the columns in B, $x_B = B^{-1}b$ and $x_N = 0$. A basic solution is primal feasible if $x_B \geq 0$, and is dual feasible if $c_B^T - c_N^T A_B^{-1} A_N \leq 0$. It is optimal if it is primal and dual feasible.

For medium-scale linear programs, we succeeded in checking (exactly !!!) in reasonable time whether a given basis is primal or dual feasible [DFK$^+$02]. This suggests the following approach. Use an inexact LP solver to determine an "optimal" basis B. Check the basis for primal and/or dual feasibility. If so, declare it optimal. If not and the basis is X-feasible ($X \in \{$primal,dual$\}$), use an exact X-simplex algorithm starting at B to find the true optimum. If B is neither primal nor dual feasible, I do not really know how to proceed; one can first use the primal simplex method to find the optimum of the subproblem defined by the satisfied constraints and then use the dual simplex method to add the remaining constraints, but this is not really satisfactory. The hope is that the inexact method find a basis B that is close enough to the optimum, so that even a slow exact algorithm can find it. Table 1 supports this hope. The last column shows the time required to check whether the basis returned by CPLEX is optimal, and if not, to obtain the optimal basis from it. Column V shows the number of constraints violated by the basis returned by CPLEX.

The general question for optimization problems is the following. Design (exact) algorithms that start from a given solution x_0 towards an optimal solution. The running time should depend on some natural distance measure between the given and the optimal solution.

5   Conclusion

Reliability (trustworthiness) is a desirable feature of humans and also programs. I love to use TeXand one of the reasons is that it never crashes. Many users of LEDA tell me that the reliability of its programs is important to them and that they do not really care about the speed.

The strive for reliability poses many hard and relevant scientific questions. Part of the answers can be found within combinatorial algorithmics, but for many of them we have to reach out and make contact with other areas of computer science: numerical analysis, computer algebra, and even semantics and program verification.

References

[BEH+02]   E. Berberich, A. Eigenwillig, M. Hemmer, S. Hert, K. Mehlhorn, and E. Schömer. A computational basis for conic arcs and boolean operations on conic polygons. to appear in ESA 2002, http://www.mpi-sb.mpg.de/~mehlhorn/ftp/ConicPolygons.ps, 2002.

[BFM+01]   C. Burnikel, S. Funke, K. Mehlhorn, S. Schirra, and S. Schmitt. A separation bound for real algebraic expressions. In *ESA 2001*, Lecture Notes in Computer Science, pages 254–265, 2001.   http://www.mpi-sb.mpg.de/~mehlhorn/ftp/ImprovedSepBounds.ps.gz.

[BK89]   M. Blum and S. Kannan. Designing programs that check their work. In *Proceedings of the 21th Annual ACM Symposium on Theory of Computing (STOC'89)*, pages 86–97, 1989.

[BW96]   M. Blum and H. Wasserman. Reflections on the pentium division bug. *IEEE Transaction on Computing*, 45(4):385–393, 1996.

[CCPS98]   W.J. Cook, W.H. Cunningham, W.R. Pulleyblank, and A. Schrijver. *Combinatorial Optimization*. John Wiley & Sons, Inc, 1998.

[CGA]   CGAL (Computational Geometry Algorithms Library). www.cgal.org.

[CNAO85]   N. Chiba, T. Nishizeki, S. Abe, and T. Ozawa. A linear algorithm for embedding planar graphs using PQ-trees. *Journal of Computer and System Sciences*, 30(1):54–76, 1985.

[DFK⁺02]   M. Dhiflaoui, S. Funke, C. Kwappik, K. Mehlhorn, M. Seel, E. Schömer, R. Schulte, and D. Weber. Certifying and repairing solutions to large LPs, How good are LP-solvers? to appear in SODA 2003, http://www.mpi-sb.mpg.de/~mehlhorn/ftp/LPExactShort.ps, 2002.

[HT74]   J.E. Hopcroft and R.E. Tarjan. Efficient planarity testing. *Journal of the ACM*, 21:549–568, 1974.

[KCF⁺02]   J. Keyser, T. Culver, M. Foskey, S. Krishnan, and D. Manocha. ESOLID: A system for exact boundary evaluation. In *7th ACM Symposium on Solid Modelling and Applications*, pages 23–34, 2002.

[KLPY99]   V. Karamcheti, C. Li, I. Pechtchanski, and Chee Yap. A core library for robust numeric and geometric computation. In *Proceedings of the 15th Annual ACM Symposium on Computational Geometry*, pages 351–359, Miami, Florida, 1999.

[KMMS02]   D. Kratsch, R. McConnell, K. Mehlhorn, and J.P. Spinrad. Certifying algorithms for recognizing interval graphs and permutation graphs. SODA 2003 to appear, http://www.mpi-sb.mpg.de/~mehlhorn/ftp/intervalgraph.ps, 2002.

[LEC67]   A. Lempel, S. Even, and I. Cederbaum. An algorithm for planarity testing of graphs. In P. Rosenstiehl, editor, *Theory of Graphs, International Symposium, Rome*, pages 215–232, 1967.

[LED]   LEDA (Library of Efficient Data Types and Algorithms). www.algorithmic-solutions.com.

[MN99]   K. Mehlhorn and S. Näher. *The LEDA Platform for Combinatorial and Geometric Computing*. Cambridge University Press, 1999. 1018 pages.

[MNS⁺96]   K. Mehlhorn, S. Näher, T. Schilz, S. Schirra, M. Seel, R. Seidel, and C. Uhrig. Checking geometric programs or verification of geometric structures. In *Proceedings of the 12th Annual Symposium on Computational Geometry (SCG'96)*, pages 159–165, 1996.

[OTU87]   T. Ottmann, G. Thiemt, and C. Ullrich. Numerical stability of geometric algorithms. In Derick Wood, editor, *Proceedings of the 3rd Annual Symposium on Computational Geometry (SCG '87)*, pages 119–125, Waterloo, ON, Canada, June 1987. ACM Press.

[OW96]   T. Ottmann and P. Widmayer. *Algorithmen und Datenstrukturen*. Spektrum Akademischer Verlag, 1996.

[SM90]   G.F. Sullivan and G.M. Masson. Using certification trails to achieve software fault tolerance. In Brian Randell, editor, *Proceedings of the 20th Annual International Symposium on Fault-Tolerant Computing (FTCS '90)*, pages 423–433. IEEE, 1990.

[Smi70]   Brian T. Smith. Error bounds for zeros of a polynomial based upon Gerschgorin's theorems. *Journal of the ACM*, 17(4):661–674, October 1970.

[VG99]   Joachim Von zur Gathen and Jürgen Gerhard. *Modern Computer Algebra*. Cambridge University Press, New York, NY, USA, 1999. Chapters 1 and 21 cover cryptography and public key cryptography.

[WB97]   H. Wasserman and M. Blum. Software reliability via run-time result-checking. *Journal of the ACM*, 44(6):826–849, 1997.

[Yap99]   C.K. Yap. *Fundamental Problems in Algorithmic Algebra*. Oxford University Press, 1999.

A Lower Bound for Randomized Searching on m Rays

Sven Schuierer

Novartis Pharma AG, Lichtstr. 35, CH-4002 Basel, Switzerland
sven.schuierer@pharma.novartis.com

(In Honor of Prof. Th. Ottmann's 60th Birthday)

Abstract. We consider the problem of on-line searching on m rays. A point robot is assumed to stand at the origin of m concurrent rays one of which contains a goal g that the point robot has to find. Neither the ray containing g nor the distance to g are known to the robot. The only way the robot can detect g is by reaching its location. We use the *competitive ratio* as a measure of the performance of a search strategy, that is, the worst case ratio of the total distance D_R traveled by the robot to find g to the distance D from the origin to g.

We present a new proof of a tight lower bound of the competitive ratio for randomized strategies to search on m rays. Our proof allows us to obtain a lower bound on the optimal competitive ratio for a fixed m even if the distance of the goal to the origin is bounded from above.

Finally, we show that the optimal competitive ratio converges to $1 + 2(e^\alpha - 1)/\alpha^2 \, m \sim 1 + 2 \cdot 1.544 \, m$, for large m where α minimizes the function $(e^x - 1)/x^2$.

1 Introduction

Searching for a goal is an important and well studied problem in robotics. In many realistic situations the robot does not possess complete knowledge about its environment, for instance, the robot may not have a map of its surroundings or the location of the goal may be unknown [2; 3; 4; 6; 8; 11; 12; 15].

Since the robot has to make decisions about the search based only on the part of its environment that it has explored before, the search of the robot can be viewed as an *on-line* problem. This invites the application of *competitive analysis* to judge the performance of an on-line search strategy. The *competitive ratio* of a search strategy is defined as the worst case ratio of the distance traveled by the robot to the distance from the origin to the goal, maximized over all possible locations of the goal [17].

Consider the following problem in this context. A point robot is imagined to stand at the origin s of m rays one of which contains the goal g whose distance to s is unknown. The robot can only detect g if it stands on top of it. A simple strategy for the robot to find g is to visit the rays in cyclic order. On each ray the robot travels for some distance (further than on the previous visit of the same ray) and then returns to the origin. In fact, the optimal strategy is of this kind.

If only deterministic strategies are considered, Baeza-Yates *et al.* [1] and Gal [7] present an optimal search strategy that achieves a competitive ratio of $1 + 2m^m/(m-1)^{m-1} (\sim 1 + 2em$ for large m).

The simplicity of its description and the elegance of its solution make searching on m rays one of the most fundamental problems in on-line searching; furthermore, many

R. Klein et al. (Eds.): Comp. Sci. in Perspective (Ottmann Festschrift), LNCS 2598, pp. 264–277, 2003.
© Springer-Verlag Berlin Heidelberg 2003

problems of searching in more complex geometries can be directly reduced to it [4; 5; 13; 14; 16].

In the randomized case Kao, Reif, and Tate [10] present an optimal randomized algorithm to search on m rays and prove its optimality in the special case $m = 2$ (see also [7]). The algorithm achieves a competitive ratio of

$$c_m^* = 1 + \frac{2}{m} \min_{a>1} \frac{a^m - 1}{(a-1)\ln a}.$$

For general m, Kao et al. prove a matching lower bound [9]. It should be noted that the proof by Kao et al. as well as the results presented here apply only to *periodic* strategies, that is, strategies that visit the rays in a fixed cyclic order [9]. In this paper we partially reprove and extend the results by Kao et al. using a different and simpler approach. More precisely, the contribution of this paper is threefold.

1. We present a tight lower bound on the competitive ratio of randomized strategies to search on m rays using a proof that is simpler and more elementary than that by Kao et al. [9].
2. The lower bound we prove also applies to searching on bounded rays; that is, our proof provides a lower bound even if we assume that a search algorithm is given an upper bound on the maximal distance to the goal in advance.
3. In order to estimate the competitive ratio of the optimal randomized strategy for large m it is useful to know its constant of proportionality. In the deterministic case it can be easily seen to be $1 + 2em$. We show that the competitive ratio c_m^* of randomized searching on m rays converges to $1 + 2(e^\alpha - 1)/\alpha^2 m \sim 1 + 2 \cdot 1.544 m$, for large m where α minimizes the function $(e^x - 1)/x^2$.

The paper is organized as follows. In the next section we introduce some definitions and present the optimal randomized algorithm to search on m rays. In Section 3 we prove a lower bound for randomized strategies for searching on m rays. In particular, we show how to adapt our lower bound proof to the case when an upper bound D on the distance to goal is known. Finally, in Section 4 we show that c_m^* approaches $1 + 2 \cdot 1.544 m$ as m goes to infinity.

2 Randomized m-Way Ray Searching

Consider a randomized search strategy X to search on m rays. The distance $dist_X(g)$ traveled by a robot using X to find the goal g depends on the random choices of X and is a random variable. We define the competitive ratio c_X of X to be the worst case ratio of the expected value of $dist_X(g)$ over the distance of g to the origin, maximised over all possible positions of g. In other words, the strategy X has competitive ratio c_X if c_X is the smallest value such that, for all goal positions g,

$$E[dist_X(g)] \leq c_X dist(g). \tag{1}$$

Sometimes an additive constant on the right hand side of Equation 1 is allowed to avoid degeneracies if the goal is very close. We choose a different solution by assuming that the distance to the goal is at least one unit.

2.1   An Optimal Algorithm

The following algorithm to search on m rays is presented and analyzed by Kao *et al.* [9; 10]. A slight variant of it was first suggested by Gal [7].

Algorithm Randomized m-way Ray-Search
let σ be a random permutation of $\{0, \ldots, m-1\}$
let a_m^* be the number that minimizes the function $(a^m - 1)/((a-1)\ln a)$
let ε be a random real uniformly chosen from $[0, 1)$
let $d_0 \leftarrow (a_m^*)^\varepsilon$; $i \leftarrow 0$
while the goal is not found **do**
     let $r \leftarrow \sigma(i \bmod m)$
     explore ray r up to distance d_i
     $d_{i+1} \leftarrow d_i \cdot a_m^*$
     $i \leftarrow i+1$
end while

It can be shown that Algorithm *Randomized m-way Ray-Search* has a competitive ratio of

$$c_m^* = 1 + \frac{2}{m} \frac{(a_m^*)^m - 1}{(a_m^* - 1)\ln a_m^*}. \tag{2}$$

Note that Algorithm *Randomized m-way Ray-Search* is an example of a *periodic* strategy, that is, for every $i \geq 0$, the ray visited in Step $i+m$ is the same ray as visited in Step i. In next section we prove a matching lower bound for the competitive ratio of any periodic randomized search strategy.

3   Proving a Lower Bound

In order to prove a lower bound for randomized searching on m-rays we make use of Yao's corollary to the well-known von Neumann minimax principle [18]. It states that a lower bound for *deterministic* strategies on a given input distribution is also a lower bound for a *randomized* strategy on its worst case input.

So in the following we deal with deterministic strategies and probability distributions on m rays which we call *hiding strategies*. We show that, for every $\varepsilon > 0$, there is a hiding strategy $\mathcal{H}$, such that, for each deterministic search strategy X, the expected competitive ratio $E_{\mathcal{H}}(X)$ of X under $\mathcal{H}$ is at least $(1-\varepsilon)^3 c_m^*$. By Yao's lemma this implies that the competitive ratio of a randomized strategy X on its worst case input is at least c_m^*.

3.1   Searching and Hiding Strategies

We represent a deterministic search strategy as an infinite sequence X of pairs (x_k, J_k) where x_k is the distance that the robot travels in Step k along a given ray until it returns and J_k is the index of the next time when the robot travels along the same ray again [7]. Note that this representation is unique up to an ordering of the sequence in which the rays are visited the first time. Since in our case the rays are indistinguishable in the beginning, we leave the sequence in which the rays are visited the first time unspecified. We define $J^{-1}(k)$ to be the inverse of the function J; that is, if there is an index i with $J_i = k$, then we define $J^{-1}(k) = i$; if the robots visits the ray for the first time in Step k, then we define $J^{-1}(k) = -1$ and $x_{-1} = 0$. If X is a periodic strategy, then $J_k = k + m$ and $J^{-1}(k) = k - m$.

We model a hiding strategy $\mathcal{H}$ as a collection of m measures $H_r : [0, \infty) \longrightarrow \mathbb{R}[+]$, $0 \le r \le m - 1$, with

$$\sum_{r=0}^{m-1} \int_0^\infty dH_r(h) = 1.$$

where $\int_0^{h_0} dH_r(h)$ is the probability that the goal is hidden in ray r up to a distance of at most h_0. Hence, the additional contribution to the expected competitive ratio when the robot explores the ray r_k in Step k up to distance x_k is

$$\int_{x_{J-1}(k)}^{x_k} \left(2 \sum_{i=0}^{k-1} x_i + h \right) \frac{dH_{r_k}(h)}{h}$$

since ray r_k has been explored before up to distance $x_{J-1}(k)$ and the robot has traveled a distance of $2\sum_{i=0}^{k-1} x_k$ before Step k. Therefore, the expected competitive ratio $E_{\mathcal{H}}(X)$ of search strategy X is given by

$$E_{\mathcal{H}}(X) = \sum_{k=0}^\infty \int_{x_{J-1}(k)}^{x_k} \left(2 \sum_{i=0}^{k-1} x_i + h \right) \frac{dH_{r_k}(h)}{h}$$

$$= 1 + 2 \sum_{k=0}^\infty \sum_{i=0}^{k-1} x_i \int_{x_{J-1}(k)}^{x_k} \frac{dH_{r_k}(h)}{h}. \qquad (3)$$

3.2 The Hiding Strategy $\mathcal{H}_D$

In order to prove a lower bound for searching on m rays we need to define a specific hiding strategy. Let $D > 0$. We use the hiding strategy $\mathcal{H}_D$ which is defined by the measure

$$H_D(h) = \begin{cases} 0 & \text{for } 0 \le h \le 1 \\ \dfrac{\delta_D}{m} \ln(h) & \text{for } 1 \le h \le D \\ \dfrac{1}{m} & \text{for } h \ge D, \end{cases}$$

for each of the m rays[1] where $\delta_D = 1/(1 + \ln D)$. $\mathcal{H}_D$ is a straightforward generalization of the hiding strategy used in the lower bound proof by Gal for the case $m = 2$ [7]. The probability measure H_D has a density of $\delta_D/(mh)$, for $1 \le h < D$, and a probability

---

[1] The measure H of the hiding strategy $\mathcal{H}$ used by Kao et al. [9; 10] is given by $H(h) = -h^{-\varepsilon}/m$, for $h \ge 1$. This leads to a much more complicated optimization problem than the problem $(*)$ below.

atom of mass δ_D/m at $h = D$. The function H_D has the following very useful property for $1 \leq x \leq D$.

$$\int_x^\infty \frac{dH_D(h)}{h} = \int_x^D \frac{dH_D(h)}{h} = \frac{\delta_D}{m}\int_x^D \frac{1}{h^2}dh + \frac{\delta_D}{m}\frac{1}{D}$$

$$= \frac{\delta_D}{m}\left(\frac{1}{x} - \frac{1}{D} + \frac{1}{D}\right) = \frac{\delta_D}{m}\frac{1}{x}.$$

The main technical result in proving a lower bound is the following lemma.

Lemma 1. *If X is a periodic strategy with $E_{\mathcal{H}_D}(X) \leq c_m^*$, for all $D > 0$ then, for every $\varepsilon > 0$, there is a $D > 0$ such that*

$$E_{\mathcal{H}_D}(X) \geq (1-\varepsilon)^3 \left(1 + 2\frac{(a_m^*)^m - 1}{m(a_m^* - 1)\ln a_m^*}\right).$$

The remainder of this section is devoted to the proof of Lemma 1.

3.3 A Minimization Problem

In the following let X be an optimal, periodic, deterministic strategy to search on m rays under the hiding strategy $\mathcal{H}_D$. Let n be the first index such that $x_n \geq D$. After reordering the terms of Equation 3 and integrating along each ray r it is easy to see that the expected cost of X is given by

$$E_{\mathcal{H}_D}(X) = 1 + 2\sum_{k=0}^\infty x_k \sum_{r=0}^{m-1} \int_{x_{k-r}}^\infty \frac{dH_D(h)}{h}$$

$$\geq 1 + 2\sum_{k=0}^n x_k \sum_{r=0}^{m-1} \int_{x_{k-r}}^D \frac{dH_D(h)}{h}$$

$$\geq 1 + 2\frac{\delta_D}{m}\left(\sum_{k=0}^n \sum_{r=0}^{m-1} \frac{x_k}{x_{k-r}} - 1\right) \qquad (4)$$

$$\geq 1 + 2\frac{\sum_{k=0}^n \sum_{r=0}^{m-1} x_k/x_{k-r} - 1}{m(1 + \ln x_n)}$$

where we define $x_{-m+1} = \cdots = x_{-1} = 1$. Note that we need to subtract 1 in Equation 4 as x_n is not integrated along the ray $r_{n \bmod m}$ with a positive measure anymore. Hence, we are interested in a lower bound for the ratio

$$\frac{-1 + \sum_{k=0}^n \sum_{r=0}^{m-1} x_k/x_{k-r}}{m(1 + \ln x_n)}. \qquad (5)$$

Let $a_i = x_i/x_{i-1}$, for $-m+2 \leq i \leq n$ which implies that $x_n = \prod_{i=-m+2}^n a_i$. Since $a_i = 1$, for $i < 0$, we have $x_n = \prod_{i=0}^n a_i$. Hence,

$$E_{\mathcal{H}_D}(X) \geq 1 + 2\frac{\sum_{k=0}^n \sum_{r=0}^{m-1} x_k/x_{k-r} - 1}{m(1 + \ln x_n)}$$

$$= \frac{\sum_{k=0}^n \sum_{r=0}^{m-1} \prod_{i=k-r+1}^k a_i - 1}{m(1 + \ln \prod_{i=0}^n a_i)}, \qquad (6)$$

where we define $\prod_{i=k-r+1}^k a_i = 1$, if $k - r + 1 > k$.

If we fix x_n, then a lower bound for the numerator of Expression 5 is given by a solution to the following minimization problem

$$(*) \text{ minimize } -1 + \sum_{k=0}^{n} \left(1 + a_k + a_k a_{k-1} + \cdots + \prod_{i=k-m+2}^{k} a_i \right)$$

$$\text{where } \prod_{i=0}^{n} a_i = x_n, \quad a_i > 0, \quad \text{and} \quad a_{-m+1} = \cdots = a_{-1} = 1. \tag{7}$$

The optimal value of the objective function of $(*)$ only depends on x_n (and n and m).

Instead of solving $(*)$ directly we split the objective function into the $m-1$ sums $S_r = \sum_{k=0}^{n} a_k a_{k-1} \cdots a_{k-r+1}$, for $1 \le r \le m-1$. We minimize each of the sums S_r separately, under Constraint 7. Clearly, the sum of the $m-1$ minimal values of S_r, for $1 \le r \le m-1$, is no larger than the minimal value for $(*)$ and, therefore, a lower bound for the numerator of Expression 5.

3.4   Minimizing the Sum S_r

In the following let r be a fixed integer between 1 and $m-1$. We define $A_k = a_{k-r+1} \cdots a_k$, that is, $S_r = \sum_{k=0}^{n} A_k$. The main idea of the proof is to show that if the sum S_r of the products A_k, $0 \le k \le n$, is minimized, then the numbers a_i are approximately equal, for $0 \le i \le n$.

If the a_i are exactly equal, then $a_i = \alpha_n$ where α_n is the $(n+1)$st root of x_n (since $x_n = \prod_{i=0}^{n} a_i$) and $A_k = \alpha_n^r$, for all $0 \le k \le n$, and the value of the objective function of $(*)$ is given by

$$\sum_{k=0}^{n} \sum_{i=0}^{m-1} \alpha_n^i - 1 = (n+1)(\alpha_n^m - 1)/(\alpha_n - 1) - 1. \tag{8}$$

If we divide the right hand side of Equation 8 by $m(1 + \ln x_n) = m(1 + (n+1)\ln\alpha_n)$—and multiply by two and add one, then we have an expression that converges to Expression 2 as D (and with it n) goes to infinity.

In the following we show that the a_i are approximately equal if S_r is minimized and that the minimal value of $(*)$ converges to Expression 8 as D goes to infinity.

We distribute the products A_k into $r+1$ sets $\mathcal{A}_{-1}, \mathcal{A}_0, \ldots, \mathcal{A}_{r-1}$ where the product A_k belongs to set $\mathcal{A}_s$ if $k \bmod r = s$ and $k \ge r$, that is,

$$\mathcal{A}_s = \left\{ a_{s+1} \cdots a_{s+r}, a_{s+r+1} \cdots a_{s+2r}, \ldots, a_{s+(c_s-1)r+1} \cdots a_{s+c_s r} \right\}$$

where $c_s = \lfloor (n-s)/r \rfloor$ is the cardinality of $\mathcal{A}_s$. The remaining products A_k belong to $\mathcal{A}_{-1}$. Note that all a_i with $s+1 \le i \le s+c_s r$ a part of exactly one of the products A_k in $\mathcal{A}_s$. In the following we show that if $S_r = \sum_{k=0}^{n} A_k$ is minimized, then all the products in a set A_s have all the same value.

Lemma 2. *Let $A_k, A_{k+r} \in \mathcal{A}_s$. If $A_k \ne A_{k+r}$, for some $r-1 \le k \le n-r$, then there is a sequence $(\bar{a}_{-m+1}, \bar{a}_{-m+2}, \ldots, \bar{a}_n)$ that satisfies Constraint 7 and, additionally,*

$$\bar{S}_r = \sum_{k=0}^{n} \bar{A}_k < \sum_{k=0}^{n} A_k = S_r$$

where $\bar{A}_k = \prod_{i=k-r+1}^{k} \bar{a}_i$, for $0 \le k \le n$.

Proof: Let $r-1 \leq k_0 \leq n-r$ with $A_{k_0} \neq A_{k_0+r}$, say, $A_{k_0} < A_{k_0+r}$. Then, there is a $\alpha > 0$ such that $A_{k_0}(1+\alpha) = A_{k_0+r}/(1+\alpha)$. If we define a new sequence by

$$\bar{a}_i = \begin{cases} a_i & \text{if } i \neq k_0, \\ (1+\alpha)a_i & \text{if } i = k_0, \text{ and} \\ a_i/(1+\alpha) & \text{if } i = k_0+1, \end{cases}$$

then $x_n = \prod_{i=0}^n a_i$ and

$$\bar{S}_r = \sum_{k=0}^n \bar{A}_k = \sum_{\substack{k=0 \\ k \neq k_0, k_0+r}}^n A_k + \frac{A_{k_0+r}}{1+\alpha} + (1+\alpha)A_{k_0} < \sum_{k=0}^n A_k = S_r,$$

where the last inequality follows from the fact that $x/(1+\alpha) + y(1+\alpha)$ achieves its minimum for $x/(1+\alpha) = y(1+\alpha)$. Hence, we can replace a_{k_0} and a_{k_0+1} by $\bar{a}_{k_0}$ and $\bar{a}_{k_0+1}$ and obtain a sequence where the product of the elements is still x_n but the sum of the products $\bar{A}_k$ is smaller. □

By a simple induction it can now be seen that all the products in each $\mathcal{A}_s$ have the same value. If $r = 1$, then we have the following corollary of Lemma 2.

Corollary 1.

$$\sum_{k=0}^n a_k \geq (n+1)\alpha_n.$$

Proof: By Lemma 2 $\sum_{k=0}^n a_k$ is minimized under Constraint 7 if $a_0 = a_1 = \cdots = a_n$. Since $\prod_{k=0}^n a_k = x_n$, we obtain $a_k = \alpha_n$, for $0 \leq k \leq n$. □

If we consider the product $\prod_{k \in \mathcal{A}_s} A_k$, then this product contains almost all a_k (except for the at most r values $a_0, \ldots, a_s$ at the beginning and the at most r values $a_{s+c_s r+1}, \ldots, a_n$ at the end). In the following we show that the equality of the values of A_k, for $k \in \mathcal{A}_s$, implies that A_k converges to α_n^r as D goes to infinity.

In order to do this we need to bound the product of the remaining at most $2r \leq 2m$ values a_i that are not part of the product $\prod_{k \in \mathcal{A}_s} A_k$. We first give an upper bound on the size of α_n and a lower bound on the size of n before we bound a_i from above, with $0 \leq i \leq n$. In the following let γ_m^* be the unique solution of the equation $c/(1+\ln c) = mc_m^*$.

Lemma 3. *If X is a strategy with $E_{\mathcal{H}_D}(X) \leq c_m^*$, then*

1. $\alpha_n \leq \gamma_m^*$,
2. $\ln D / \ln \gamma_m^* \leq n+1$, *and*
3. $a_i \leq mc_m^*(1+\ln D)$, *for $0 \leq i \leq n$.*

Proof: We first prove that $\alpha_n \leq \gamma_m^*$. We have

$$\frac{\gamma_m^*}{m(1+\ln\gamma_m^*)} = c_m^* \geq E_{\mathcal{H}_D}(X) \geq \frac{\sum_{i=0}^n a_i}{m(1+\ln x_n)}$$

$$\overset{\text{(Corollary 1)}}{\geq} \frac{(n+1)\alpha_n}{m(1+(n+1)\ln\alpha_n)} \geq \frac{\alpha_n}{m(1+\ln\alpha_n)}.$$

Therefore, $\alpha_n \leq \gamma_m^*$.

The second inequality follows from the first by $D \leq x_n \leq (\gamma_m^*)^{n+1}$, that is, $\ln D \leq (n+1)\ln\gamma_m^*$. Finally, we have by Equation 4

$$c_m^* \geq E_{\mathcal{H}_D}(X) \geq \frac{a_i}{m(1+\ln D)}$$

and the third claim follows. $\square$

We now can give a lower bound on S_r.

Lemma 4. *If D is large enough, then*

$$S_r \geq (1-\varepsilon)(n-2m)\alpha_n^r.$$

Proof: Let $0 \leq s \leq r-1$. Since $A_k = A_{k'}$, for all $k,k' \in \mathcal{A}_s$ by Lemma 2, we can define $A_s^* \overset{def}{=} A_k$, with $k \in \mathcal{A}_s$, and A_s^* only depends on s and not on k. Recall that c_s is the size of $\mathcal{A}_s$. The sum S_r is now given by

$$S_r = \sum_{k=0}^{n} A_k \geq \sum_{s=0}^{r-1}\sum_{k\in\mathcal{A}_s} A_k = \sum_{s=0}^{r-1} c_s A_s^*. \tag{9}$$

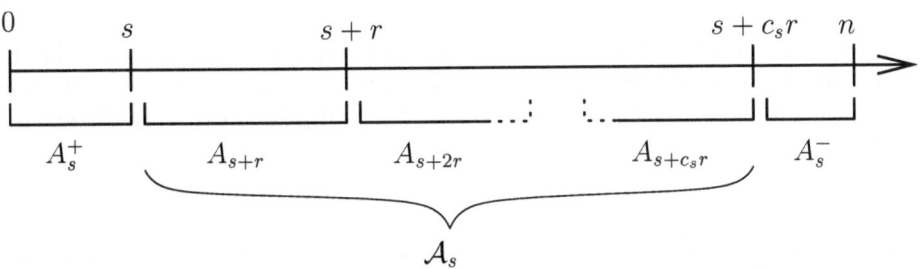

Figure 1. The products that belong to set $\mathcal{A}_s$.

Let $A_s^+ = a_{-m+1}\cdots a_s$, and $A_s^- = a_{s+c_s r+1}\cdots a_n$ (see Figure 1). We have, for $0 \leq s \leq r-1$,

$$A_s^+(A_s^*)^{c_s}A_s^- = A_s^+ \prod_{k=1}^{c_s}\left(\prod_{i=s+(k-1)r+1}^{s+kr} a_i\right)A_s^- = A_s^+\left(\prod_{i=s+1}^{s+c_s r} a_i\right)A_s^- = x_n.$$

By Lemma 3 is easy to see that the values A_s^+ and A_s^- which both are the product at most $m-1$ elements a_i are bounded from above by $(mc_m^*(1+\ln D))^{m-1}$. Since n grows with $\ln D$ by Lemma 3, we can choose D large enough such that

$$\frac{1}{(A_s^+ A_s^-)^{(m+1)/n}} \geq \frac{1}{(mc_m^*(1+\ln D))^{2(m^2-1)/n}} = \frac{1}{O(m^2\ln D)^{2(m^2-1)/n}} \geq (1-\varepsilon).$$

since $c_m^* = \Theta(m)$ by Theorem 3. Hence,

$$A_s^* = \frac{x_n^{1/c_s}}{(A_s^+ A_s^-)^{1/c_s}} \geq \frac{\alpha_n^{(n+1)/c_s}}{(A_s^+ A_s^-)^{(m+1)/n}} \geq (1-\varepsilon)\alpha_n^r \tag{10}$$

since $(m+1)c_s \geq n \geq c_s r$. Combining the results with Equation 9 we obtain

$$S_r = \sum_{k=0}^{n} A_k \geq \sum_{s=1}^{r} c_s A_s^* \overset{(10)}{\geq} \sum_{s=1}^{r} c_s(1-\varepsilon)\alpha_n^r \geq (1-\varepsilon)\alpha_n^r(n-2m)$$

as claimed. □

3.5   Establishing the Lower Bound

We now make use of the lower bounds for the sums S_r that we have obtained above.

$$\frac{-1+\sum_{k=0}^{n}\sum_{r=0}^{m-1}\prod_{i=k-r+1}^{k} a_j}{m(1+\ln x_n)} \geq$$

$$\geq \frac{\sum_{r=0}^{m-1}(1-\varepsilon)\alpha_n^r(n-2m)}{m(1+(n+1)\ln\alpha_n)} - \frac{1}{m(1+\ln D)}$$

$$= \frac{1-\varepsilon}{m}\frac{n-2m}{n+1}\frac{\sum_{r=0}^{m-1}\alpha_n^r}{\ln\alpha_n + 1/(n+1)} - \frac{1}{m(1+\ln D)}.$$

It is easy to see that the above expression converges to Expression 8 if D and with it n goes to infinity. In fact, a more careful analysis show that if we choose D large enough so that $(n-2m)/(n+1) \geq (1-\varepsilon)$ and $\ln D \geq 1/\varepsilon$, then the above expression is bounded from below by

$$\frac{(1-\varepsilon)^3}{m}\frac{\sum_{r=0}^{m-1}\alpha_n^r}{\ln\alpha_n} - \frac{\varepsilon}{m}.$$

Thus, we have proven that, for all $\varepsilon > 0$, there exists a $D > 0$ such that, for all periodic deterministic strategies X that search m rays

$$E_{\mathcal{H}_D}(X) \geq (1-\varepsilon) + (1-\varepsilon)^3\frac{2}{m}\min_{1<a<\infty}\frac{a^m-1}{(a-1)\ln a}.$$

This completes the proof of Lemma 1.

Let $E_g(X)$ denote competitive ratio of strategy X if the goal is placed at position g. By Yao's lemma the above result implies the following theorem.

Theorem 1. *For every $\varepsilon > 0$ and for every periodic randomized strategy X that searches on m rays, there is a goal position g such that*

$$E_g(X) \geq (1-\varepsilon) + (1-\varepsilon)^3\frac{2}{m}\frac{(a_m^*)^m-1}{(a_m^*-1)\ln a_m^*}.$$

Theorem 1 and implies the following corollary as already shown by Kao *et al.* [9].

Corollary 2. *The Strategy Randomized m-way Ray-Search is an optimal periodic randomized search strategy.*

3.6 Bounded Searching

In fact, our analysis also yields a bound on the convergence rate for optimal randomized strategies if an upper bound on the distance to the goal is known.

Theorem 2. *Let X be an optimal randomized search strategy to search m rays. If the maximum distance of the origin to the goal is bounded by D, then there is a goal position g such that*

$$E_g(X) \geq \left(1 - O\left(\frac{m^2 \ln m \ln \ln D}{\ln D}\right)\right)\left(1 + \frac{2}{m}\frac{(a_m^*)^m - 1}{(a_m^* - 1)\ln a_m^*}\right), \tag{11}$$

for large enough D.

Proof: Assume that we are given an upper bound D on the distance to the goal. We again make use of Yao's lemma. Hence, we have to provide a hiding strategy $\mathcal{H}$ that is bounded by D such that any deterministic strategy to search on m rays has a competitive ratio as in Equation 11.

Of course, the hiding strategy $\mathcal{H}_D$ is well suited. Since we assumed above that $\mathcal{H}_D$ is known to the strategy, all the bounds we have shown remain valid.

So let X be an optimal deterministic strategy to search on m rays if we are given $\mathcal{H}_D$. By our previous analysis we have

$$E_{\mathcal{H}_D}(X) \geq$$
$$1 + \frac{1}{O(m^2 \ln D)^{2(m^2-1)/n}}\frac{n-2m}{n+1}\frac{1}{1+\frac{1}{\ln D}}\frac{2(\alpha_n^m - 1)}{m(\alpha_n - 1)\ln \alpha_n} - \frac{1}{m(1+\ln D)}$$

since

$$\left(1 + \frac{1}{\ln D}\right)\ln \alpha_n = \ln \alpha_n + \frac{\ln x_n}{\ln D}\frac{1}{n+1} \geq \ln \alpha_n + \frac{1}{n+1}.$$

Furthermore, since $n + 1 \geq \ln D / \ln \gamma_m^*$ by Lemma 3 and $\gamma_m^* = \Theta(m^2 \log m)$ by Theorem 3, we obtain

$$\frac{n-2m}{n+1} = 1 - O\left(\frac{m \log m}{\ln D}\right).$$

Now

$$\ln O(m^2 \ln D)^{2(m^2-1)/n} = \frac{2(m^2-1)}{n}\ln O(m^2 \ln D)$$
$$\leq \frac{2(m^2-1)\ln \gamma_m^*}{\ln D - \ln \gamma_m^*}\ln O(m^2 \ln D)$$
$$= O\left(\frac{m^2 \log m \ln \ln D}{\ln D}\right).$$

If D is large enough, the above expression is close to zero and, thus,

$$\ln O(m^2 \ln D)^{2(m^2-1)/n} \leq 2\ln\left(1 + O\left(\frac{m^2 \log m \ln \ln D}{\ln D}\right)\right).$$

This implies

$$\frac{1}{O(m^2 \ln D)^{2(m^2-1)/n}} \geq \left(1 - O\left(\frac{m^2 \log m \ln \ln D}{\ln D}\right)\right),$$

for large enough D, and the claim follows.   □

4   The Asymptotic Competitive Ratio

In this section we investigate the growth rate of the competitive ratio for randomized search strategies on m rays. As Table 1 shows it seems that the competitive ratio grows linearly with the number of rays. In the following we are going to compute the exact constant of proportionality. Knowing this constant allows us to make fairly accurate estimates for the competitive ratio of randomized searching for arbitrary values of m. It also provides a basis for assessing the maximal gain of a randomized strategy over the best deterministic strategy. As a side result we also obtain an estimate for the step length.

m	Opt. rand. step length	Rand. comp. ratio c_m^*	$\dfrac{c_m^*-1}{2m}$	Opt. det. step length	Det. comp. ratio d_m^*	$\dfrac{c_m^*-1}{d_m^*-1}$
2	3.591	4.591	0.898	2.0	9.0	0.449
3	2.011	7.732	1.122	1.5	14.5	0.499
4	1.622	10.841	1.23	1.333	19.963	0.519
5	1.448	13.942	1.294	1.25	25.414	0.53
6	1.350	17.037	1.336	1.167	30.860	0.537
7	1.287	20.130	1.366	1.143	36.302	0.542
10	1.186	29.404	1.42	1.111	52.623	0.55
20	1.086	60.296	1.482	1.053	107.001	0.559
50	1.033	152.95	1.519	1.02	270.105	0.565
100	1.016	307.365	1.532	1.01	541.936	0.566

Table 1. A comparison of the optimal randomized and deterministic strategy to search on m rays.

In the deterministic case it is easy to see that the constant of proportionality is $2e$ since we have an explicit solution. The competitive ratio is given by

$$1 + 2\frac{m^m}{(m-1)^{m-1}} = 1 + 2m\left(1 + \frac{1}{m-1}\right)^{m-1} \longrightarrow 1 + 2em.$$

Also the step length is given explicitly by $1 + 1/(m-1)$.

In the randomized case the step length a_m^* is given implicitly as the value $a > 1$ that minimizes the expression

$$\frac{a^m - 1}{(a-1)\ln a}$$

[10]; once having computed a_m^* the competitive ratio c_m^* is obtained by

$$c_m^* = 1 + 2\frac{(a_m^*)^m - 1}{m(a_m^* - 1)\ln a_m^*}.$$

It is not clear how c_m^* behaves as m goes to infinity. In the following we make use of the Lambert function W which is given by the functional equation $W(x)e^{W(x)} = x$.

Theorem 3. *The optimal step length of a randomized strategy to search on m rays is approximately $1 + \alpha/m$, where $\alpha = 2 + W(-2e^{-2}) \sim 1.59$ minimizes the function $(e^x - 1)/x^2$ and the optimal competitive ratio converges to*

$$1 + 2\frac{e^\alpha - 1}{\alpha^2}m \sim 1 + 2 \cdot 1.544\,m$$

as m goes to infinity. Moreover, for every $m \geq 2$, $1 + 2 \cdot 1.545\,m$ is an upper bound on the competitive ratio c_m^*.

Proof: In order to prove the claim we write a_m^* as $a_m^* = 1 + \alpha_m/m$ and consider the function

$$f_m(x) = 1 + 2\frac{(1 + x/m)^m - 1}{m((1 + x/m) - 1)\ln(1 + x/m)} = 1 + 2\frac{(1 + x/m)^m - 1}{xm\ln(1 + x/m)}.$$

Since $(1 + x/m)^m$ converges uniformly to e^x and $m\ln(1 + x/m)$ converges uniformly to x on the interval $[1,2]$ as m goes to infinity, we obtain that f_m converges uniformly to

$$f_\infty(x) = 1 + 2\frac{e^x - 1}{x^2}$$

on the interval $[1,2]$. The unique minimum α of f_∞ is given by the equation

$$(2 - \alpha)e^\alpha - 2 = 0$$

and we obtain $\alpha = 2 + W(-2e^{-2}) \sim 1.59$. Since $\alpha \in (1,2)$ and f_m converges uniformly to f_∞ on $[1,2]$, there is an $m_0 \geq 1$ such f_m has a minimum in $(1,2)$, for all $m \geq m_0$. Since f_m has exactly one minimum α_m, α_m converges to α as m goes to infinity.

In order to see that $1 + 2 \cdot 1.545\,m$ is an upper bound on the competitive ratio c_m^* we observe that the derivative of f_m w.r.t. m is positive. Hence, $f_{m+1}(x) \geq f_m(x)$, for $x \geq 1$ and, therefore, the minimum of $f_{m+1}(x)$ is at least as large as the minimum of $f_m(x)$. This implies that the limit of the minima is an upper bound.   □

The above theorem implies that the time needed by the optimal randomized strategy is only $(e^\alpha - 1)/\alpha^2/e \sim 0.568$ times the time needed by the optimal deterministic strategy to find the goal on worst case inputs as m goes to infinity.

5 Conclusions

We present a new proof that

$$c_m^* = 1 + \frac{2}{m} \min_{a>1} \frac{a^m - 1}{(a-1)\ln a}$$

is a lower bound for randomized search strategies on m rays. We also give a lower bound on the convergence rate of the competitive ratio to c_m^* if an upper D on the distance to the goal is known. Finally, we compute the constant of proportionality for c_m^* and show that c_m^* converges to $1 + 2(e^\alpha - 1)/\alpha^2 m \sim 1 + 2 \cdot 1.544\, m$ for large m where α minimizes the function $(e^x - 1)/x^2$.

References

1. R. Baeza-Yates, J. Culberson, and G. Rawlins. Searching in the plane. *Information and Computation*, 106:234–252, 1993.
2. Margrit Betke, Ronald L. Rivest, and Mona Singh. Piecemeal learning of an unknown environment. In *Sixth ACM Conference on Computational Learning Theory (COLT 93)*, pages 277–286, July 1993.
3. K-F. Chan and T. W. Lam. An on-line algorithm for navigating in an unknown environment. *International Journal of Computational Geometry & Applications*, 3:227–244, 1993.
4. A. Datta, Ch. Hipke, and S. Schuierer. Competitive searching in polygons—beyond generalized streets. In *Proc. Sixth Annual International Symposium on Algorithms and Computation*, pages 32–41. LNCS 1004, 1995.
5. A. Datta and Ch. Icking. Competitive searching in a generalized street. In *Proc. 10th Annu. ACM Sympos. Comput. Geom.*, pages 175–182, 1994.
6. A. Datta and Ch. Icking. Competitive searching in a generalized street. *Comput. Geom. Theory Appl*, 13:109–120, 1999.
7. S. Gal. *Search Games*. Academic Press, 1980.
8. Christian Icking and Rolf Klein. Searching for the kernel of a polygon: A competitive strategy. In *Proc. 11th Annu. ACM Sympos. Comput. Geom.*, pages 258–266, 1995.
9. M. Y. Kao, Y. Ma, M. Sipser, and Y. Yin. Optimal constructions of hybrid algorithms. *J. of Algorithms*, 29:142–164, 1998.
10. M. Y. Kao, J. H. Reif, and S. R. Tate. Searching in an unknown environment: An optimal randomized algorithm for the cow-path problem. *Information and Computation*, 131(1):63–80, 1997.
11. R. Klein. Walking an unknown street with bounded detour. *Comput. Geom. Theory Appl.*, 1:325–351, 1992.
12. J. M. Kleinberg. On-line search in a simple polygon. In *Proc. of 5th ACM-SIAM Symp. on Discrete Algorithms*, pages 8–15, 1994.
13. A. López-Ortiz and S. Schuierer. Position-independent near optimal searching and on-line recognition in star polygons. In *Proc. 4th Workshop on Algorithms and Data Structures*, pages 284–296. LNCS, 1997.
14. A. López-Ortiz und S. Schuierer. Lower bounds for streets and generalized streets. *Intl. Journal of Computational Geometry & Applications*, 11(4):401–422, 2001.
15. C. H. Papadimitriou and M. Yannakakis. Shortest paths without a map. In *Proc. 16th Internat. Colloq. Automata Lang. Program.*, volume 372 of *Lecture Notes in Computer Science*, pages 610–620. Springer-Verlag, 1989.

16. S. Schuierer. Efficient robot self-localization in simple polygons. In H. Christensen, H. Bunke, and H. Noltemeier, editors, *Sensor Based Intelligent Robots*, volume 1724, pages 220–239. LNAI, 1999.

17. D. D. Sleator and R. E. Tarjan. Amortized efficiency of list update and paging rules. *Communications of the ACM*, 28:202–208, 1985.

18. A. Yao. Probabilistic computations: Towards a unified measure of complexity. In *Proc. 18th IEEE Symp. on Foundations of Comp. Sci.*, pages 222–227, 1977.

Single and Bulk Updates in Stratified Trees: An Amortized and Worst-Case Analysis

Eljas Soisalon-Soininen[1] and Peter Widmayer[2]

[1] Department of Computer Science and Engineering, Helsinki University of Technology, P.O.Box 5400, FIN-02015 HUT, Finland
ess@cs.hut.fi
[2] Institut für Theoretische Informatik, ETH Zentrum/CLW, CH-8092 Zürich, Switzerland
widmayer@inf.ethz.ch

Abstract. Stratified trees form a family of classes of search trees of special interest because of their generality: they include symmetric binary B-trees, half-balanced trees, and red-black trees, among others. Moreover, stratified trees can be used as a basis for relaxed rebalancing in a very elegant way. The purpose of this paper is to study the rebalancing cost of stratified trees after update operations. The operations considered are the usual insert and delete operations and also bulk insertion, in which a number of keys are inserted into the same place in the tree. Our results indicate that when insertions, deletions, and bulk insertions are applied in an arbitrary order, the amortized rebalancing cost for single insertions and deletions is constant, and for bulk insertions $O(\log m)$, where m is the size of the bulk. The latter is also a bound on the structural changes due to a bulk insertion in the worst case.

Prologue

Search tree rebalancing technology is one of the cornerstones of the information society. Without it, no fast web searches would be possible, and no database system could operate efficiently. Ever since their inception in 1962 [1], balanced search trees have inspired researchers to make them ever faster and more powerful. This paper is devoted to Thomas Ottmann, a world champion in search tree rebalancing, on the occasion of his 60th birthday. He invented and co-invented a number of classes of balanced search trees with a variety of desirable features, such as brother trees early in his career (it goes without saying that today he would call them sibling trees) and stratified trees later on. The latter enjoy the fine property that insert and delete operations have only a constant cost for link changes. The authors of this paper met for the first time when the first author was a visiting scholar and the second author a doctoral student in Thomas Ottmann's innovative balancing studio. They have continued their cooperation ever since, and for this occasion joined forces to point out the utility of stratified trees—for the theory of algorithms, for a number of applications, and for the professional paths of scientists.

1 Introduction

In the area of research on search trees there have been some initiatives to develop a general framework from which special tree classes can be drawn. The idea here is to

R. Klein et al. (Eds.): Comp. Sci. in Perspective (Ottmann Festschrift), LNCS 2598, pp. 278–292, 2003.
© Springer-Verlag Berlin Heidelberg 2003

define a general family of classes of trees, and provide it with access and update operations, so that certain important properties are achieved. One notable example is the class of stratified trees defined by Ottmann and Wood [18]. When defining this class Ottmann and Wood had in mind what the desirable properties of dynamic search trees are, apart from the obligatory logarithmic access cost. On the basis of the need for trees that can be restructured (or rebalanced) by only a constant number of link changes (this is important when binary trees have adjunct secondary structures such as in priority-search trees [4], segment-range trees [22], and persistent search trees [21]) the idea was to find out the reasons for the constant linkage cost and define a most general framework for binary trees with this property. This framework of stratified trees created by Ottmann and Wood [18] is related to both the stratified trees of van Leeuwen and Overmars [11] and the dichromatic trees of Guibas and Sedgewick [5]. Designing update operations with constant linkage cost for stratified trees implies such operations for any class included in the family. These are, among others, symmetric binary B-trees [2], half-balanced trees [15], red-black trees [5], and red-h-black trees [6]. In the context of self-organizing binary search trees [23] constant linkage cost has been studied by Lai and Wood [9].

Constant linkage cost is also important when search trees are used in highly concurrent systems. This is because link changes require exclusive locking of nodes in the tree thus preventing search operations from going through. It should be noted that other types of changes, such as colour flips or changes in the balance factor, do not require the prevention of searches because such changes cannot cause a search to choose a wrong way. Notably, stratified trees have been shown ([16; 17]) to work well and extremely simply in the context of relaxed balancing designed for improving concurrency in search trees (see e.g. [14]). In [16; 17] it is shown that relaxed stratified trees also have a constant linkage cost.

Relaxed rebalancing is a valuable concept for applications in which a large number of updates has to be processed rapidly, and there is more time to rebalance the tree later. Whenever the updates are confined to a small part of the tree, relaxed rebalancing may even profit from the chance that rebalancing requirements from several updates may cancel each other [16; 17]. Furthermore, if all keys in a large number of insertions (or at least a large fraction) happen to fall into the same location in the search tree (a bulk insertion), one should exploit this locality so as to perform rebalancing more efficiently. The complexity of bulk insertions has been studied e.g. in [12; 10; 19]. Applications of bulk insertions vary from document databases [19; 20] and data warehousing [7] to real-time databases [8].

In this paper we will propose an algorithm for bulk insertion into stratified trees and analyze its worst case number of pointer changes (cf. Lemmas 3 and 4). Our result is that for a bulk of size m there are at most $O(\log m)$ pointer changes, in contrast to $O(m)$, which would be needed if the keys in the bulk were inserted individually. It should be noted that it is indeed of importance to have a decrease in the number of pointer changes, even though in the process of bulk insertion a bulk tree is constructed, which takes linear time. Pointer changes are not local operations, as are the steps in constructing the bulk tree, and may thus cause cache misses, or, if the search tree is stored externally, disk accesses. In applications it is often necessary to construct several bulk trees from a sequence of keys arrived at as a burst [12; 19], but the pointer changes remain the most important non-local operations.

For stratified trees with constant linkage cost, it is interesting to have a closer look at the total cost of rebalancing. The worst case cost is $O(\log n)$, where n is the size of the tree, because changes in strata must perhaps be made up to the root, but what about the amortized cost? There exist amortized results for specific tree classes. For (a,b)-trees with $b \geq 2a$ [13] and for red-black trees [3] (p. 428) the amortized cost of rebalancing is constant for mixed insertions and deletions. Even for bulk insertions, $O(\log m)$ amortized cost algorithms for (a,b)-trees are available [10; 19], when bulk insertions appear in arbitrary order together with singleton insertions and deletions. We will show in the present paper that for stratified trees similar amortized results are obtained (cf. Lemmas 8, 9, and 10). Altogether, we will prove the following theorem:

Theorem 1. *For the insertion of a bulk tree of m keys into a given stratified tree of n keys, the total number of pointer changes is bounded from above in the worst case by $O(min\{\log m, \log n\})$. The bulk tree can be constructed in time $O(m)$. Over an arbitrarily mixed sequence of insertions, deletions and bulk insertions, starting with an initially empty stratified tree, each insertion and each deletion has only a constant amortized cost, and each bulk insertion has an amortized cost bounded by $O(min\{\log m, \log n\})$.*

2   Stratified Trees

Roughly speaking, stratified trees consist of component trees that are arranged in layers (strata). For the purpose of simplicity of explanation, we have chosen a particular class of stratified trees, although our approach is not limited to this class, either in terms of the bulk insertion operation or in terms of the potential function for the amortized analysis. Similarly, for the sake of presentation we have limited ourselves to the case of leaf search trees, where a router at an inner node stores the maximum key in its left subtree, but our approach also works for trees where the keys themselves are stored at the internal nodes, as well as for other routing schemes. The class of stratified trees that we have chosen is structurally the same as the symmetric binary B-trees [2]; we define it inductively as follows, based on the set of component trees shown in Fig. 1:

(1) Any single one of the component trees is a stratified tree.

(2) Given a stratified tree, replace each leaf by a component tree; the result is a stratified tree.

(3) Nothing else is a stratified tree.

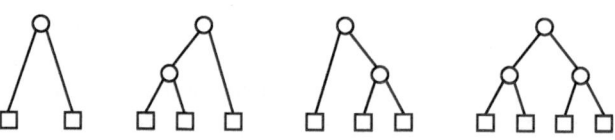

Figure 1. The set of component trees.

We call a component with i leaves a degree-i-component or i-component for short, for $i = 2, 3, 4$. The partition of the tree into layers is significant; we will therefore take it into account explicitly by colouring the root of each component black and all other nodes white (suitable for implementation). Alternatively (perhaps better for viewing), we can draw a horizontal line just above the component roots and above the leaves. For an example of a stratified tree marked in both ways, see Fig. 2.

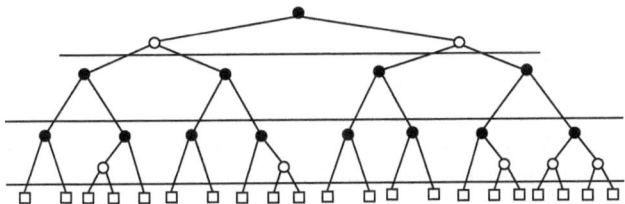

Figure 2. A stratified tree, with component roots marked black, and layers represented by horizontal lines.

2.1 Insertion

In the preparation of the bulk insertion operation and the amortized analysis, let us briefly recall the insertion of an individual key [16]. First, a search operation starting at the root of the stratified tree locates a leaf node, say q (see Fig. 3).

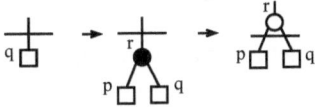

Figure 3. Insertion of a single key.

Leaf q is replaced by a new interior node r with two children, q and a new leaf p containing the new key. Since the resulting tree is no longer a stratified tree, a structural repair process starts at r and propagates upwards in the tree. First, node r is *pushed up* one layer, by simply changing the layer boundary. Note that this does not involve a pointer change. The temporary tree to which r belongs now has within its layer 3, 4, or 5 leaves (see Fig. 4).

It is not necessarily a legitimate component, but in the case of 3 or 4 leaves it can easily be changed into such a component, with at most 3 pointer changes. In this case, the insertion terminates. If, however, the temporary tree has 5 leaves, its root node t is pushed up to the next layer above. Note that in this case, t changes its colour from black to white, but there is no pointer change. If the push-up occurs at the root layer, t becomes a 2-component and causes the tree to grow by adding a new layer at the top.

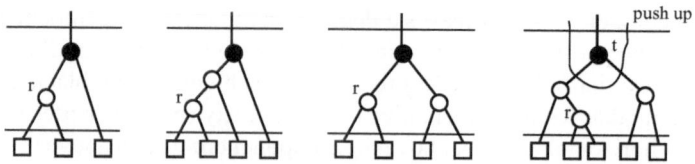

Figure 4. The effect of a push-up on the next higher layer.

The analysis of the insertion process has been given in the above considerations and is expressed in the following Lemma, due to [16].

Lemma 1. *The number of pointer changes for the insertion of a single key into a stratified tree is bounded from above by a constant in the worst case.*

2.2 Deletion

The deletion of an individual leaf proceeds in a similar manner. After the search for the leaf, p, to be deleted, the component above p is inspected. In the case that it is not a 2-component before the deletion, a local change within the component will repair it with a constant number of link changes, and the deletion terminates. If, however, it is a 2-component before the deletion, it will lose its only interior node r (see Fig. 5).

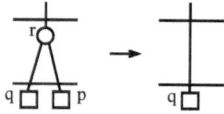

Figure 5. Deletion of a single key: the critical case.

A first repair attempt tries to find a node in the neighbouring child component with root s of r's parent component with root t to create at least a 2-component as parent of q (see Fig. 6).

Figure 6. Find a spare node among the siblings of the deleted node.

Note that if this attempt succeeds, the deletion terminates and uses only a constant number of pointer changes. If this fails, however, the second attempt tries to rearrange the nodes of all sibling components of r and of its parent component as much as necessary in order to get a stratified tree. Again, if this succeeds, the deletion terminates with a constant number of pointer changes. Otherwise, the parent component of r is a 2-component, and so is its only sibling (see Fig. 7).

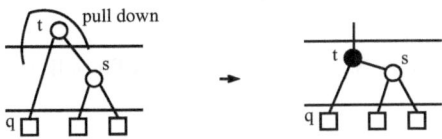

Figure 7. The initiation of a pull-down operation.

Now, a *pull-down* operation moves t to the parent layer of q and propagates the deletion problem one layer up in the tree. Note that in this case there are no pointer changes.

The analysis of the deletion process has been given in the above considerations and is expressed in the following Lemma, due to [16].

Lemma 2. *The number of pointer changes for the deletion of a single key from a stratified tree is bounded from above by a constant in the worst case.*

3   Bulk Insertion

We are interested in the insertion of an entire (possibly large) set of keys (the *bulk*) all together. We assume that all the keys in the bulk fall in between two adjacent keys in the given search tree T, and we call this a *bulk insertion*. Roughly speaking, a bulk insertion consists of two phases. In the first phase, we construct a suitable, fairly specific stratified tree from the bulk of keys, the *bulk tree B*. In the second phase, we insert the bulk tree B into the given tree T by cutting T at the position where B belongs. The difficulty is now that whenever we cut a link within T for this purpose, we must attach a dangling subtree of T elsewhere. Similarly, a component of T that has lost a subtree by a cut should get a replacement. To show how this can be done, let us first describe the specific bulk trees that we construct. Then we will show how to insert the bulk tree into the given tree.

3.1   Bulk Tree Construction

To prepare for the insertion of the bulk into a given stratified tree, we build the bulk tree as a very special stratified tree. The reason why the bulk tree must be of this specific form will become clear from the method of bulk insertion. The general shape of a bulk tree is shown in Fig. 8.

The root layer is a 3-component. Any other layer consists of a 2-component on each of its two boundaries, and a 4-component adjacent to it. In between, the components may differ from layer to layer. For the layer below the root layer, both "adjacent 4-components" coincide, i.e., we have a 2-component, a 4-component, and a

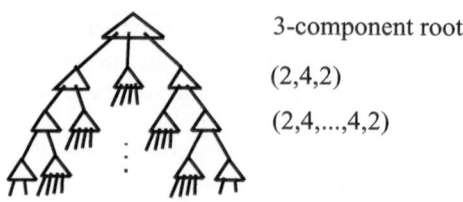

3-component root

(2,4,2)

(2,4,...,4,2)

Figure 8. The general shape of a bulk tree.

2-component, from left to right. In addition to these requirements on the boundary, we allow at most a constant number of 2-components and 4-components on each layer of a bulk tree, with the possible exception of the first six layers where the in-between components are allowed to be arbitrary. The question is now whether this limited class of stratified trees allows for an arbitrary number of leaves, for a sufficiently large number. To make things simpler, let us limit the freedom in choosing the degree of components to the four nodes that are children of the 4-components adjacent to the boundary, for each layer. All other components must be 3-components. This immediately implies that below a layer of z components, for any $z \geq 12$, the next layer must have between $3z - 8$ and $3z + 8$ components. Hence it is sufficient to allow for an "initial" set of bulk trees whose numbers of leaves span an interval from k to at least $3k - 7$, for some value k, since the next higher values of $3k - 8$ and more can be achieved on the next layer of the tree, and therefore all numbers of leaves will be achievable. For this initial set of bulk trees, an easy calculation shows that under the given conditions, the number of nodes on the successive layers (starting with the root node) is 1, 3, 8, 20..28, 44..108, and 92..428, where $i..j$ denotes any value in the interval from i to j, depending on the choice of components' degrees. The last of these intervals is the first in the sequence that fits our needs, and hence all numbers of leaves from 92 up can be realized according to our strict rule. Altogether, all numbers of leaves from 44 up can be realized, since 44..108 overlaps 92..428, and smaller numbers are treated separately at constant cost.

The result of the construction can be expressed in the following Lemma.

Lemma 3. *For a given bulk of m keys in sorted order, a bulk tree of height $O(\log m)$ can be constructed in time $O(m)$.*

3.2   Bulk Tree Insertion

The bulk tree insertion progresses in layers, from bottom to top. The root layers of the given tree T or the bulk tree B will be treated differently and described in a moment. First, we will focus on intermediate (non-root) layers. The bottom layers of B and T are aligned so as to be the same. Starting with both bottom layers (i.e., the layers that yield the new bottom layer), within a layer the left and right boundary component trees of B are considered in connection with their adjacent components of T. If B does not cut a link of T, the insertion at the current layer is complete, and the process continues on the next layer up (see Fig. 9).

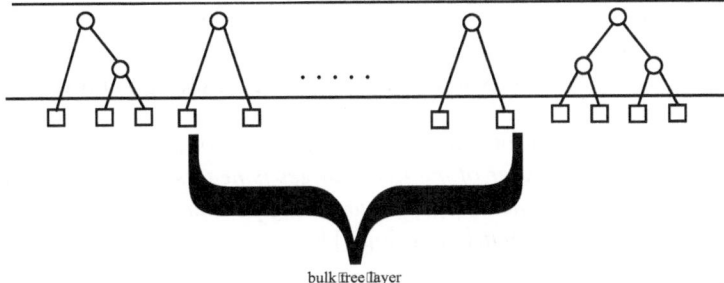

Figure 9. Bulk tree insertion: no cut.

If, however, the bulk tree layer cuts a link (or maybe more than one), pointer changes become necessary to preserve the search tree property. Fig. 10 illustrates by means of dashed vertical lines the possible positions in which a component of T can be cut.

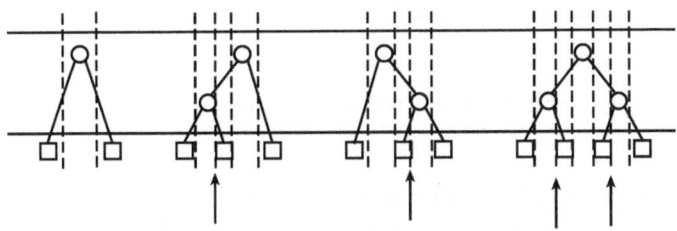

Figure 10. Possible cut positions in components.

Positions that are marked with an arrow indicate that not just one, but two links are cut. In these cases, a local pointer change as indicated in Fig. 11 will lead to the situation that only one link within a component layer needs to be taken care of: It is a dangling link and can be viewed as one (low) half of a cut link, with the other half non-existent.

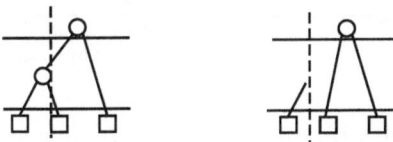

Figure 11. Two cuts lead to one dangling link.

Let us now look at a cut in an intermediate layer in detail, and then consider the situation at a root layer where the bulk insertion will terminate. The subsequent observations, as exposed in Lemmas 5 and 6, will prove the following Lemma:

Lemma 4. *For the insertion of a bulk tree of m keys into a given stratified tree of n keys, at most a constant number of pointer changes is necessary for each layer that is present in both trees, and hence the total number of pointer changes is bounded from above in the worst case by $O(min\{\log m, \log n\})$.*

Cuts in Intermediate Layers

In an intermediate layer, i.e. neither the root layer of T nor of B (see Fig. 10 with the modification of Fig. 11), we are left with one cut link per component, either with both halves present or the low half only. The low half of a link may lead directly to the next layer down (or a leaf), or it may lead to a single node on the current layer with two links down (see Fig. 12).

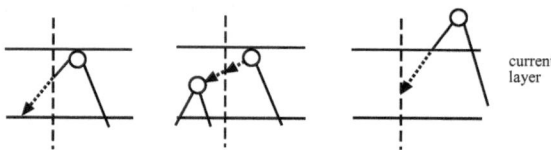

Figure 12. Two low and high half-links.

Similarly, the high half of a link can be viewed as coming from the next layer up or from a single node on the current layer (again, see Fig. 12). It can be seen from Fig. 10 and Fig. 11 that this covers all cases. Not surprisingly, we constructed B in such a way that these cases can easily be taken care of. We describe this separately for low half links and high half links.

Low Half Links A low half link, leading either to no node on the current layer or to one node, can be simply attached to the boundary 2-component of B, as shown in Fig. 13 for the left boundary of B, with a constant number of pointer changes.

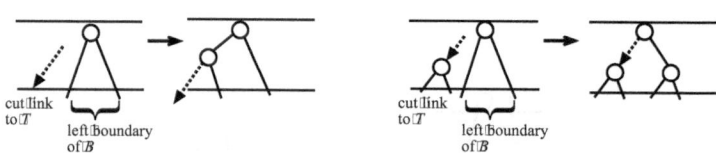

Figure 13. Attach cut link at B's boundary.

High Half Links A high half link, coming either from the next layer up or from a single node on the current layer, can get the 2-component from B's boundary as its new child (see Fig. 14 for the right boundary of B).

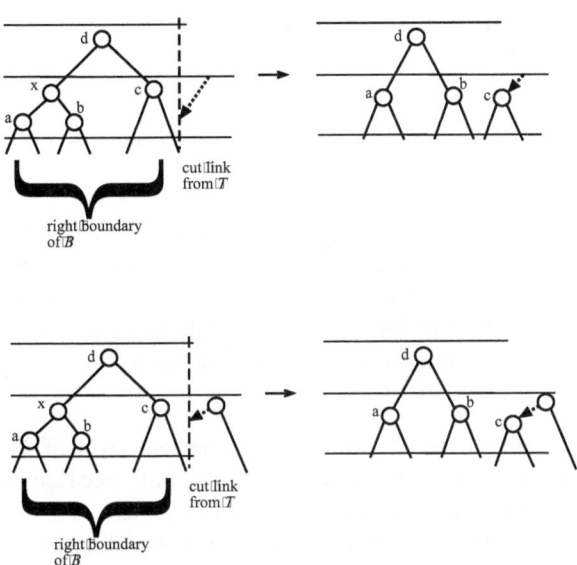

Figure 14. Attach B's boundary component to T's link and restructure within B (node x disappears).

This cures the situation for T's link, but makes a change within B necessary. The parent of the child that was lost for B gets a new child from the 4-component adjacent to the lost child in B (again, see Fig. 14). Again, we only need a constant number of pointer changes within the layer.

We summarize this in the following Lemma:

Lemma 5. *The insertion of an intermediate (non-root) bulk tree layer into an intermediate (non-root) given stratified tree layer can be handled with a constant number of pointer changes.*

The Root Layer

For the root layer, we distinguish whether the root of T is on a lower layer than the root of B. Both cases are simpler than the intermediate layer considerations.

The Root of T Is Lower In the case where the root of T is on a lower layer than the root of B, we progress up to and including that layer, treating it just like an intermediate layer. As a result, the root of T will be adjacent to the boundary of B, with no cut link (see Fig. 15).

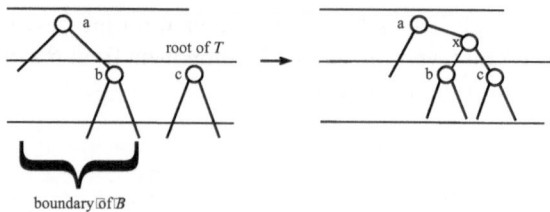

Figure 15. The root of T is attached to B (node x appears).

Now, the root of T is attached as a new child to a new interior node created on the next layer up from B. This next layer up from B exists, by the assumption that the root of T is lower than the root of B, but could be the root layer of B. Since the root component of B is a 3-component, the attachment will still be possible locally. This completes the bulk insertion with a constant number of pointer changes.

The Root of T Is Not Lower   In this case, at the root layer of B, we cannot apply the procedure for intermediate layers to cure a cut link, because we only have a 3-component for B at hand. Instead, we choose a component of T with which we will combine the root component R of B in a different way (see Fig. 16).

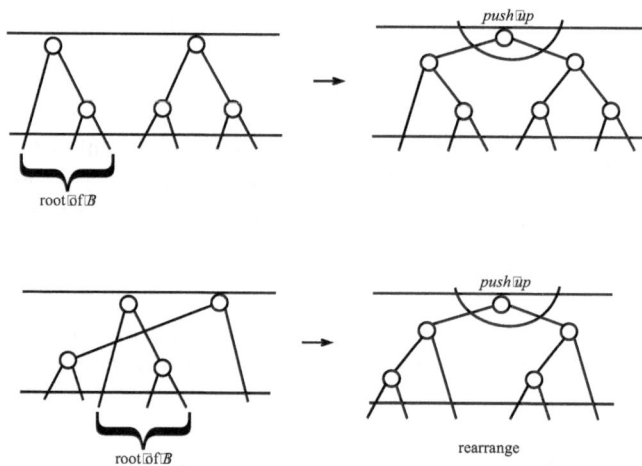

Figure 16. Combine the root layer of B with T.

The component of T is chosen as follows. If R cuts a link of a component C of T, then C is chosen, and otherwise a component of T adjacent to R is chosen. Now, together R and the chosen component of T have between 5 and 7 links to the next layer down. These links are rearranged into one root node r and two legitimate components within a layer; if there is no cut link of T, the rearrangement is minimal, since we merely add a new root and attach both components. In any case, a constant number of pointer changes suffices. Then, r initiates a push-up operation exactly as in a single insertion. If both trees, T and B have the same number of layers, the push-up will immediately cause the tree to grow by one layer and terminate. Otherwise, it will progress upwards exactly like a single insertion and hence lead to at most another constant number of pointer changes.

We summarize the effect of this restructuring in the following Lemma:

Lemma 6. *The insertion of a bulk tree layer into a given stratified tree layer, where at least for one of both trees, the layer is the root layer, can be handled with a constant number of pointer changes, including the possible propagation effect of a push-up operation.*

4   Amortized Complexity Analysis

We have shown in the previous section that our bulk insertion algorithm for stratified trees needs no more than $O(\log m)$ pointer changes in the worst case for a bulk of size m. In addition to pointer changes, however, the modification of layer information (colour changes of nodes) should also be accounted for. In the worst case, we cannot avoid the fact that colour changes progress up to the root of the given tree. That is the reason for our interest in an amortized analysis. We will show the following lemma by proving lemmas 8, 9, and 10.

Lemma 7. *Over an arbitrarily mixed sequence of insertions, deletions and bulk insertions, starting with an initially empty stratified tree, each insertion and each deletion has only a constant amortized cost, and each bulk insertion has only a logarithmic amortized cost.*

We show our amortized bounds by means of a potential function (or, equivalently, bank account balance), exactly in the same way as Tarjan [24] and Mehlhorn [13] pioneered the amortized analysis. Let the number of i-components in a stratified tree T be n_i, for $i = 2, 3, 4$. Then the potential $p(T)$ of T is defined as $p(T) = n_2 + 2n_4$. The amortized cost a_j of the j-th operation in a sequence is its true cost t_j for performing the steps of the operation, plus the difference in potential it induces. In order for T to be understood, let p_j denote the potential of T after the j-th operation, and in particular, let p_0 be the initial potential, with $p_0 = 0$ by adding the empty tree to the class of stratified trees for the sake of this analysis. Hence, more formally, we have $a_j = t_j + p_j - p_{j-1}$. Summing up across a sequence of k operations, we get $\sum_{j=1}^{k} a_j = \sum_{j=1}^{k} t_j + p_k - p_0$, since all the intermediate p_j cancel. With the initial $p_0 = 0$ and the invariant that always $p_j \geq 0$ by definition, the sum of all amortized costs is an upper bound on the sum of all true costs.

Let us now check the change in potential and the true costs for each operation in turn, step by step.

Amortized Insertion Cost

In the first step, the insertion increases the potential by 1, due to the creation of a 2-component on the leaf level, and triggers a push-up. Let h be the number of layers that the push-up climbs, until it finally stops. The true cost for the insertion is then $h + 1$, if we measure the cost of climbing and changing colour within a layer as unity (we consider h layers plus the leaf layer). How does the potential change? To see this, let us study what happens to the tree structure. A push-up can only continue to the next layer up when it enters from below into a 4-component. It turns this 4-component into a 2-component and a 3-component (recall Fig. 4). Therefore, each individual push-up decreases the potential by 1, totalling to a potential decrease of h along the push-up path. In total, we get an amortized cost of $h + 1 - h$ for a single insertion, i.e., an amortized cost of 1. For easier future reference, we summarize this in the following Lemma:

Lemma 8. *The insertion cost of a single key into a given stratified tree is amortized constant.*

Amortized Deletion Cost

Let us be rather more brief with regard to the amortized analysis of deletions. Whenever a deletion terminates on a layer, the true cost on that layer can only be constant, and the change in potential can only be constant. When a deletion propagates up to the next layer by means of a pull-down operation, a 2-component turns into a 3-component, and hence the potential decreases by 1. On the next layer up, the pull-down can create a 2-component from a 3-component, but if it does, the operation terminates. Therefore, if h is the number of layers that the pull-down climbs, and unity is the cost of changing colour and climbing within a layer, then the amortized cost of a deletion is $h + 1 - h$, which is again 1. Summarizing, we get the following Lemma:

Lemma 9. *The deletion cost of a single key from a given stratified tree is amortized constant.*

Amortized Bulk Insertion Cost

In a bulk insertion, there are three parts that together determine the potential of the resulting tree, namely the potential of the given tree, the potential of the bulk tree, and the potential changes that processing on the boundary of the bulk tree induces. In calculating the difference in potential before and after the bulk insertion, the potential of the given tree cancels.

For the bulk tree, the potential is $O(\log m)$, where m is the number of keys in the bulk. This upper bound on the potential is guaranteed by construction: On each of the $O(\log m)$ layers of the bulk tree, we allow for at most a constant number of 2-components and 4-components (all others must be 3-components), with the exception of a constant number of layers at the top of the bulk tree which together can have a constant extra number of 2- and 4-components.

It remains for us to show that restructuring at the boundary increases the potential by at most $O(\log m)$ overall, but this is again easy to see. The reason is that on each layer and on each side of the bulk tree in a bulk insertion, combining the bulk tree and the given tree can increase the potential of the resulting tree at most by a constant. Hence, the total increase in potential in a bulk insertion is $O(\log m)$.

The true operational cost in a bulk insertion is at most constant on each layer, and hence the true total cost is $O(\log m)$ as well. This leads to the following Lemma:

Lemma 10. *The insertion cost for a bulk tree with m keys into a given stratified tree with n keys is amortized $O(min\{\log m, \log n\})$.*

5   Discussion and Conclusion

The stratified trees of Ottmann and Wood [18] form an interesting family of search tree classes, because they are defined so as to capture the constant linkage cost and present the essential features required to achieve this property. Our purpose in this paper was to prove other important properties related to the constant linkage cost of updates. We have shown that stratified trees have constant amortized cost for total rebalancing when insertions and deletions are applied in an arbitrary order into an initially empty stratified tree. This extends the previously-known results on (a, b)-trees and red-black trees to stratified trees. Note that for height-balanced trees such as AVL-trees a corresponding result is not true; the amortized constant rebalancing cost is obtained for insertions and deletions separately, not for mixed operations.

Dynamic bulk operations for search trees have been the focus of some recent interest, and we wanted to extend some old results to stratified trees and also obtain some new results. A result of the latter type states that if a bulk contains m keys to be inserted, then the total linkage cost is $O(\log m)$. A result of the former type is that bulk insertions have amortized cost $O(\log m)$ when applied in combination with singleton insertions and deletions. The same result has previously been proved for (a, b)-trees [19] and for relaxed (a, b)-trees [10], but in [10] for operations with linear worst case time.

References

1. G.M. Adel'son-Vel'skii and E.M. Landis. An algorithm for the organisation of information. *Dokl. Akad. Nauk SSSR* **146** (1962), 263–266 (in Russian); English Translation in *Soviet. Math.* **3**, 1259–1262.
2. R. Bayer: Symmetric binary B-trees: Data structure and maintenance algorithms. *Acta Informatica* **1** (1972), 290–306.
3. T.H. Cormen, C.E. Leiserson, R.L. Rivest, and C. Stein: *Introduction to Algorithms, Second Edition*, The MIT Press, Cambridge, Massachusetts, 2001.
4. E.M. McCreight: Priority search trees. *SIAM Journal on Computing* **14** (1985), 257–276.
5. L.J. Guibas and R. Sedgewick: A dichromatic framework for balanaced trees. In: *Proceedings of the 19th Annual IEEE Synposium on Foundations of Computer Science*, Ann Arbor, pp. 8–21, 1978.
6. Chr. Icking, R. Klein, and Th. Ottmann: Priority search trees in secondary memory. In: *Graphtheoretic Concepts in Computer Science (WG '87)*. Lecture Notes in Computer Science 314 (Springer-Verlag), pp. 84–93.
7. C. Jermaine, A. Datta, and E. Omiecinski. A novel index supporting high volume data warehouse insertion. In: *Proceedings of the 25th International Conference on Very Large Databases*. Morgan Kaufmann Publishers, 1999, pp. 235–246.

8. T.-W. Kuo, C-H. Wei, and K.-Y. Lam. Real-time data access control on B-tree index structures. In: *Proceedings of the 15th International Conference on Data Engineering*. IEEE Computer Society, 1999, pp. 458–467.
9. T.W. Lai and D. Wood: Adaptive heuristics for binary search trees and constant linkage cost. *SIAM Journal on Computing* **27**:6 (1998), 1564–1591.
10. K.S. Larsen. Relaxed multi-way trees with group updates. In: *Proceedings of the 20th ACM SIGMOD-SIGACT-SIGART Symposium on principles of Database Systems*. ACM Press, 2001, pp. 93–101. To appear in *Journal of Computer and System Sciences*.
11. J. van Leeuwen and M.H. Overmars: Stratified balanced search trees. *Acta Informatica* **18** (1983), 345–359.
12. L. Malmi and E. Soisalon-Soininen. Group updates for relaxed height-balanced trees. In: *Proceedings of the 18th ACM SIGMOD-SIGACT-SIGART Symposium on Principles of Database Systems*. ACM Press, 1999, pp. 358–367.
13. K. Mehlhorn: *Data Strucures and Algorithms, Vol. 1: Sorting and Searching*, Springer-Verlag, 1986.
14. O. Nurmi, E. Soisalon-Soininen, and D. Wood: Concurrency control in database structures with relaxed balance. In: *Proceedings of the Sixth ACM Symposium on Principles of Database Systems*, pp. 170–176, 1987.
15. H.J. Olivie: A new class of balanced search trees: Half-balanced search trees. *RAIRO Informatique Theorique* **16** (1982), 51–71.
16. Th. Ottmann, and E. Soisalon-Soininen: Relaxed balancing made simple. Institut für Informatik, Universität Freiburg, Technical Report 71, 1995.
17. Th. Ottmann and P. Widmayer: *Algorithmen und Datenstrukturen, 4. Auflage*, Spektrum-Verlag, Heidelberg, 2002.
18. Th. Ottmann and D. Wood: Updating binary trees with constant linkage cost. *International Journal of Foundations of Computer Science* **3**:4 (1992), 479–501.
19. K. Pollari-Malmi: Batch updates and concurrency control in in B-trees. Ph.D.Thesis, Helsinki University of Technology, Department of Computer Science and Engineering, Report A38/02, 2002.
20. K. Pollari-Malmi, E. Soisalon-Soininen, and T. Ylönen: Concurrency control in B-trees with batch updates. *IEEE Transactions on Knowledge and Data Engineering* **8** (1996), 975–984.
21. N. Sarnak and R.E. Tarjan: Planar point location using persistent search trees. *Communications of the ACM* **29** (1986), 669–679.
22. H.-W. Six and D. Wood: Counting and reporting insersections of *d*-ranges. *IEEE Transactions on Computers*, **C-31** (1982), 181–187.
23. D.D. Sleator and R.E. Tarjan: Self-adjusting binary search trees. *Journal of the ACM* **32** (1985), 652–686.
24. R.E. Tarjan. Amortized computational complexity. *SIAM Journal on Algebraic and Discrete Methods* **6** (1985), 306–318.

Meta-information for Multimedia eLearning

Ralf Steinmetz and Cornelia Seeberg

KOM, Darmstadt University of Technology, Merckstr. 25, 64283 Darmstadt, Germany
{Ralf.Steinmetz, Cornelia.Seeberg}@kom.tu-darmstadt.de
http://www.kom.e-technik.tu-darmstadt.de

Abstract. The generation of multimedia learning objects are expansive and time consuming. Only by reusing them in many contexts, the effort becomes worth while. Reusing requires a flexible layout. Therefore content and format should be separated. Means for searching relevant learning object are of highest importance. Metainformation has to be added to the learning object. Metadata and especially learning metadata are a powerful instrument. Complemented by an ontology and relations between the learning object, the learning material become findable. For supporting the authors, tools like a metadata editor are necessary.

1   Introduction

The advantages of multimedia learning are widely discussed:
- complex procedures and algorithm can be depicted nicely by animations,
- dangerous experiences can be shown in a video,
- huge or tiny objects of consideration can be "tamed" by simulations,
- expansive procedures can be exercised virtually before doing it in reality.
- the motivating aspect of good and colorful learning material improves the performance of the learners and so on.

The price of these features is a high effort to generate this kind of learning material. Not only expert knowledge, but also expertise to record videos or to handle applications for creating animations etc. is needed. Skills in graphical designing are as necessary as ability to deal with a image processing programs. Teamwork with multimedia experts has to be practiced. New pedagogical concepts have to be developed.
In this paper some technical approaches to solve the problem of high effort are presented.

2   Reusability

To compensate the above listed expenditures, the multimedia learning objects have to be used more than just once, either by the same author or by others. If the learning

R. Klein et al. (Eds.): Comp. Sci. in Perspective (Ottmann Festschrift), LNCS 2598, pp. 293-303, 2003.

objects can be reused, the extended effort at the generation becomes reasonable. There are several features of reusability:

- **Learning Scenario:** A learning object is normally used to teach the information for the first time. But it can be reused – probably slightly modified - for the repetition before exams or for looking up like in an encyclopedia. It can also serve as a test item either for self-controlling or for an exam.

Raw learning   Generated   Example:
object   learning objects

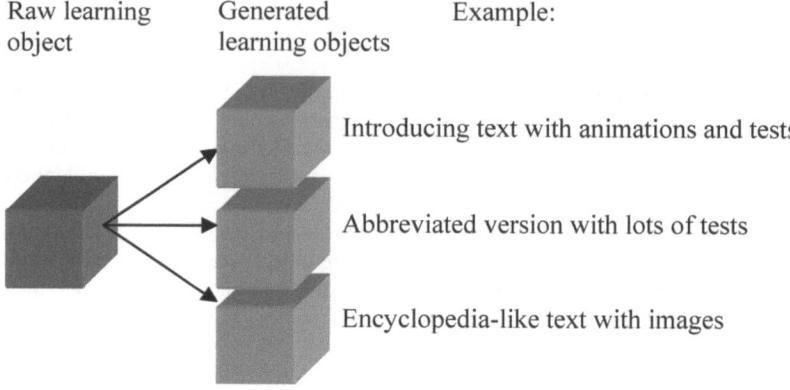

Introducing text with animations and tests

Abbreviated version with lots of tests

Encyclopedia-like text with images

Fig. 1. Generating several learning objects out of a raw learning object

- **Output Medium:** A learning object can be viewed via a computer or – if it is not of a dynamic format – as a print out.
- **Context:** Learning objects can serve in different contexts. A learning object can be an example in one course and an introduction in another.
- **Technical Aspects:** The learning objects should be optimally used in all technical environments, no matter which operating system the learner prefers, which internet connections and screen size are available.
- **Personal Preferences:** Users have different personal preferences for using an electronic document. They differ in the font size, the number of windows in use etc. The learning objects should meet any of such preferences.

Reusability means that learning object can be easily found and used by other authors and learners in different pedagogical contexts. To facilitate these aspects of reusability, some requirements have to be fulfilled. Two of them are discussed in the following sections. The requirement of a formal description of the learning objects is in more details discussed in chapter 3.

2.1 Separation of Content and Format

The separation of content and format as it is possible by storing the learning objects as XML files and presenting them via XSL, allows for different characteristics of a raw learning object. Personal preferences, output medium and technical aspects can be adapted to users. Also different layout preferences of the authors can be considered. A raw learning module stored as XML can be presented as a PDF print out with print layout in a 12 pt Times Roman font or as a HTML page with web layout in a 14 pt Arial font. This flexible environment provides for different presentation forms of the learning objects.

2.2 Parametrizeable Modules

One possibility to generate different content from one raw multimedia learning object such as a simulation or a movie, is to parameterize the learning object. That means that a simulation has a set of starting parameters from which the suitable one for the specific learning context can be chosen (see [7]). Or the relevant section of a movie or an audio file can be selected.

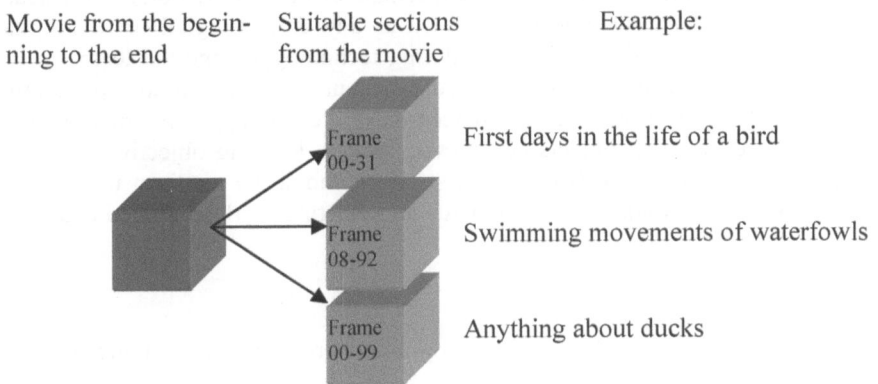

Movie from the beginning to the end   Suitable sections from the movie   Example:

Frame 00-31   First days in the life of a bird

Frame 08-92   Swimming movements of waterfowls

Frame 00-99   Anything about ducks

Fig. 2. Virtual learning objects by different starting parameters

So, from one raw learning object several virtual learning objects can be generated without much effort.

3 Metainformation

For an author to reuse his/her own learning objects or even learning objects of some other author, there has to be a powerful means for searching. Otherwise, the learning object cannot be found. Search engines usually offer thousands of hits to a user as a result of a query, if the keywords the user provided are very popular or generic. The problem is that it is not possible to describe the content of HTML pages, videos or animations in an adequate way. What is needed is information about information; also called metadata - labeling, cataloging and descriptive information structured in such a way that allows learning objects to be properly searched and processed [18]. With metadata users can describe much more accurately what kind of information they actually want to find.

3.1 Learning Object Metadata

One approach for metadata describing learning resources is the "Learning Object Metadata (LOM)" [5] scheme by the IEEE Working Group P1484.12. It is mainly influenced by the work of the IMS (Educom's Instructional Management Systems) [3] project and the ARIADNE Consortium (Alliance of Remote Instructional Authoring and Distribution Networks for Europe) [1]. The LOM scheme uses almost every category of the metadata scheme Dublin Core [2], which is used in the bibliographic world, and extends it with categories and attributes tailored to the need of learners and authors searching the web for material.

The LOM approach specifies the syntax and semantics of learning object's metadata. In this standard, a learning object is defined as any entity, digital or non-digital, which can be used, reused or referenced during technology-supported learning. Examples of learning objects include multimedia content, instructional content, instructional software and software tools, referenced during technology supported learning. In a wider sense, learning objects could even include learning objectives, persons, organizations, or events. The IEEE LOM standard should be conform to, integrate with, or reference to existing open standards and existing work in related areas (see [17]).

Purpose

In the LOM specification [13], the following points are mentioned among others as the purpose of this standard:

"To enable learners or instructors to search, evaluate, acquire, and utilize learning objects.

To enable the sharing and exchange of learning objects across any technology supported learning system.

To enable the development of learning objects in units that can be combined and decomposed in meaningful ways.

..."

The standard provides for extensions of the below listed categories. So, it is possible and LOM conform to add a category for parameters (see [7]). This way, to one physical raw learning object. may exist several metadata records with different starting parameters. And thus, another purpose of LOM can be to enable generating virtual learning object

Structure

The definition of LOM divides the descriptors of a learning object into nine categories:

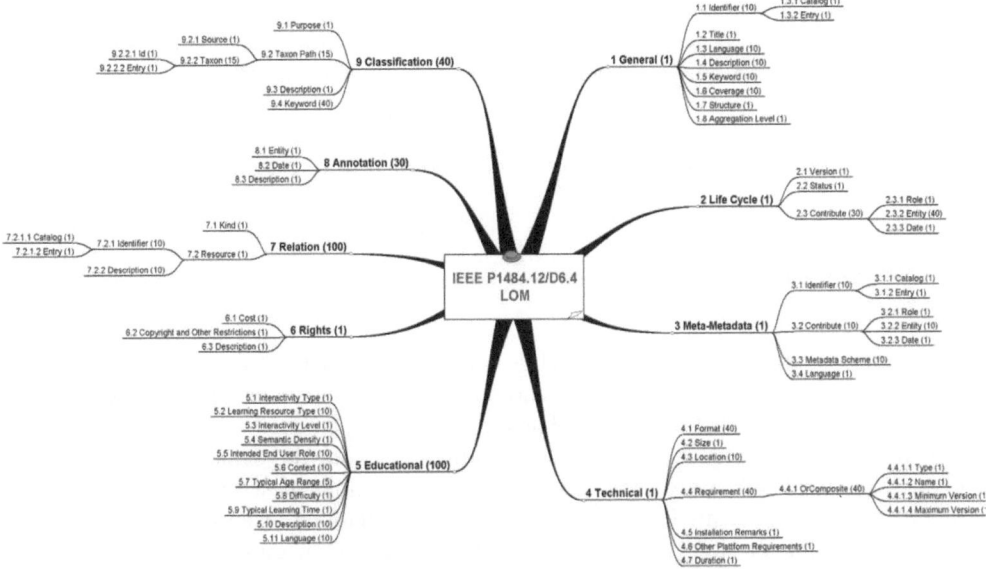

Fig. 3. The nine categories of LOM

Category 1: General, regroups all context-independent features of the resource.

Category 2: Lifecycle, regroups the features linked to the lifecycle of the resource.

Category 3: Meta-metadata, regroups the features of the description itself (rather than those of the resource being described).

Category 4: Technical, regroups the technical features of the resource.

Category 5: Educational, regroups the educational and pedagogic features of the resource.

Category 6: Rights, regroups the legal conditions of use of the resource.

Category 7: Relation, regroups features of the resource that link it to other resources.

Category 8: Annotation, allows for comments on the educational use of the resource.

Category 9: Classifications, allows for description of a characteristic of the resource by entries in classifications

Taken all together, these categories form what is called the "Base Scheme". Some elements like the description element of the general category allow free text as values, while for other elements the values are restricted to a limited vocabulary.

Following Dublin Core, all categories are optional in the LOM scheme. The reason for this is simple. If someone wants to use all categories and attributes from LOM, she/he has to fill out at least 60 fields. Entries like author, creation date or to some extent keywords can be filled automatically by an authoring system. But then there are still many entries left, which the author has to fill her-/himself. The time effort to describe all properties of a resource is considered as a hindrance to a wide distribution and usage of a metadata scheme. Using learning objects to build courses requires more information than the description of a single resource can provide. All categories are optional and the base scheme can easily be extended to fit particular needs (for how to use LOM for building courses, i.e. compositions of several learning objects, see [11]).

The values for the relation category of LOM are taken from Dublin Core. The values are.

```
{isPartOf, HasPart, IsVersionOf, HasVersion, IsFormatOf, HasFor-
mat, References IsReferencedBy, IsBasedOn, IsBasisFor, Requires,
IsRequiredBy}
```

Unfortunately, the bibliographical background of these relations is obvious. Furthermore the relations mix content-based and conceptual connections between the learning objects. The fact that a learning object is referencing an another one, is an indication that both learning objects contain information about the same topic. It is not enough information for a course author to decide, whether these connected learning modules can be presented in a certain order. The relations "isPartOf/hasPart" and "isVersionOf/hasVersion" are useful for organizing and managing generated lessons. To help assembling lessons they are not helpful. Adding two layers of metainformation to the metadata, the drawbacks of LOM can be compensated.

3.2 Ontology

As in traditional, printed books, an index of the keywords of the domain to be learnt, is very helpful to find quickly the wanted topic. An ontology contains the relevant keywords of the knowledge domain and it offers also relations between the keywords, both hierarchical and non-hierarchical.

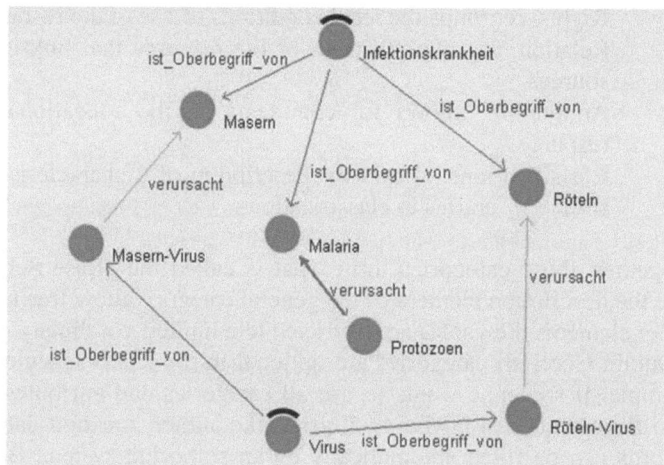

Fig. 4. An example for an ontology

The course author or the learner can get information about the available topics. The learners get an overview how the topics are connected. Together with the LOM information, learners can search for learning objects, they know nothing about. Example: learners can search in the ontology for bacteria causing diarrhea and the search the metadata files for a movie described by the result of the ontology search.

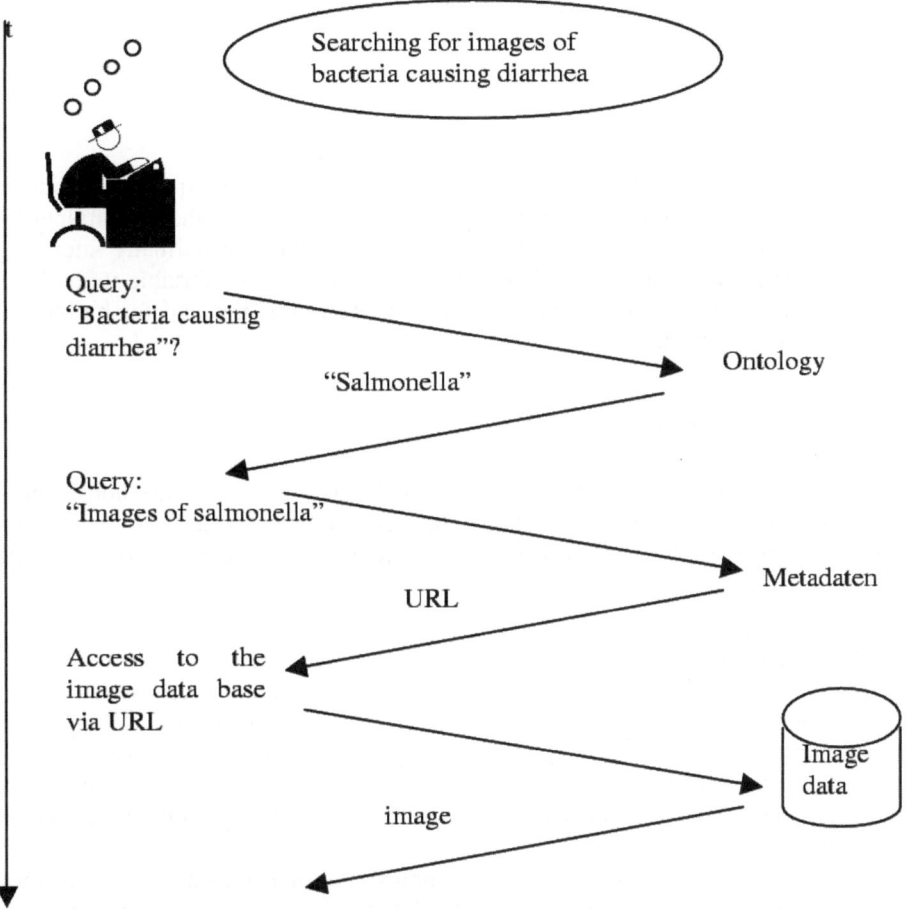

Fig. 5. Procedure of searching for learning objects with an ontology and metadata

3.3 Rhetorical-Didactic Relations

The second additional kind of metainformation are relations between learning objects. These relations can help the course authors or the learners to find clusters of learning modules.

The relations between single learning objects should be restricted to didactic relations. These are for both the course author and the learner useful to gain additional, more profound or explaining material.

Based on the Rhetorical Structure Theory of Mann and Thompson [14], a set of so-called rhetorical-didactic relations can be used to connect learning objects. Examples are "example" or "deepens". Also tests and exams can be added to informing learning objects by the relation "exercises" (see [15]).

4   Tools

For the authors, to describe their learning object is additional expenses, beside the fact that generating multimedia content is as such more costly than traditional text content. Additionally, the authors are domain experts and not normally knowledge engineers. Therefore, the authors have to be supported by comfortable tools. In the following sections, tools for composing and editing LOM files and for building up ontologies are sketched.

4.1 New Kinds of Tools

An optimal environment for authors is an integrative authoring suite consisting of both content and metainformation editors and a course builder (see [12]). An example for this scenario is the project k-MED [4]. Here, the medical ontology is called Concept-Space.

ConceptSpace-Editor

With the ConceptSpace-Editor concepts of the knowledge domain can be set, deleted, renamed and modified. There are several applications for managing and enhancing the ontology and for navigating on it for searching concepts. The user interface has to be intuitive since in this project the content experts are medical specialist and not computer scientists. These tools can be operated cooperatively, both synchronously and asynchronously.

Since the modeling of the ontology is a complex and time consuming task, methods for enriching ontologies at least semi-automatically [8], [9] are extremely helpful.

LOM-Editor

The LOM-Editor is the tool to generate the metadata record of a learning object, connect them to the concepts of the ontology and export the metadata as an XML description. The LOM base scheme consists of 60 fields. The willingness to describe the learning object by at least a minimum of metadata is related to the comfort of the tool. The LOM editor used in the k-MED project [16] compiles as much information as possible from the learning object themselves, like seize, creation date and format. Additionally, authors can employ personal templates.

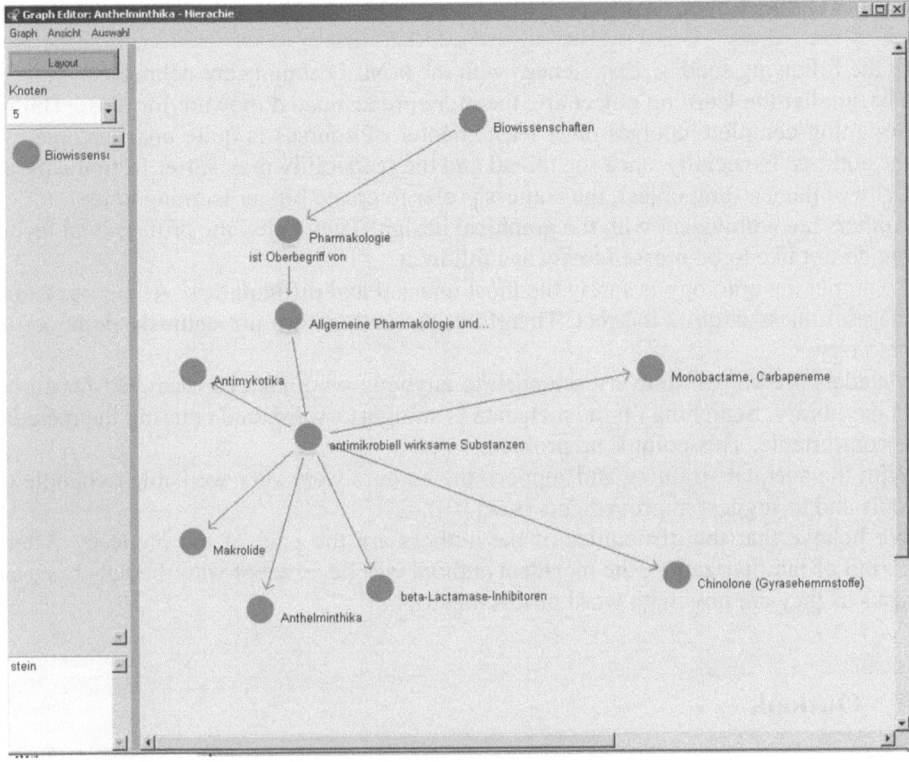

Fig. 6. ConceptSpace Editor (*k*-MED)

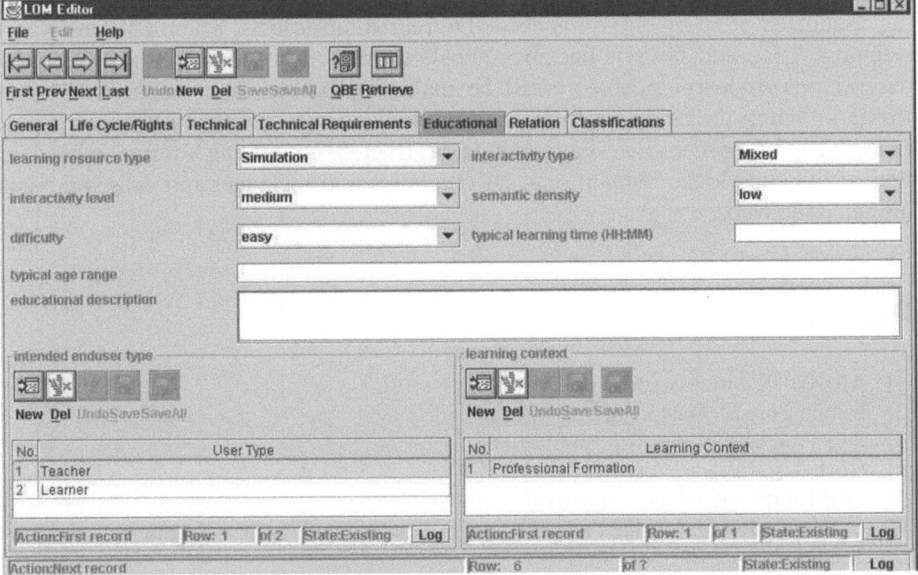

Fig. 7. KOMs LOMEditor

4.2   Feedback from Authors

In the following section, experiences with the k-MED authors are delineated.

The smaller the learning object are, the more predestinated they are for reuse. But not designing complete courses or at least chapter of courses is quite unaccustomed for the authors. Especially since the thread and the readability may suffer from the modularity of the learning object, the authors prefer to create bigger learning units.

Authors are ambivalent with the graphical design. They enjoy the professional layout, but do not like to be pressed to always follow it.

To model the ontology is surely the most unusual and difficult task. At the same time, the usefulness is quite indirect. Therefore, the authors are not enthusiastic about the ontology.

Metadata are known to every scientist, to anybody who has ever searched for a book in the library. Searching on the metadata is straightforward, and entering the metadata is comfortable. This point is no problem.

With the adequate training and support, the authors were very well able to handle the tools and to suggest improvements (see [10]).

We believe that the difficulties of the authors are the pain of the pioneers. After a period of familiarization, the metadata authors will be as adept with the new tools and tasks as they are now with word processing tools.

5   Outlook

The modularization of the learning material allows for more individualized learning. Due to the rapid development in scientific areas, the half-life of knowledge decreases rather fast. A permanent process of learning is required. That means, life-long learning conducted often by oneself is needed to remain up-to-date. The traditional way of teaching "once and for all" becomes obsolete. As it is planned by the LOM draft, descriptions of learning object shall be make it possible for machines to support learners finding the needed information. We have shown that metadata alone are not sufficient. Tim Berners-Lee suggests in [6] a structure combining a formal representation of the knowledge domain (e.g. an ontology) and metadata and calls this vision the semantic web.

References

[1]   ARIADNE: http://ariadne.unil.ch
[2]   Dublin Core Metadata: http://purl.org/dc
[3]   IMS Project; http://www.imsproject.org
[4]   k-MED: www.k-med.org
[5]   LOM: http://ltsc.ieee.org/wg12/
[6]   T. Berners-Lee, J. Hendler, and O. Lassila: The Semantic Web. Scientific American 284, 5, 2001.
[7]   A. El Saddik: Interactive Multimedia Learning. Springer-Verlag, Heidelberg. 2001.

[8] A. Faatz, S. Hörmann, C. Seeberg, and R. Steinmetz: Conceptual Enrichment of Ontologies by means of a generic and configurable approach. In In Proceedings of the ESSLLI 2001 Workshop on Semantic Knowledge Acquisition and Categorisation.2001.

[9] A. Faatz, C. Seeberg, and . Steinmetz: Ontology Enrichment with Texts from the WWW . In Proceedings of ECML-Semantic Web Mining 2002. Springer-Verlag, Heidelberg. 2002.

[10] S. Hörmann, C. Seeberg, and A. Faatz: Verwendung von LOM in k-MED. Technical Report, KOM, DarmstadtUnivresity of Technology. TR-KOM-2002-02. Darmstadt. 2002.

[11] S. Hörmann, A. Faatz, O. Merkel, A. Hugo, and R. Steinmetz: Ein Kurseditor für modularisierte Lernressourcen auf der Basis von Learning Objects Metadata zur Erstellung von adaptierbaren Kursen. In LLWA 01 - Tagungsband der GI-Workshopwoche "Lernen-Lehren-Wissen-Adaptivität". Ralf Klinkenberg, Stefan Rueping, Andreas Fick, Nicola Henze, Christian Herzog, Ralf Molitor, Olaf Schroeder (Eds), 2001. Research Report #763.

[12] S. Hörmann and R. Steinmetz:. k-MED: Kurse gestalten und adaptieren mit rhetorisch-didaktischen Relationen. In Rechnergestützte Lehr- und Lernsysteme in der Medizin. J. Bernauer and M. R. Fischer and F. J. Leven and F. Puppe and M. Weber. 2002.

[13] IEEE WG 12: http://ltsc.ieee.org/doc/wg12/LOM_1484_12_1_v1_Final_Draft.pdf

[14] W.C. Mann, and S.A. Thomson: Rhetorical Structure Theory: A Theory of Text Organization. Technical Report RS-87-190, Information Science Institute, USC ISI, USA. 1987.

[15] C.Seeberg: Life Long Learning – Modulare Wissensbasen für elektronische lernumgebungen. Springer-Verlag, Heidelberg. 2002.

[16] A: Steinacker: Medienbausteine für web-basierte Lernsysteme. Dissertation, D17 Darmstadt University of Technology. Darmstadt, 2001.

[17] A. Steinacker, A. Ghavam, and R. Steinmetz: Metadata Standards for Web-based Resources. IEEE Multimedia, 8(1):70-76, 2001.

[18] W3C: "Metadata Activity Statement"; http://www.w3.org/Metadata/Activity.html

Information Technology Practitioner Skills in Europe: Current Status and Challenges for the Future

Wolffried Stucky[1], Matthew Dixon[2], Peter Bumann[3], Andreas Oberweis[4]

[1] Universität Karlsruhe (TH)
Institut AIFB
76128 Karlsruhe, Germany
stucky@aifb.uni-karlsruhe.de

[2] Birkbeck, University of London
School of Economics, Mathematics and Statistics
Malet Street, Bloomsbury
London WC1E 7HX, United Kingdom
MatthewD@iisfairfield.demon.co.uk

[3] Council of European Professional Informatics Societies – CEPIS
Stresemannallee 15
60596 Frankfurt a.M., Germany
secretary@cepis.org

[4] J.W. Goethe-Universität Frankfurt
Lehrstuhl für Entwicklung betrieblicher Informationssysteme
60054 Frankfurt a.M., Germany
oberweis@computer.org

Abstract. The current state of the IT practitioner labour market in Europe is considered, together with the role and contribution of the Council of European Professional Informatics Societies (CEPIS). An important recent CEPIS study is reported, and a new approach to tracking the progress of people towards full IT professional status is introduced. The growing contribution of the new European Certification for Informatics Professionals is flagged, and the paper concludes with some suggestions for policy priorities at the European level.

1 Introduction and Motivation

In spite of an extended period of discussions in various bodies about the IT "skills gap" in Europe, there remain many questions about the exact nature of the problem, and how to tackle it, whether from the political or economic perspective. Attempts at policy measures to effectively improve the situation for the benefit of European employment and the economy are still in their early stages in a number of countries. In response to this situation, the EU Commission established an "ICT Skills Monitoring Group" in 2001, and this reported to the recent "e-Skills Summit" in Copenhagen (in October 2002). Since the IT profession needs to contribute to effective policy development as one of the strong drivers of national competitiveness, CEPIS Member

R. Klein et al. (Eds.): Comp. Sci. in Perspective (Ottmann Festschrift), LNCS 2598, pp. 304–317, 2003.
© Springer-Verlag Berlin Heidelberg 2003

Societies agreed through their Council, also in 2001, to commission a serious study from an acknowledged expert. The results of the study are surveyed in this paper.

The paper is structured as follows: After defining the relevant terms in the field of IT skills, Section 2 introduces CEPIS together with its goals. Section 3 presents results from the empirical study on the labour market in Europe. Section 4 discusses several proposals of CEPIS concerning IT skill challenges within Europe. Section 5 summarises the paper and gives an outlook on future tasks.

1.1 Usage of Terms

The terms used in discussing IT skills questions are often understood in different ways by different people. The following terms are key concepts, with the following broad meanings.

Skills: The set of requirements needed by employers from those who are capable of satisfactorily carrying out each relevant occupation. In the context of labour market work, the word is generally used to refer to the overall market parameters.

Occupations: The set of separate broad roles carried out within a particular working area. There are many occupational frameworks in IT, these are discussed in more detail in Section 1.2.

Competencies: The set of capabilities that people in a particular occupation need to have, in order to reliably and consistently perform that role to an adequate level of performance (the term is therefore close to, and often used in this context interchangeably with, *Skills*)

Education: The instilling of the underlying principles that are taught (generally to young people) in broad preparation for life, including working life. Publicly-funded education in most EU Member States consists of three broad levels: 1) Primary, 2) Secondary, and 3) Tertiary (generally in universities). Generally, education provides the underpinning knowledge and understanding required for achieving workplace competencies.

Training: Focused learning directly related to capabilities required in specific jobs. Much (although not all) of training for IT practitioners is in the use of specific software tools: it is largely commercially delivered, and often *certified* by the supplier of this software.

Continuing Professional Development (often referred to generally as *Lifelong Learning*):
The often-regular learning required to maintain employability and effective performance as occupational requirements change over a career.

Figure 1 illustrates the relationships between the different terms.

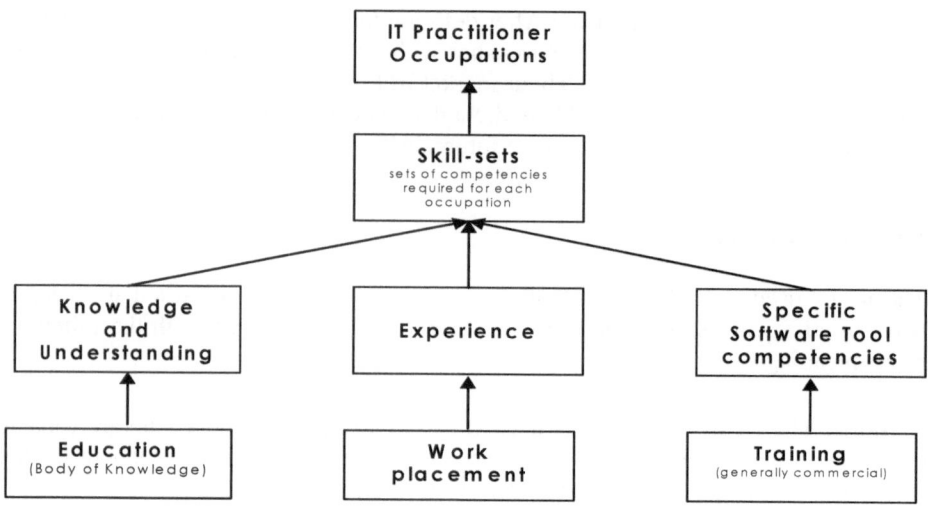

Fig. 1. Relationships between Occupations, Skills and components of Skill Development

1.2 What Are IT Practitioners and Where Do They Work?

There are probably as many views as to what IT practitioners are and do as there are people who take an interest in this area. It is inevitable that readers of this paper from different EU Member States will have a range of particular perspectives. The different kind of work in (and around) labour markets, including the work of those considering and developing:

- careers material,
- education and training approaches and programmes,
- professional formation and career development requirements, and
- salary surveys,

all require the assumption of a set of occupational definitions (with a greater or lesser detail of skill/competence requirements) in the broad area of interest. These different focuses inevitably have different requirements as to the amount of detail of skill requirements and the "granularity" of the classification. For example, salary surveys generally require information about a comparatively large number of occupational categories, so that individual employers will learn about market conditions for the categories they happen to use – this can result in *many dozens* of "job titles" being used for different kinds of IT practitioners. On the other hand, careers material needs to paint a "broad brush" picture of general areas of activity in an attractive way – here the number of types of IT practitioner would generally be much lower – generally *less than 10*.

In addition to the range of application areas for occupational classifications, the speed of development of the technology, of the tools, systems and methodologies arising from this and their penetration into the marketplace have also played a role. As IT management and occupational analysts in different organizations, contexts and countries have tried – over the last few decades - to distil structure from the fast-changing picture, it is understandable that a number of different occupational frameworks have emerged.

But from the point of view of sound analysis of the IT practitioner labour market – of tracking developments and understanding the labour market well enough that sensible decisions can be made, both at the enterprise and at the public policy levels – the most important set of occupational frameworks are those for which *surveys* (whether of employers or of individuals) have been carried out. Even within a single country (e.g. the United Kingdom) a range of different frameworks have been used for surveys. While there are broad similarities, the fact that the frameworks used are not the same means that data gathered from the different surveys cannot generally be usefully compared, and thus contribute to building up an overall picture enriched by evidence from all the different surveys.

The great desirability of the use of a single ("unifying") framework for all surveys (and ideally, for as many other applications as possible) led to an initiative over the late 1990s within the UK to win wide support for such a framework, built largely from the BCS's *Industry Structure Model*. The resulting framework – "Skills Framework for the Information Age" (SFIA) - developed with active steer from industry, has considerable merit, but needs even greater buy-in. Details can be found at www. e-skillsnto.org.uk/sfia/. The only framework of an international character developed thus far arose from an initiative by a number of large European ICT companies, which led to the development of the *"career-space"* framework. Since this is intended first and foremost for careers promotion purposes, it is a relatively "coarse-grained" framework (with thirteen generic job profiles, including communications specialisms), and it is only 1-dimensional – i.e. it does not directly reflect the fact that different occupations exist at different levels of technical complexity and responsibility. More details can be found at www.career-space.com.

It is important to recognise that – for a fast moving set of industrial sectors and occupations – the task of tracking development in a robust quantitative way will always be very difficult. All occupational frameworks in such an environment need to have review and updating built in. Even more difficult in that situation is the task of forecasting future skill needs, both in terms of "competence-content" and of development of numbers required in the labour market as a whole, since trend data is lost if a single occupational framework cannot be sustained.

Probably the most important principle to be understood when trying to develop a sound view of IT practitioner skills is to recognise the *fundamental distinction between the **sector** and **occupational** perspective*. IT (supplier) companies arise from, and are strongly influenced by, the abilities of the "technical people". But such companies generally also employ people with a range of other skills, in support operations of various kinds (e.g. accounts people, admin. people, marketing and sales people, personnel staff, general managers, office cleaners, etc.). In addition, organizations in most other parts of the economy – e.g. banks, manufacturing companies, local authorities, hospitals (and other health care operations), airlines, retail businesses, government departments, etc. - all make considerable use of Information Technology, and in doing so, generally have "IT departments" that employ teams of IT practitioners. In many countries, the number of IT practitioners employed in these "IT user organizations" is greater than the numbers employed in IT (supplier) companies. Table 1 shows the structure in its simplest form: a 2 x 2 matrix. The numbers of people employed in the 4 boxes show the basic profile of a country's IT practitioner community.

	IT (Supply) Companies	IT User Organisations
IT Practitioner Occupations		
Other Occupations		

Table 1. Partitioning the IT Practitioner Community

2 CEPIS

The Council of European Professional Informatics Societies, CEPIS, is a non-profit organisation seeking to improve and promote high standards among informatics professionals in recognition of the impact that informatics has on employment, business and society. CEPIS unites:
- 34 European national member societies of IT professionals,
- from 29 countries of Europe,
- representing more than 200,000 individual IT professionals.

CEPIS Goals can be stated as follows: Recognising the potential of information and communication technologies to generate employment, to improve the competitiveness of business and the quality of life of all citizens, the members of CEPIS commit themselves so far as it lies within their power to work to achieve high levels of:
- Informatics education and training,
- competence and ethical behaviour among European informatics professionals throughout their careers,
- user skills in the European workforce,
- awareness among all Europeans.

The above duties imply particular requirements that need to be fulfilled by professional informatics practitioners, as indicated below.

Objectives for Professionals

- Networking of professionals across Europe,
- common interests and free movement of European informatics professionals,
- mutual recognition of professional informatics qualifications and the acceptance of professional standards,
- encouragement of compliance with the CEPIS Code of Professional Conduct,
- scientific and technical cooperation in the field of information and its exploitation,
- sponsorship of activities which inform and illuminate matters of public interest.

Objectives in the European Union

- To inform European Institutions and recognised European bodies of the views of the informatics professionals and to liaise with them to provide opportunities for convergence and cooperation,
- to advise on and contribute to the development of European legislation in the light of the continuing evolution of informatics,
- to promote the coordination of regulations, legislation and standards of relevance to the informatics profession
- to stimulate the use of informatics in ways that will improve the quality of life and provide benefits throughout European society.

For more information, see the CEPIS website (www.cepis.org).

3 The Labour Market in Europe: Results from an Empirical Study

The report "Information Technology Skills in Europe" [1], commissioned by CEPIS in 2001, surveyed the current state of IT practitioner skills within the European Union. It presents an overview of the IT practitioner labour market und summarises in more detail recent trends in employment in four different countries (Germany, Ireland, Sweden, and the United Kingdom). The future development of the size of the IT practitioner workforce was then explored using different plausible employment growth scenarios, with annual increases of 2% to 15%, following an initial downturn. For the full report, see www.cepis.org/prof/eucip/cepis_report.pdf.

The study sets the analysis in its economic and policy contexts, and then shows a number of comparative statistics of employment levels and the characteristics of the workforce, all based on the official national statistics arising from regular surveys of the national workforce. The position of those in IT practitioner occupations is found by retrieving the figures for the two IT-related occupations within the internationally agreed ISCO occupational framework: ISCO 213 (Computing Professionals) and ISCO 312 (Computer Associate Professionals). These "official statistics" for the size of the national IT practitioner workforce were checked against the "prevailing consensus" within each of the four countries as viewed from industry and labour market experts, and the differences analysed.

The economic context of IT skills was analysed in relation to the scale and importance of IT-related activity and this workforce, as well as in relation to labour market performance (e.g. in the effectiveness of price rise responses to shortage and the consequent increase of supply), national characteristics, and employers' and individual roles.

The policy context was reviewed in relation to:

- Academic and vocational education and training infrastructures,
- approaches to labour market policies,
- policy responses to skills shortages,
- experience with relevant initiatives, and
- the roles of national and European policy-makers.

The study examines in particular recent trends in the relevant indicators, which show important differences between EU member states, as well important developments over time, not least in relation to activity in tackling the "Y2K" date-change problem. The report shows interesting comparisons – some similarities, some differences – between the situation of the computing professional workforce in the four countries examined in more detail, in particular in relation to:

- The fraction of female employment,
- the age distribution of IT practitioners,
- self-employment within the workforce,
- the supply (IT) industry share of employment,
- the computing professionals' "Highest Academic Achievement", and
- the amount of training recently received.

In addition, the report shows, for all member states, the development of the *IT practitioner share of total national employment*, and how this developed over recent years, and the role of migration.

The report then elaborates the issues involved in estimating future skill shortages, reviews the two main previous EU level studies, and adopts an innovative approach to forecasting employment levels, using four "scenarios" which could prove useful in policy analysis.

4 CEPIS Proposals

Over recent years, CEPIS has made important contributions to the understanding and tackling of the IT skills challenges within Europe. In particular, CEPIS work is notable in the following areas:

- European Computer Driving Licence (ECDL),
- European Informatics Skills Structure (EISS),
- Information Technology Practitioner Skills in Europe (Labour Market Study),
- European Certificate for Informatics Professionals (EUCIP).

In the following we will have a closer look at ECDL (Section 4.1) and EUCIP (4.2).

4.1 European Computer Driving Licence (ECDL)

There is quite a bit of progress being made until now in the work of promoting IT professionalism in Europe, however, it all started within a fairly modest framework. The engagement of CEPIS in promoting IT Skills has begun with IT literacy. In order to promote IT literacy CEPIS has in the year 1996 developed the European Computer Driving License (ECDL) for which the basic concept came from the Finnish Information Processing Association (FIPA). The ECDL mainly addresses the user community for which CEPIS encourages greater levels of ICT literacy competence through a modular learning programme. The ECDL is an internationally recognised standard which certifies that an individual has achieved the necessary knowledge and skills needed to use the most common computer applications efficiently and productively. It sets a base standard for the IT skills that are necessary for most people seeking employment, or working, in the Information Society. For the dissemination of the ECDL CEPIS installed the ECDL Foundation, located in Dublin, Ireland. The members of the Foundation are CEPIS and its member societies. The Foundation has the right to give licenses to CEPIS member societies who in turn accredit test centres and certify learning providers in their territories according to commonly valid quality assurance procedures. The ECDL has until now attracted more than two million students in Europe and abroad. Due to the open and industry neutral concept the ECDL has received the recognition of the "Employment and Social Dimension of the Information Society" (ESDIS) from the European Commission. Based upon the success of the ECDL CEPIS and its member societies became motivated to apply a similar concept to the education of IT practitioners.

The further promotion of professionalism in the Information and Communication Technology (ICT) Industry was performed by means of developing two major documents namely the EISS, the European Informatics Skills Structure, and the EICL, the European Informatics Continuous Learning Programme which are deemed to serve the development of Lifelong Learning.

4.2 European Certificate for Informatics Professionals (EUCIP)

As a natural follow-up product of the European Computer Driving License which was developed by CEPIS under the EC funded LEONARDO programme being aimed at educating a very broad audience, a new product was developed being addressed to help filling the so-called IT skills gap in Europe. In the year 2000 CEPIS has commenced to work on the preparation of a certification programme for IT professionals: called the "European Professional Informatics Competence Service" (EPICS). This project was renamed in the year 2002 into EUropean Certificate for Informatics Professionals" (EUCIP). It defines a core level of knowledge that all people wishing to become IT professionals should have; this will be extended into specialised areas where an individual needs more extensive and detailed knowledge in order to carry out a particular job. EUCIP is targeted at new entrants into the profession and towards those who wish to formalise their knowledge. The EUCIP product is ready in a prototype version to go on to the market in the year 2003. The service to be delivered to the market consists of a Europe-wide IT certification scheme based on the EUCIP syllabus. EUCIP offers a unique and open set of services and a portal to a community which alleviates re-education and competence enhancement for the ICT profession. EUCIP is designed to be complementary to public education and certification and will also rely on the acceptance of public educators as well as on private training. As there is competition from national programmes which may never reach the necessary international level of acceptance, EUCIP is ideally suited to overcome the lack of Europe-wide harmonisation in the recognition of certification for IT professionals. The product shall ideally be brought to the market via the CEPIS member societies as licensees who have already knowledge of the market and experience from the ECDL, thus representing the maximum chance for success and a minimum of risk

It is expected that the results from an EC-funded market validation phase initiated in 2002 will bring enough information about what will actually be required by the market in order to commence with the deployment phase. As experience has shown until now, there will be a continuous requirement for up-dating and for modification to the syllabus according to changing market demand. It is anticipated that at the outcome of the market validation phase there will be a fully developed product which can then be brought to the market in European countries and abroad having the advantage of being demanded because of its unique concept. This will be a certification scheme based upon the EUCIP framework harmonised throughout Europe.

The ultimate objective is to link EUCIP into lifelong learning for IT professionals. Linkages are being built not only with training providers but also with vendor certification schemes and learning providers (such as third level educational establishments).

One of the areas of growing importance in skills policy in many countries is the strengthening of opportunities for *lifelong learning*. Growth of public investment in such provision is generally triggered by a combination of:

- industrial structural change (with demand for employment in certain sectors falling, and major re-training programmes therefore needed for skills for which demand is growing),
- the introduction of new technologies and tools in the workplace, and
- the growing competitive pressure on businesses of most kinds arising from the slow but steady movement towards the globalisation of markets.

Increasingly such pressures produce a shift of employment realities that result in the need for people to recognise that they cannot rely on earning their living in a particular occupation throughout their life, and must increasingly take responsibility for acquiring updated, and often quite different, skills as their career progresses. Governments have increasingly urged and supported individuals taking responsibility for their continued employability via an accepted commitment to lifelong learning. However, also all efforts from public educators like universities and others have not been the overall solution to close the IT skills gap and to solve the lack of qualified practitioners in the IT profession in Europe.

The big debate around IT practitioner skills is to do with precisely what education, training, qualifications, competencies, and thus "skills", practitioners actually need to perform cost-effectively in each occupational role. In this situation, while the "supply channels" from education still have a most important role to play, the "supply model" into these occupations is not a simple "linear" one (with a straightforward career path from IT specialisation within education, an IT degree and a career as an IT professional).

EUCIP sets out to address the current and anticipated skills shortages in the ICT industry in Europe. The shortage is currently estimated at over 1 million people. It will provide a range of services to assist in raising the skill level of those already in the ICT industry and to also provide an attractive means of entry into the ICT industry for new entrants of any age. The EUCIP concept as developed by CEPIS will do this by providing a set of services including information about courses, diagnostic testing of current abilities, certification of skill levels attained and access to educational material about state of the art technological developments. EUCIP is developed to create a Trans-European Network for Certification through its member societies leading to the provision of education and training using telecommunications for web based delivery. EUCIP will stimulate the commercial deployment of multi-lingual, interactive, and multimedia educational and training tools and services across the EU through the provision of a standard agreed syllabus, which is accepted by the market place. Further objectives are to raise the ICT competence level, attract new practitioners to the professional ICT field, define ICT professional entry levels, offer entry-level certification, provide other professionals with the ICT knowledge that they need from a business perspective through a web based learning management system. Provide a framework for ICT professionals to keep up-to-date, and contribute to closing Europe's ICT skills gap.

The plans are that a delivery mechanism which relies on the CEPIS member societies will eventually make the EUCIP product available to citizens across Europe by means of the Internet needs to be validated within the project starting with a market validation phase, being followed by a preparatory market deployment phase, focussing on the roll-out of pilot projects in several European countries. This will then provide the basis for the successful deployment of EUCIP in the CEPIS partner countries and subsequently throughout Europe.

EUCIP will provide for services which are particularly suited to the needs of practitioners and also for people working at home. EUCIP has a syllabus which has been validated by acknowledged training providers and the higher level colleges. It will provide an automated means of testing the skills and knowledge acquired. It will set up a quality assurance scheme to ensure that the certification process is valid and accepted throughout Europe. Through the CEPIS member societies the venture will have easy access to the national markets, i.e. learning providers and students. The physical content of the training will be delivered through local learning providers and test centres and will be supported by the local informatics societies. The education services will, in general, be in the local language.

4.3 Competence Maturity Model

The CEPIS contribution to EUCIP can, perhaps, be most effectively understood by reference to the new framework being considered within CEPIS, under the title of a *competence maturity model.*

The proposed competence maturity model, developed by the authors of this paper and presented at European e-Skills Summit in Copenhagen 2002, shows five distinct stages in the development of full professional competence in relation to IT practitioner work:

- *IT Awareness* (basic knowledge),
- *IT Literacy* (knowledge to operate a PC),
- *Expert User* (Special Knowledge – expertise with application software, helping other users),
- *Professional Entry Level* (Professional knowledge), and full
- *Professional Level* (Advanced Professional Knowledge).

In order to demonstrate how the model can work we may regard the personnel in employment in each of our said categories: as can be seen in Figure 2 there are many people on the "IT Awareness" level, less people on the "IT Literacy" level and so on. The scaling in this diagram is simplified of course, but this fact is not of importance for the understanding of the model. In addition to the personnel in employment there are many people on the first level who are unemployed at the moment. And on the other hand there is a skills gap in the higher knowledge categories: there are vacant positions requiring a certain knowledge level.

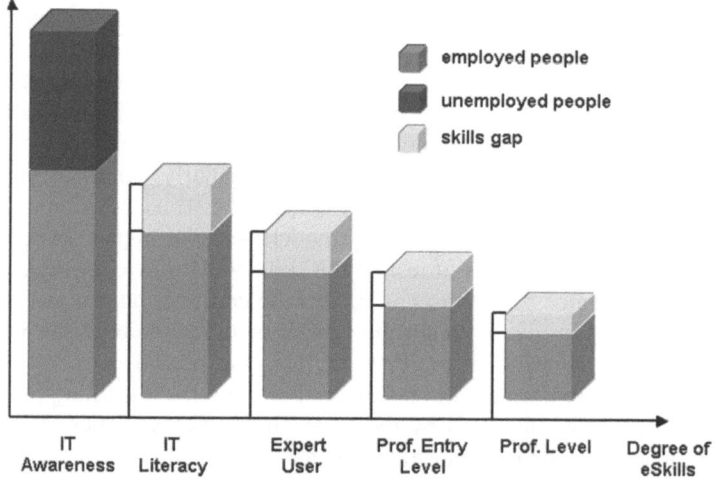

Fig. 2. Competence Maturity Model (a)

As can be seen from Figure 3, people move step-by-step along towards profes-sional level, with unemployed people starting with the first steps (ECDL), "topping up" the stack at the next level along. In effect this confirms a "cascade" process in operation as the reality of the labour market development.

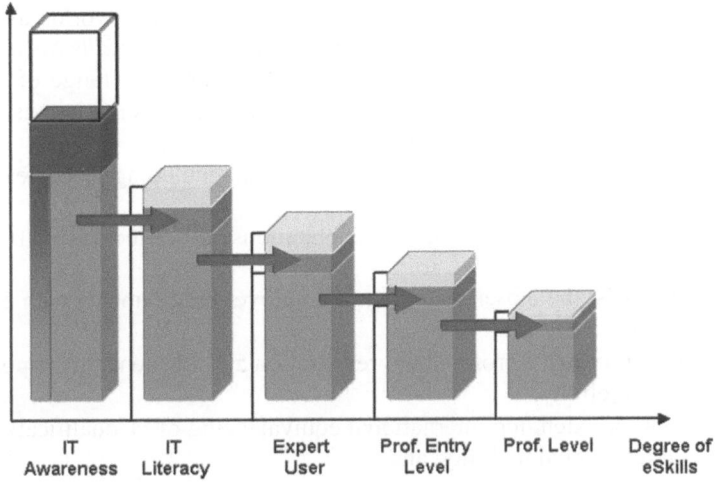

Fig. 3. Competence Maturity Model (b)

The result of this process can be seen in Figure 4: we will have less unemployed people on the "IT Awareness" level and reduced skills gap at all other higher levels.

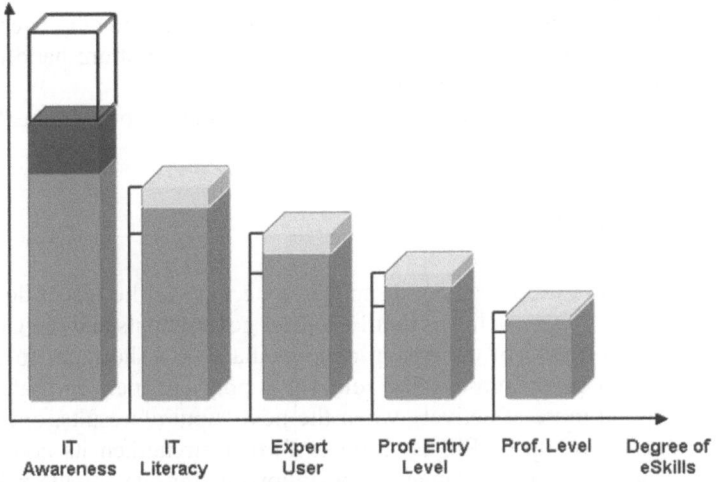

Fig. 4. Competence Maturity Model (c)

As can be seen, in terms of this *Competence Maturity Model*, CEPIS's current and planned contributions are:
- ECDL helps in the step from *Awareness* to *Literacy*,
- EUCIP helps in the step from *Expert User* to *Professional Entry Level*.

CEPIS will continue to monitor the situation, and will liaise with the European Commission on other possible development, where appropriate, and in the light of developments in the labour market and experience with ECDL and EUCIP.

A recent Conference examining these issues for the world as a whole ("IFIP/OECD/WITSA Joint Working Conference on *Global IT Skills Needs – the role of Professionalism*" – see www.globalitskills.org) elaborated a range of issues (not least in relation to the range of different occupational frameworks in existence internationally) and agreed an agenda involving:

- a high level reference model covering groups including IT professionals, IT practitioners and others:
  - to identify the differences in obligations associated with different types of work,
  - to assist the closer matching of employer requirements with educational provision,
- an inventory of IT professional registration arrangements in support of international mobility,
- options for extending international equivalencing of IT qualifications to support international mobility, and
- exploring the value of greater alignment of occupational frameworks internationally for different purposes (with different customers), including:
  - Labour market statistic-gathering (for policy-makers, planners),
  - career opportunity clarification (for prospective entrants),
  - quality assurance (for competence area identification and credential validation),
  - career progression management (for individuals and employers),
  - international mobility (use in relation to immigration; handling regulation and employer hiring),
  - clarify "IT professional", "IT profession" and "IT practitioner".

5 Summary and Outlook

The IT skills issue in Europe has not disappeared because of the recent downturns in the ICT market, and CEPIS believes that European governments and the Commission should use the opportunity of the easing of immediate skill shortages to stand back and take sensible steps to improve the relevant supply channels so that the labour market will respond more effectively when the next "upturn" results, once again, in serious shortages. Overall, CEPIS is ready to further strengthen its contribution to tackling the problems, and recommends the following European-wide priorities for the future:

- Core policy analysis (in particular on whether market is failing) to be based on sound labour market figures,
- adopt measures that recognise that not all aspiring IT practitioners have computing degrees,
- recognise that not all IT practitioners work in the IT "supply" sector,

- move towards common EU standards for assessment of professional competence,
- move to common European-wide occupational frameworks,
- EU Commission to concentrate on finding, reviewing and disseminating good practice,
- CEPIS will work with Eurostat and employer bodies to improve validity of labour market understanding.

Acknowledgement

We wish to thank Daniel Sommer for his support in preparing the final version of this paper.

References

1. M. Dixon: Information Technology Practitioner Skills in Europe, Council of European Professional Informatics Societies (CEPIS), Frankfurt/Germany, 2002 (www.cepis.org/prof/eucip/cepis_report.pdf)

Collaboration via Internet and Web

Ivan Tomek

Jodrey School of Computer Science, Acadia University
Wolfville, Nova Scotia, Canada
ivan.tomek@acadiau.ca

Abstract. Internet was originally created to make it possible to run programs on remote computers distributed over a variety of interconnected networks. In the next major advance, the Web was invented mainly to make access to documents on Internet as easy as possible. Current Web-related work - the Semantic Web - is motivated by the need for automated document processing. Although document access remains the Web's main purpose, both the Internet and the Web have also always been extensively used for various forms of social interaction, perhaps most importantly for collaboration. With the universal spread of computing and advancing globalization, support for social interaction and collaboration is becoming ever more important and research and development in this area are intense. In this paper, we classify the main approaches to social interaction support, present a model of a powerful framework for collaboration, and give examples of several existing Internet- and Web-based applications that support parts of this model. We conclude by hypothesizing that a next step in Web development might be to use accumulated experience with *ad hoc* platform-dependent collaboration tools to develop a general Web framework supporting social interaction. This framework would advance the Web from its document- and data-centric present to a technology that supports collaboration and social interaction. We call this paradigm the Inhabited Web.

Introduction

Historical Perspective

Computer networks were invented to make it possible to run programs on remote computers. As the number of independent networks started to grow, *Internet* was created to connect individual networks together and make it possible to run programs on remote computers regardless of network boundaries. Among the many Internet applications, two emerged as prominent - access to remote documents and e-mail.

Whereas e-mail and other forms of communication proved to be satisfactory for the initial needs of collaboration, support for access to documents was perceived as unsatisfactory. This inspired Berners-Lee [1] to develop a universal approach to document access which led to the establishment of the *World Wide Web*. The essence of the solution was a protocol (HTTP), an addressing scheme (URI and URL), and a document markup language (HTML) that provided a universal scheme for accessing conforming documents. This combination of standards led to the separation of document format from applications that access and render documents, and led to the develop-

R. Klein et al. (Eds.): Comp. Sci. in Perspective (Ottmann Festschrift), LNCS 2598, pp. 318-329, 2003.
© Springer-Verlag Berlin Heidelberg 2003

ment of increasingly sophisticated Web servers and browsers capable of accessing all forms of electronic information and following trails of links from one document to another.

As HTML became widespread, two of its substantial limitations became an obstacle to further Web progress. The first was that although HTML documents are easy to read by humans, they don't have enough information to allow sophisticated computer processing because the meta-information contained in them includes only layout-related cues but nothing related to the document's semantic structure. The second limitation of HTML is that it is essentially a fixed standard with no built-in mechanism for extension.

This realization led to a new standardization effort and the current phase of Web evolution called the *Semantic Web*. The work centers on the Extended Markup Language or XML, and XML Schema [2], both of them addressing description of semantic content of data. Using XML Schema, every XML user can define new document structures and create XML instances - XML documents whose semantic content is defined by their reference to the corresponding Schema. Any application that can access the schema can convert the implied hierarchical structure into a tree object and process the document by accessing its structural elements. The schema can also be used to create new schemata that extend the original one or combine it with other schemata.

XML is also proving to be a very powerful foundation for a variety of other protocols and languages and most current Web protocols and languages are now defined in terms of XML. They include XHMTL - the new version of HTML [3], SOAP - a messaging protocol, WSDL - a specification language for describing Web Services, UDDI - a language for describing and discovering Web services [4], and many others. All these protocols and languages are undergoing intense development and testing, mainly in the context of Web Services, the anticipated basis of future electronic commerce. Whatever their particular use, most of these standards address the needs of document - and data-representation and their purpose is to allow automated processing of data by computer programs with minimal human intervention. This should allow sophisticated data-oriented interactions between businesses (B2B), businesses and clients (B2C), and clients and clients (C2C).

Beyond the Document- and Data-Centric Web

The previous section shows that the history of Internet and the Web has been, and continues to be, *document*-and *data-centric*. However, we have seen that computer networks, Internet, and the Web, have always been intensely used for *social interaction and collaboration*. In fact, evolution of the Internet and Web has itself fundamentally depended on electronic communication among its geographically distributed developers. Although several protocols, such as SMTP [5], supporting this alternative use of Internet have been developed and are massively used, truly powerful applications supporting social interaction in general, and collaboration in particular, have generally been developed outside the main stream of Internet and Web normative activity, in isolation, and without any underlying common foundation. Existing collaborative environments thus remain isolated *ad hoc* and platform-dependent with application-defined data formats, just as documents used to be isolated and application- and platform-defined before the advent of the Web. In this article, we hope to present a convincing argument that the universal importance of social interaction on the Web requires exploration of unifying frameworks. We hope that our vision will stimulate research into standards supporting social interaction, raising it to the same level of importance as electronic processing of distributed documents. For reasons that will be explained, we call this alternative Web paradigm the *Inhabited Web*.

In the rest of this article, we first attempt to deduce the general features of Internet- and Web-based collaborative applications from existing practice and relate them to existing models. We then briefly describe several such environments and their conceptual foundation and architecture. The last section describes our goals and our current work on the Inhabited Web.

Social and Collaborative Uses of Internet and Web

What Is Required for Social Interaction?

The major tool for social interaction on Internet has always been e-mail. However, judging by the progressive emergence of alternative forms of electronic communication, the asynchronous nature of e-mail has not satisfied all communication needs. *Talk* programs made it possible to communicate synchronously, and *chat* programs allowed synchronous communication with context restricted to user-defined virtual 'rooms'. Chat programs and newsgroups, which could be considered asynchronous versions of chat, demonstrated the need for private virtual communication spaces. Eventually, emergence of new media made it possible to complement or replace text communication with graphical user interfaces (GUIs), graphics, audio, and video.

Although e-mail, talk programs, newsgroups, and chat are powerful means of communication, they are isolated from one another and from other applications and do not provide a complete collaboration framework approaching real-world interaction. A truly powerful environment for collaboration requires conscious emulation of collaboration-supporting facilities offered by the real world, extended by facilities provided by modern technology including, but not limited to, information technology. An ideal collaborative software environment thus has the form of a *virtual world* offering the following features:

- Persistence. Like the real world, a virtual world with all its contents exists continuously even when it is undergoing changes, extensions, and customization.
- Sufficiently complete representations of its human inhabitants (agents, for short).
- Ability for agents to form groups that meet and interact in dedicated spaces/places [6] modeled on the topology of the real world.
- Ability to create, customize, and remove such places.
- Ability to create, access, share, modify, copy, and destroy documents and other information objects.
- Provide shared access to tools.
- Support mobility, allowing agents to move from one place to another and move objects and tools from one place to another as needed.
- Ability to define and assume roles delimiting agents' modes of operation.

- Security.
- Most importantly, the ability to not only instantiate current tool and object templates, but also to customize or extend them and create new ones, and integrate new functionality and new technologies.

Existing Models

Social interaction in the real world suggest that only a virtual environment providing the affordances described above and capable of continuous evolution can fully satisfy present and future needs of social interaction and collaboration. Such an environment should obviously contain all the features of e-mail, talk, and chat, but extend and unify them into an integrated whole capable of responding to new needs and integrating new technologies.

Historically, work on environments of this kind started with the emergence of MUDs [7]. MUDs (Multi-User Dungeons or Multi-User Dialogs) were originally created in the late 1970s as networked versions of the Dungeons and Dragons fantasy game and emulated fairytale fantasy worlds. Eventually, they evolved from a rather rigid 'hard coded' framework to include most of the features enumerated above. Their essential attributes include the following:

- Avatars - software proxies of human and software agents playing assigned roles and interacting with other avatars and the environment.
- A universe of interconnected virtual places inhabited by avatars and containing objects and tools.
- Communication within a group inhabiting a shared location, as well as between individuals, possibly across the boundaries of individual places.
- Mobility - ability of avatars to move and transport objects and tools from one place to another.
- Ability to instantiate templates of objects, places, and tools, for example by creating new locations of a particular type or new documents and objects.
- Extendibility - ability to create new templates of tools and objects and to extend or customize the existing ones without shutting the environment down.
- Persistence - uninterrupted existence over time.

MUDs have, of course, developed substantially from their initial design. Originally implemented as client-server applications in procedural languages and used via obscure commands over Telnet, they evolved into object-oriented versions (MOO - MUD Object-Oriented) implemented mostly in the object-oriented Lambda MOO language [8] developed for this purpose. Eventually, they acquired graphical user interfaces (GUIs) and representation taking advantage of Web browsers.

As a parallel development, the concept of virtual worlds evolved into 2D and 3D implementations, sometimes based on VRML (Virtual Reality Modeling Language) [9], a 3D-representation language, which is currently being replaced by XML-based X3D [10].

At present, there is a rapidly growing commercial interest in the development of both text-based and graphics-based Virtual Environments (VEs) and Collaborative Virtual Environments (CVEs) in applications including e-business, e-government, knowledge management, workplace collaboration, education, and recreational uses. This is demonstrated, for example, by several recent issues of the Communications of ACM largely or completely dedicated to this subject.

A Classification of Virtual Environments

To delineate the scope of the rest of this article, this section presents a classification of virtual environments and briefly discusses its implications.

A recent issue of the Communications of ACM dedicated to virtual environments [11] allows us to classify Virtual Environments into three categories:

1. VEs using a universe compartmentalized into modules and textual or graphical user interfaces (GUIs).
2. VEs that use 3D graphics in a universe compartmentalized into modules.
3. VEs that use graphics and model the universe as a 'continuum'.

Each of these categories is suitable for particular uses, each is amenable to a different conceptual model, and each has its own set of technical and social issues.

The major *technical issue* of any virtual environment is its scalability with respect to its complexity and the number of participants that it can support. In GUI-based VEs, this is restricted largely to the number of concurrent participants and is not a major problem. For graphics-based VEs, scene complexity becomes a major issue, especially if the participants are to be provided with acceptable response time and a consistent view of the universe. It is interesting to note that research shows that graphical detail is not a major usability requirement because even in text-based VEs, 70% of users experience the feeling of 'being there' with a sufficient immersion into the group activity [12]. This suggests that achieving high visual realism is not an essential parameter of most VE applications.

In terms of *use*, the major distinction between the three categories is again between text-based and graphics-based VEs. For most applications, text- and GUI-based VEs (we will call them MOOs for simplicity) are the best solutions: Besides their scalability, their main advantage is that if the message is not fundamentally spatial, it is carried by pure text more succinctly and without a cognitive overhead. MOOs also eliminate the need to manipulate the graphical environment, and are much easier to extend. Graphics-based VEs should thus be restricted to certain kinds of social environments [13], spatially oriented engineering design [14], architectural design, simulations such as some kinds of war games, shape design [15], and similar uses.

In terms of the conceptual model, MOOs and 3D environments with modular structure such as SPLINE [16] can be considered as two manifestations of space partitioned into disjoint modules. As such they can share the same conceptual model that has been referred to as *place-based collaborative environment* or PBCE. The 3D continuum model, on the other hand, does not easily fit into the same framework. In this article, we focus on PCBEs in general and MOO-like environments in particular.

The next section presents three examples of MOO-based applications, their conceptual models, and architectural design. We do not discuss 3D PBCE but many of the presented concepts can be extended to include it and the shared framework that we are developing must consider it as one possible manifestation.

Three Examples: Jersey MOO, MUM, and CVW

This section describes three CVEs. The first two are examples of our own work - pilot projects developed to test the suitability of CVEs in support of geographically distributed software development teams. The third example is an environment used at MITRE Corporation as an underlying component of a commercial environment supporting knowledge management.

Jersey MOO

Jersey MOO was our first MOO project. Its detailed description is contained in [17] and the following is a brief summary of its main features.

Jersey was designed to explore the suitability of the MOO concept to support teams of geographically dispersed Smalltalk programmers. Its initial user interface - Telnet text - was later converted to a Web browser interface whose main window is shown in Figure 1. The top right-hand portion of the window is a list containing information about the user's current location. It shows the present objects and human and software agents, as well as exits to adjacent rooms. When the user selects an item, the list below shows information about it and its executable commands. The two radio buttons between the lists allow the user to switch between executing a command and displaying information about it. In the 'execute' mode, clicking a command displays its Smalltalk [18] form in the input field (the anticipated users were Smalltalk programmers and communication with the Smalltalk server was via ASCII Smalltalk messages) and to execute it, the user only needs to add arguments and send the command. Clicking the command in the 'help' mode displays command information and examples of its typical uses.

Jersey MOO supports human and software agents, and a single level of 'rooms'. It also provides access to documents handled by the Web browser and treats them as MOO objects, allowing agents to move them from one place to another. URL links are also treated as MOO objects. Other types of preprogrammed objects can also be created via the Smalltalk interface, and because the interface provides direct control over the server, new types of objects and new functionality can be defined at runtime without disrupting the operation of the universe.

The major improvements of Jersey over existing MOOs are its user interface, the fact that the user does not have to remember MOO commands because most of the activity can be accomplished by clicking commands associated with a selected object, and runtime extendibility. As we started using Jersey in the context of its own development we found this last asset very important. As an example, when our group meetings in Jersey became confusing due to the delays in creating and transmitting textual communication, one of the developers programmed a meeting tool while a meeting was in progress and the group immediately started using it.

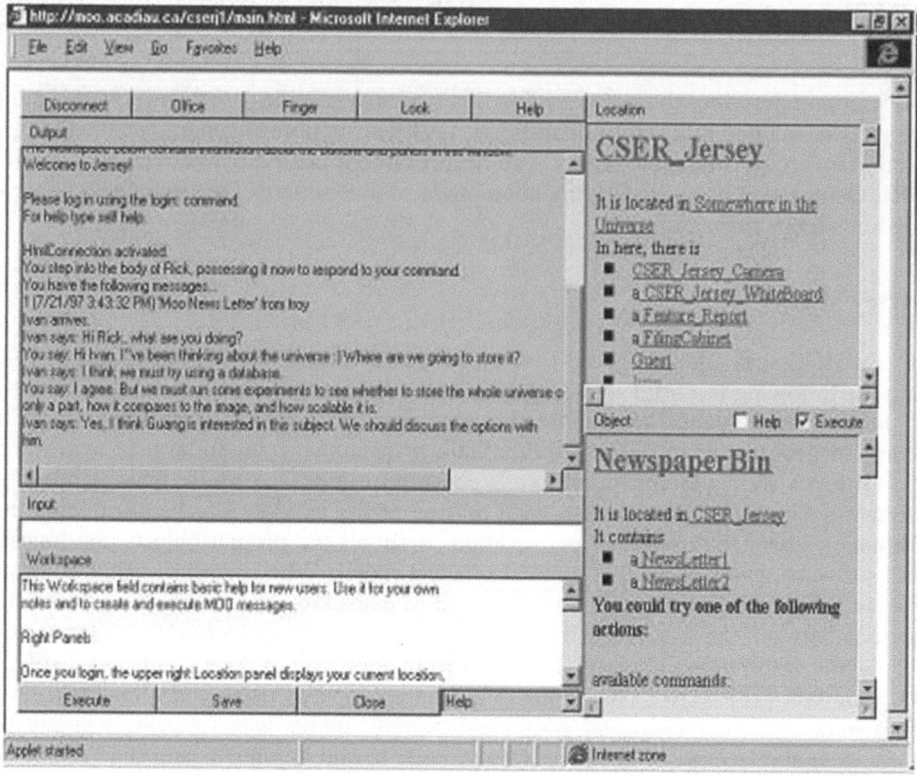

Fig. 1. The main window of Jersey MOO.

The limitations of Jersey include its Smalltalk-only interface (but other MOOs are similar in this respect because they typically require the use of the Lambda MOO language), its isolation from other applications, its relatively rigid architecture, and the fact that it was not designed to be event-aware. As a consequence, Jersey users cannot subscribe to important events such as the release of a new version of a software module. Other shortcomings include the lack of advanced MOO features, for example, roles and groups. Most of these gaps could be corrected, but our experiences led us to design another MOO based on a different architecture and conceptual framework. This MOO is described in the following section.

MUM

The acronym MUM stands for Multi-Universe MOO and derives from the fact that MUM allows users to create an arbitrary number of *interconnected universes* [19,20]. The implementation of this feature relies on a multi-layered architecture in which universes are registered in meta-servers that allow discovery of active universes and transition from one universe to another (Figure 2).

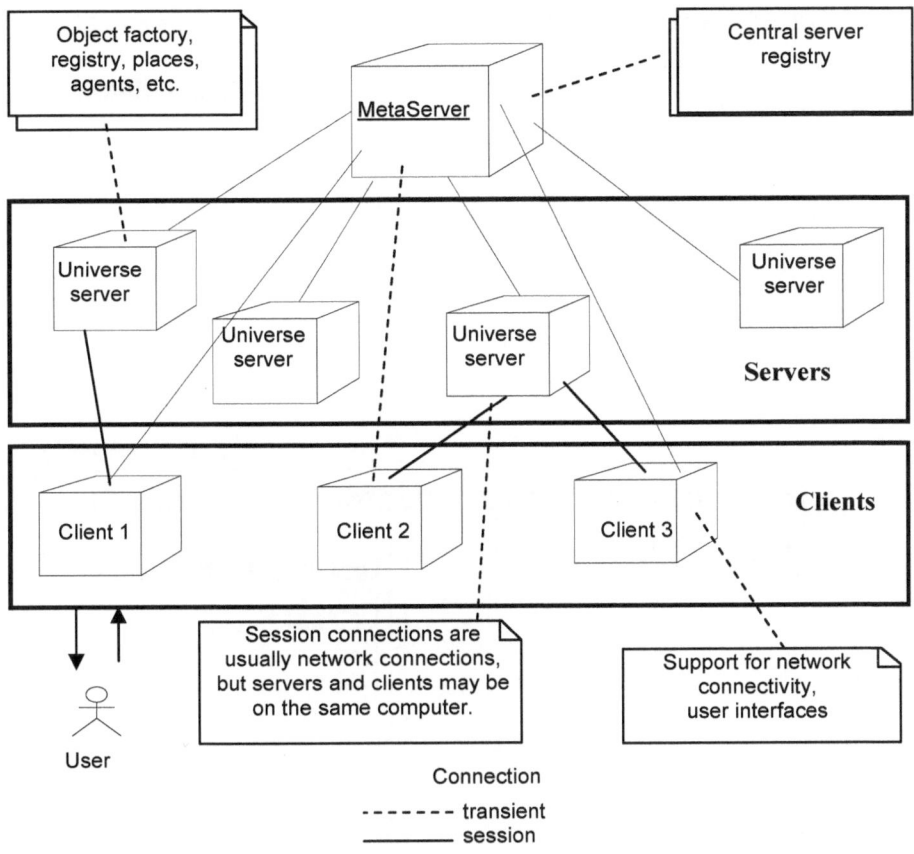

Fig. 2. Essential MUM architecture.

Besides its multi-universe nature, the main innovations of MUM are its event-awareness and its pluggable user interfaces. Implementation of events, whose lack is a fundamental gap in Jersey, is based on the use of a finite state machine (FSM) combined with an event queue containing event objects. This allows agents to subscribe to arbitrary events and obtain executable notification objects when the event occurs.

MUM user interfaces are implemented as MUM objects that can be downloaded from a repository (a place in a MUM universe) either manually or automatically when the interface is needed. The interface itself is GUI-based and does not use a Web browser. As an example, the system window of MUM and its main communication window are shown in Figure 3.

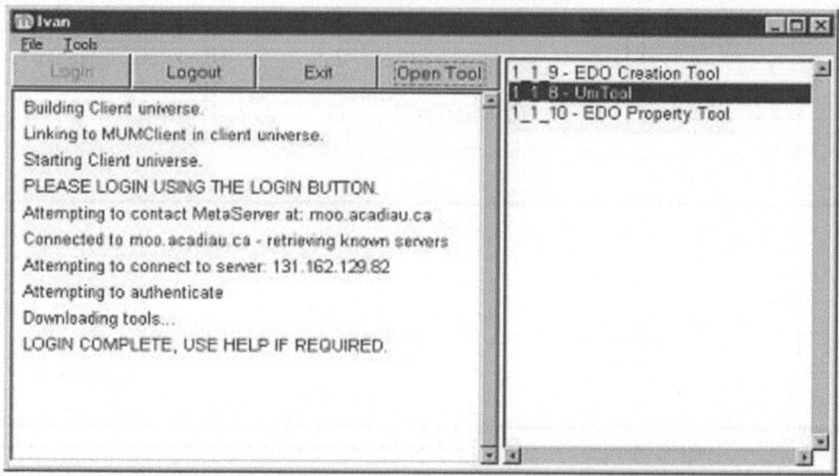

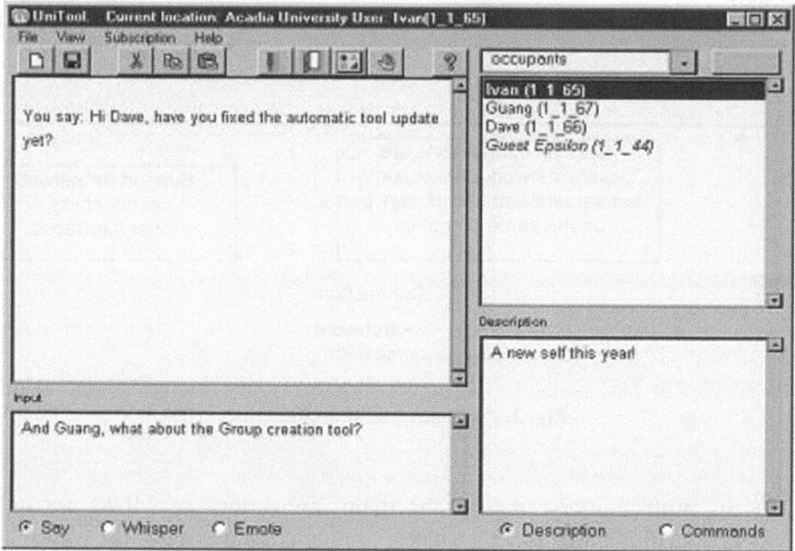

Fig. 3. Top: MUM system window with system log on the left and a library of user interfaces on the right. Bottom: The main communication window.

MUM provides a number of other features such as groups and roles and is, like its predecessor Jersey, implemented in Smalltalk [21].

CVW

CVW is an acronym for Collaborative Virtual Workplace. It is a MOO-like environment developed and used at MITRE Corporation and in several military and government applications [22, 23]. CVW is a typical PBCE and one of the references describes it as essentially 'a chat room application coupled with a document repository'.

It is, however, more than that and its features comprise standard MOO features including several GUIs (Figure 4), rooms located on floors contained in buildings, text-
based communication (person to room occupants, person to person in or outside the
room), audio and video conferencing, a shared whiteboard, ability to handle documents in several formats, and event logging. Users can also form groups and assume
roles such as those required in meetings. Unlike other MOOs, CVW does not seem to
allow its users to extend the environment with new features.

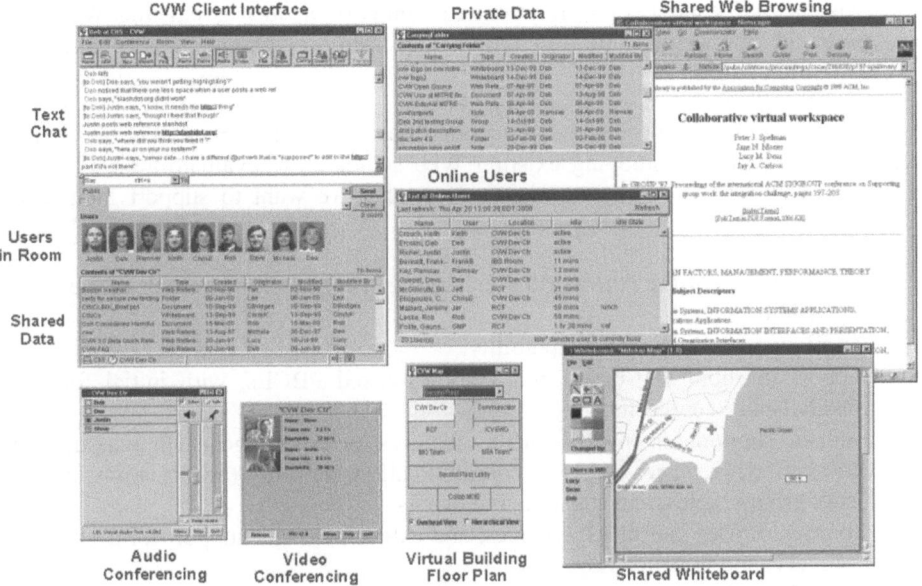

Fig. 4. Examples of CVW user interfaces [23].

An interesting aspect of CVW is its conceptual model, which consists of three
main components - participants, contexts, and conferences. *Participants* are human or
software agents with roles, such as administrator or facilitator. *Conferences* correspond to meetings and contain a roster of participants, their roles, and means of communication. *Contexts* define contexts in which the participants are meeting and contain a list of participants, a manager, a virtual room, and objects and tools. Maybury
[24] describes CVW as an essential component of a larger knowledge management
system.

Conclusions and Current Work

Our experience with Jersey and MUM, as well as study of recent work such as CVW,
led us to the conclusion that the conceptual model represented by PBCE environments
is a sound basis for powerful extendible environments supporting social interaction
and collaboration that should be supported by a combination of Web standards. We
decided to call this vision an *Inhabited Web* (IW) because it can be visualized as a

Web universe inhabited by agents, objects, and tools. Defining IW through a set of standards would provide the following advantages:

- If widely accepted, universal standards could open the Web to a new public and new uses, just as HTTP and HTML and then XTML-based standards did.
- Separating communication an representation formats from applications would enable interoperability among different IW applications just as HTML and the latter XML-based protocols and languages did.
- Existence of standards would stimulate competition just as existing Web standards did, and at the same time reduce duplication and simplify development of new applications.

We are currently exploring the conceptual basis of an Inhabited Web to create a conceptual model, and studying existing Web standards to evaluate their suitability for its implementation. The essential features that we want to support include the following:

- Consistent use of the real-world metaphor as the basis for as much of the model as possible.
- Persistence with run-time adaptability.
- Support for both text-based and graphics-based PBCEs, with initial focus on text/GUI-based environments.
- 'Built-in' support for universal MOO features including agents (human and software varieties), places (nested), communication, mobility (agents, tools, and objects), groups, roles, and subscribable events.
- Libraries of user interfaces, services, and extensions with support for description, discovery, and downloading.
- Semantic Web inclusion.
- Support for multiple interconnected universes.
- Support for extension of functionality, standards, and underlying technology.
- Support for IW-aware applications and tools.
- Ease of creation of new universes by expert and non-expert users.
- Support for a command-based interface.

We hope that the vision that we described will stimulate research that may lead to a new phase in Web development, a Web that builds on the document-centric Semantic Web and provides an environment organized on the basis of the familiar real-world paradigm and inhabited not only by documents but also by human and software agents, arbitrary objects, and IW-aware tools.

References

1. Berners-Lee T.: Weaving the Web. HarperCollins Publishers, New York (2000).
2. Skonnard A., Gudgin M.: Essential XML Quick Reference, Addison-Wesley, 2002.
3. XHTML™ 1.0 The Extensible HyperText Markup Language,
   http://www.w3.org/TR/xhtml1/

4. Newcomer A.: Understanding Web Services. Addison-Wesley, 2002.
5. SMTP: Simple Mail Transfer Protocol, http://cr.yp.to/smtp.html
6. Harrison, S., Dourish, P. Re-Place-ing Space: The roles of place and space in collaborative systems, , 1996.
7. Haynes, C., Holmevik, J. R. High Wired: On the Design, Use, and Theory of Educational MOOs. University of Michigan Press, 1998.
8. MOO Home Page, http://www.moo.mud.org/
9. Andrea L. Ames, et al.: VRML Source Book. John Wiley & Sons (1996).
10. X3D ™ - Extensible 3D. http://www.web3d.org/x3d/
11. Communications of the ACM, December 2201.
12. Towell, J.F., Towell, E.R. Presence in text-based networked virtual environments or "MUDS". Presence 6(5) 590-595, 1997.
13. Damers, B. *Avatars!* Peachpit Press, 1998.
14. Smith R. Shared Vision. Communications of the ACM, December 2201.
15. Horvath I., Rusak Z. Collaborative Shape Conceptualization in Virtual Design Environments. Communications of the ACM, December 2201.
16. Barrus, J. et al. Locales: Supporting Large Multiuser Virtual Environments. IEEE Computer Graphics Applications, 16(6), 50-57, 1997.
17. Tomek, I., Nicholl, R., Giles R., Saulnier, T., Zwicker J. A virtual environment supporting software developers. Proceedings of CVE '98, 1998.
18. Visual Age Smalltalk. IBM, http://www-3.ibm.com/software/ad/smalltalk/
19. Tomek I. The Design of a MOO, Journal of Computer and Network Applications, 23(3), Jul 2000, pp. 275-289
20. I. Tomek. In Erdogmus, H., Tanir, O., Advances in Software Engineering. Springer, 2002. .
21. VisualWorks Smalltalk, CINCOM, http://www.cincom.com.
22. Collaborative Virtual Workspace Overview;
    http://cvw.sourceforge.net/cvw/info/CVWOverview.php3
23. Maybury M.: Collaborative Virtual Environments for Analysis and Decision Support. Communications of the ACM, December 2201.
24. Maybury M. Knowledge on Demand: Knowledge and Expert Discovery. Proceedings of I-KNOW '02. J. UCS 2002, http://www.jucs.org.

Sweepline the Music!*

Esko Ukkonen, Kjell Lemström, and Veli Mäkinen

Department of Computer Science, University of Helsinki
P.O. Box 26 (Teollisuuskatu 23), FIN-00014 Helsinki, Finland
{ukkonen,klemstro,vmakinen}@cs.helsinki.fi

Abstract. The problem of matching sets of points or sets of horizontal line segments in plane under translations is considered. For finding the exact occurrences of a point set of size m within another point set of size n we give an algorithm with running time $O(mn)$, and for finding partial occurrences an algorithm with running time $O(mn \log m)$. To find the largest overlap between two line segment patterns we develop an algorithm with running time $O(mn \log(mn))$. All algorithms are based on a simple sweepline traversal of one of the patterns in the lexicographic order. The motivation for the problems studied comes from music retrieval and analysis.

1   Introduction

Computer-aided retrieval and analysis of music offers fascinating challenges for pattern matching algorithms. The standard writing of music as exemplified in Fig. 1 and Fig. 2 represents the music as notes. Each note symbol gives the pitch and duration of a tone. As the pitch levels and durations are normally limited to a relatively small set of discrete values, the representation is in fact a sequence of discrete symbols. Such sequences are a natural application domain for combinatorial pattern matching.

The so-called query-by-humming systems [10; 18; 14] are a good example of music information retrieval (MIR) systems that use pattern matching methods. A query-by-humming system has a content-based query unit and a database of symbolically encoded music such as popular melodies. A user of the database remembers somewhat fuzzily a melody and wants to know if something similar is in the database. Then the user makes a query to the database by humming (or whistling or playing by an instrument) the melody, and the query system should then find from the database the melodies that match best with the given one. The search over the database can be done very fast using advanced algorithms for approximate string matching, based on the edit distance and discrete time-warping (e.g. [13]).

Problems get more complicated if instead of simple melodies, we have polyphonic music like symphony orchestra scores. Such a music may have very complex structure with several notes simultaneously on and several musical themes developing in parallel. One might want to find similarities or other interesting patterns in it, for example, in order to make musicological comparative analysis of the style of different composers or even for copyright management purposes. Formulating various music-psychological

---

* A work supported by the Academy of Finland.

R. Klein et al. (Eds.): Comp. Sci. in Perspective (Ottmann Festschrift), LNCS 2598, pp. 330–342, 2003.

Figure 1. A melody represented in common music notation.

Figure 2. An excerpt of Einojuhani Rautavaara's opera *Thomas* (1985). Printed with the permission of the publisher Warner/Chappell Music Finland Oy.

phenomena and models such that one can work with them using combinatorial algorithms becomes a major challenge.

Returning back to our examples, the melody of Fig. 1 is given in Fig. 3 using so-called piano-roll representation. The content is now already quite explicit: each horizontal bar represents a note, its location in the y-axis gives its pitch level and the start and end points in the x-axis give the time interval when the note is on. Fig. 4 gives the piano-roll representation of the music of Fig. 2.

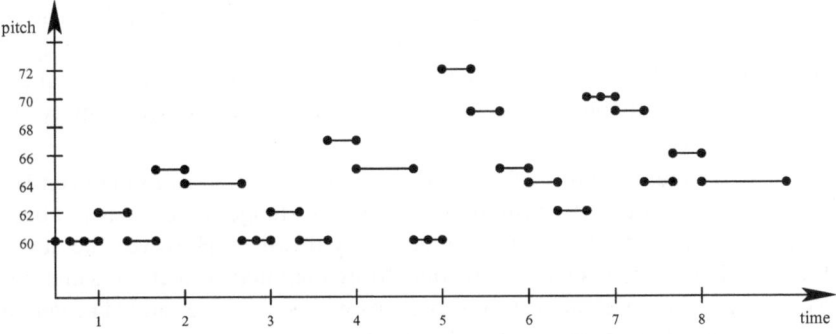

Figure 3. The example of Fig. 1 in piano-roll representation.

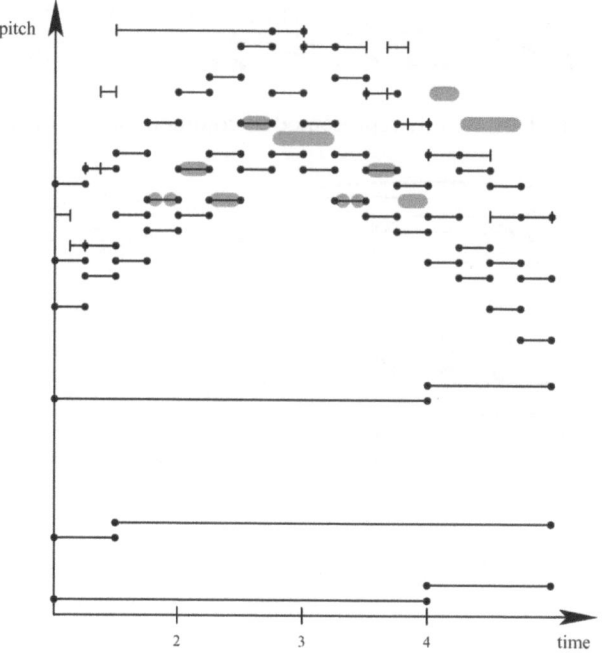

Figure 4. The example of Fig. 2 in piano-roll representation. The notes belonging to the soloist's melody are represented distinctly. Moreover, the first twelve notes of Fig. 3 are shown by shading in a translated position such that the starting points of 6 notes match and the total length of the overlap time is 6 quarter notes.

In western music, when comparing different pieces of music and melodies in particular, the absolute pitch levels of the notes are not of the primary interest. Rather, the pitch level differences between successive notes, the *intervals*, really matter. If a melody is obtained from another one by a transposition, which means adding a constant to the pitch levels of the notes, the transposed melody is still considered the same. Hence in music comparison it is customary to require invariance with respect to pitch level transpositions.

Similarly, for the durations of the notes, it is the duration ratio between successive notes that is important. Rewriting using shorter or longer notes does not basically change the melody as far as the duration ratios stay invariant. However, the number of possible rescalings is very small in practice. More common is, that the same musical theme occurs in otherwise varied forms, perhaps with some notes added or deleted, or the intervals or the duration ratios slightly changed.

This suggests that comparing musical sequences would need an appropriate form of approximate pattern matching that is invariant with respect to pitch transpositions. This could be combined with invariance with respect to rescaling of the tempo of the music but here also repeated searches with different scales can be feasible.

In this paper we will use a simple two-dimensional geometric representation of music, abstracted from the piano-roll representation. In this representation, a piece of music is a collection of horizontal line segments in the Euclidean two-dimensional space $\mathbb{R}^2$. The horizontal axis refers to the time, the vertical to the pitch values. As we do not discretize the available pitch levels nor the onset times or durations of the notes, the representation is more powerful than the standard notation of music as notes. In some cases, however, we consider the effect of the discretization on the efficiency of the algorithms.

Given two such representations, P and T, we want to find the common patterns shared by P and T when P is translated with respect to T. Obviously, the vertical component of the translation yields transposition invariance of the pattern matching while the horizontal component means shifting in time. When designing the algorithms we typically assume that T represents a large database of music while P is a shorter query piece, but it is also possible that both P and T refer to the same pattern.

Three problems will be considered.

(P1) Find translations of P such that all starting points of the line segments of P match with some starting points of the line segments in T. Hence the on-set times of all notes of P must match. We also consider a variant in which the note durations must match, too, or in which the time segment of T covered by the translated P is not allowed to contain any extra notes.

(P2) Find all translations of P that give a partial match of the on-set times of the notes of P with the notes of T.

(P3) Find translations of P that give longest common shared time with T. By the shared time we mean the total length of the line segments that are obtained as intersection of the line segments of T and the translated P.

Fig. 4 illustrates all three problems. There is no solution of (P1), but the shading shows a solution to (P2) and (P3).

For the problem (P1) we give in Sect. 3 an algorithm that needs time $O(mn)$ and working space $O(m)$. In practice the average running time is $O(n)$. Here m is the size (number of the line segments) of P and n is the size of T. For the problem (P2) we give in Sect. 4 an algorithm with running time $O(mn \log m)$ and space $O(m)$, and for the problem (P3) we describe in Sect. 5 a method that needs time $O(mn \log(mn))$ and space $O(mn)$. When the number of possible pitch levels is a finite constant (as is the case with music), the running time and working space of the algorithm for (P3) become $O(mn \log m)$ and $O(m)$. All algorithms are based on a simple *sweepline*-type [4] scanning of T. We assume that T and P are given in the lexicographic order of the starting points of the line segments. Otherwise time $O(n \log n + m \log m)$ and space $O(n + m)$ for sorting should be added to the above bounds.

Our problems are basic pattern matching questions in music when transpositions are allowed and note additions and deletions are modeled as partial matches. Problem (P3) also allows local changes in note durations. Tolerance to interval changes could be incorporated by representing the notes as narrow rectangles instead of lines.

Related Work. Our results on problems (P1) and (P2) slightly improve the recent results of Meredith et al. [19; 21] who, using similar algorithms, gave for (P1) a time bound $O(mn)$ and a space bound $O(n)$, and for (P2) a time bound $O(mn\log(mn))$ and a space bound $O(mn)$.

A popular approach to MIR problems has been to use different variants of the edit distance evaluated with the well–known dynamic programming algorithm. This line of research was initiated by Mongeau and Sankoff [20] who presented a method for the comparison of monophonic music. Edit distances with transposition invariance have been studied in [16; 13; 17]. In [8], dynamic programming is used for finding polyphonic patterns with restricted gaps. Bit–parallelism is used in the algorithm of [15] for finding transposed monophonic patterns within polyphonic music, and in the algorithm of [11] for finding approximate occurrences of patterns with group symbols.

Most of the MIR studies so far consider only monophonic music, and from the rhythmic information only the order of the notes is taken into account. An example of a more general approach is the MIR system PROMS [6] which works with polyphonic music and some durational information.

Our work has connections also to computational geometry, where point pattern matching under translations (and under more general transformations) is a well-studied field; see e.g. the survey by Alt and Guibas [1]. A problem very close to (P1) is to decide whether $A = \mathcal{T}(B)$, where $\mathcal{T}$ is an arbitrary rigid transformation, and A and B are point patterns (such as the sets of the starting points of the line segments in P and T), both of size n. This can be solved in $O(n\log n)$ time [3] by using a reduction to exact string matching. Our problem is more difficult since we are trying to match one point set with a subset of the other. A general technique, called alignment method in [12], can be used to solve problems (P1) and (P2) in time $O(mn\log n)$; we will sketch this solution in the beginning of Sect. 3. Allowing approximate point matching in problems (P1) and (P2) will make them much harder; an $O(n^6)$ algorithm was given in [2] for the case $|A| = |B| = n$, and only an improvement to $O(n^5\log n)$ [9] has been found since. However, a relaxed problem in which a point is allowed to match several points in the other pattern, leading to Hausdorff distance minimization under translations, can be solved in $O(n^2\log^2 n)$ time [5], when distances are measured by L_∞ norm (see citations in [1] for other work related to Hausdorff distance). Recently, a special case of (P1) in which points are required to be in integer coordinates was solved in $O(n\log n\log N)$ time [7] with a Las Vegas algorithm, where N is the maximum coordinate value.

2   Line Segment Patterns

A *line segment pattern* in the Euclidean space $\mathbb{R}^2$ is any finite collection of horizontal line segments. Such a segment is given as $[s, s']$ where the *starting point* $s = (s_x, s_y) \in \mathbb{R}^2$ and the *end point* $s' = (s'_x, s'_y) \in \mathbb{R}^2$ of the segment are such that $s_y = s'_y$ and $s_x \leq s'_x$. The segment consists of the points between its end points. Two segments of the same pattern may overlap.

We will consider different ways to match line segment patterns. To this end we are given two line segment patterns, P and T. Let P consist of m segments $[p_1, p'_1], \ldots, [p_m, p'_m]$ and T of n segments $[t_1, t'_1], \ldots, [t_n, t'_n]$. Pattern T may represent a large database while P is a relatively short query to the database in which case $m \ll n$. It is also possible that P and T are about of the same size, or even that they are the same pattern.

We assume that P and T are given in the lexicographic order of the starting points of their segments. The lexicographic order of points $a = (a_x, a_y)$ and $b = (b_x, b_y)$ in $\mathbb{R}^2$ is defined by setting $a < b$ iff $a_x < b_x$, or $a_x = b_x$ and $a_y < b_y$. When representing music, the lexicographic order corresponds the standard reading of the notes from left to right and from the lowest pitch to the highest.

So we assume that the lexicographic order of the starting points is $p_1 \leq p_2 \leq \cdots \leq p_m$ and $t_1 \leq t_2 \leq \cdots \leq t_n$. If this is not true, a preprocessing phase is needed, to sort the points which would take additional time $O(m \log m + n \log n)$.

A *translation* in the real plane is given by any $f \in \mathbb{R}^2$. The translated P, denoted by $P + f$, is obtained by replacing any line segment $[p_i, p_i']$ of P by $[p_i + f, p_i' + f]$. Hence $P + f$ is also a line segment pattern, and any point $v \in \mathbb{R}^2$ that belongs to some segment of P is mapped in the translation as $v \mapsto v + f$.

3   Exact Matching

Let us denote the lexicographically sorted sets of the starting points of the segments in our patterns P and T as $\overline{P} = (p_1, p_2, \ldots, p_m)$ and $\overline{T} = (t_1, t_2, \ldots, t_n)$. We now want to find all translations f such that $\overline{P} + f \subseteq \overline{T}$. Such a $\overline{P} + f$ is called an *occurrence* of $\overline{P}$ in $\overline{T}$. As all points of $\overline{P} + f$ must match some point of $\overline{T}$, $p_1 + f$ in particular must equal some t_j. Hence there are only n potential translations f that could give an occurrence, namely the translations $t_j - p_1$ where $1 \leq j \leq n$. Checking for some such translation $f = t_j - p_1$, that also the other points $p_2 + f, \ldots, p_m + f$ of $\overline{P} + f$ match, can be performed in time $O(m \log n)$ using some geometric data structure to query $\overline{T}$ in logarithmic time. This leads to total running time of $O(mn \log n)$.

We can do better by utilizing the lexicographic order. The method will be based on the following simple lemma.

Denote the potential translations as $f_j = t_j - p_1$ for $1 \leq j \leq n$. Let $p \in \overline{P}$, and let f_j and $f_{j'}$ be two potential translations such that $p + f_j = t$ and $p + f_{j'} = t'$ for some $t, t' \in \overline{T}$. That is, when p_1 matches t_j then p matches t, and when p_1 matches $t_{j'}$ then p matches t'.

Lemma 1 *If $j < j'$ then $t < t'$.*

Proof. If $j < j'$, then $t_j < t_{j'}$ by our construction. Hence also $f_j < f_{j'}$, and the lemma follows.   □

Our algorithm makes a traversal over $\overline{T}$, matching p_1 against the elements of $\overline{T}$. At element t_j we in effect are considering the translation f_j. Simultaneously we maintain for each other point p_i of $\overline{P}$ a pointer q_i that also traverses through $\overline{T}$. When q_i is at t_j, it in effect represents translation $t_j - p_i$. This translation is compared to the current f_j, and the pointer q_i will be updated to the next element of $\overline{T}$ after q_i if the translation is smaller (the step $q_i \leftarrow next(q_i)$ in the algorithm below). If it is equal, we have found a match for p_i, and we continue with updating q_{i+1}. It follows from Lemma 1, that no backtracking of the pointers is needed.

The resulting procedure is given below in Fig. 5.

```
(1)  for i ← 1,...,m do q_i ← −∞
(2)  q_{m+1} ← ∞
(3)  for j ← 1,...,n − m do
(4)      f ← t_j − p_1
(5)      i ← 1
(6)      do
(7)          i ← i + 1
(8)          q_i ← max(q_i, t_j)
(9)          while q_i < p_i + f do q_i ← next(q_i)
(10)     until q_i > p_i + f
(11)     if i = m + 1 then output(f)
(12) end for.
```

Figure 5. Algorithm 1.

Note that the main loop (line 3) of the algorithm can be stopped when $j = n − m$, i.e., when p_1 is matched against $t_{n−m}$. Matching p_1 beyond $t_{n−m}$ would not lead to a full occurrence of $\overline{P}$ because then all points of $\overline{P}$ should match beyond $t_{n−m}$, but there are not enough points left.

That Algorithm 1 correctly finds all f such that $\overline{P} + f \subseteq \overline{T}$ is easily proved by induction. The running time is $O(mn)$, which immediately follows from that each q_i traverses through $\overline{T}$ (possibly with jumps!). Also note that this bound is achieved only in the rare case that $\overline{P}$ has $\Theta(n)$ full occurrences in $\overline{T}$. More plausible is that for random $\overline{P}$ and $\overline{T}$, most of the potential occurrences checked by the algorithm are almost empty. This means that the loop 6–10 is executed only a small number of times at each check point, independently of m. Then the expected running time under reasonable probabilistic models would be $O(n)$. In this respect Algorithm 1 behaves analogously to the brute-force string matching algorithm.

It is also easy to use additional constraints in Algorithm 1. For example, one might want that the lengths of the line segments must also match. This can be tested separately once a full match of the starting points has been found. Another natural requirement can be, in particular if P and T represent music, that there should be no extra points in the time window covered by an occurrence of $\overline{P}$ in $\overline{T}$. If $\overline{P} + f$ is an occurrence, then this time window contains all members of $\overline{T}$ whose x-coordinate belongs to the interval $[(p_1 + f)_x, (p_m + f)_x]$. When an occurrence $\overline{P} + f$ has been found in Algorithm 1, the corresponding time window is easy to check for extra points. Let namely $t_j = p_1 + f$ and $t_{j'} = p_m + f$. Then the window contains just the points of $\overline{T}$ that match $\overline{P} + f$ if and only if $j' − j = m − 1$ and $t_{j−1}$ and $t_{j'+1}$ do not belong to the window.

4 Largest Common Subset

Our next problem is to find translations f such that $(\overline{P} + f) \cap \overline{T}$ is nonempty. Such a $\overline{P} + f$ is called a *partial occurrence* of $\overline{P}$ in T. In particular, we want to find f such that $(\overline{P} + f) \cap \overline{T}$ is largest possible.

There are $O(mn)$ translations f such that $(\overline{P}+f) \cap \overline{T}$ is nonempty, namely the translations $t_j - p_i$ for $1 \leq j \leq n$, $1 \leq i \leq n$. Checking the size of $(\overline{P}+f) \cap \overline{T}$ for each of them solves the problem. A brute-force algorithm would typically need time $O(m^2 n \log n)$ for this. We will give a simple algorithm that will do this during m simultaneous scans over $\overline{T}$ in time $O(mn \log m)$.

Lemma 2 *The size of* $(\overline{P}+f) \cap \overline{T}$ *equals the number of disjoint pairs* (j,i) *(i.e., pairs sharing no elements) such that* $f = t_j - p_i$.

Proof. Immediate.   □

By Lemma 2, to find the size of any non-empty $(\overline{P}+f) \cap \overline{T}$ it suffices to count the multiplicities of the translation vectors $f_{ji} = t_j - p_i$. This can be done fast by first sorting them and then counting. However, we can avoid full sorting by observing that translations $f_{1i}, f_{2i}, \ldots, f_{ni}$ are in the lexicographic order for any fixed i. This sorted sequence of translations can be generated in a traversal over $\overline{T}$. By m simultaneous traversals we get these sorted sequences for all $1 \leq i \leq m$. Merging them on-the-fly into the sorted order, and counting the multiplicities solves our problem.

The detailed implementation is very standard. As in Algorithm 1, we let $q_1, q_2, \ldots,$ q_m refer to the entries of $\overline{T}$. Initially each of q_i refers to t_1, and it is also convenient to set $t_{n+1} \leftarrow \infty$. The translations $f_i = q_i - p_i$ are kept in a priority queue F. Operation $min(F)$ gives the lexicographically smallest of translations f_i, $1 \leq i \leq m$. Operation $update(F)$ deletes the minimum element from F, let it be $f_h = q_h - p_h$, updates $q_h \leftarrow next(q_h)$, and finally inserts the new $f_h = q_h - p_h$ into F.

Then the body of the pattern matching algorithm is as given below.

(1) $f \leftarrow -\infty; c \leftarrow 0;$
　　do
(2) 　　$f' \leftarrow min(F); update(F)$
(3) 　　**if** $f' = f$ **then** $c \leftarrow c+1$
(4) 　　**else** $\{output(f,c); f \leftarrow f'; c \leftarrow 1\}$
(5) **until** $f = \infty$

The algorithm reports all (f,c) such that $|(\overline{P}+f) \cap \overline{T}| = c$ where $c > 0$.

The running time of the algorithm is $O(mn \log m)$. The m-fold traversal of $\overline{T}$ takes mn steps, and the operations on the m element priority queue F take time $O(\log m)$ at each step of the traversal.

The above method finds all partial occurrences of $\overline{P}$ independently of their size. Concentrating on large partial occurrences gives possibilities for faster practical algorithms based on *filtration*. We now sketch such a method. Let $|(\overline{P}+f) \cap \overline{T}| = c$ and $k = m - c$. Then $\overline{P}+f$ is called an occurrence with k *mismatches*.

We want to find all $\overline{P}+f$ that have at most k mismatches for some fixed value k. Then we partition $\overline{P}$ into $k+1$ disjoint subsets $\overline{P}_1, \ldots, \overline{P}_{k+1}$ of (about) equal size. The following simple fact which has been observed earlier in different variants in string matching literature will give our filter.

Lemma 3 *If $\overline{P} + f$ is an occurrence of $\overline{P}$ with at most k mismatches then $\overline{P}_h + f$ must be an occurrence of $\overline{P}_h$ with no mismatches at least for one h, $1 \le h \le k + 1$,*

Proof. By contradiction: If every $\overline{P}_h + f$ has at least 1 mismatch then $\overline{P} + f$ must have at least $k + 1$ mismatches as $\overline{P} + f$ is a union of disjoint line segment patterns $\overline{P}_h + f$. □

This gives the following filtration procedure: First (the filtration phase) find by Algorithm 1 of Section 3 all (exact) occurrences $\overline{P}_h + f$ of each P_h. Then (the checking phase) find for each f produced by the first phase, in the ascending order of the translations f, how many mismatches each $\overline{P} + f$ has.

The filtration phase takes time $O(mn)$, sorting the translations takes $O(r \log k)$ where $r \le (k+1)n$ is the number of translations the filter finds, and checking using an algorithm similar to the algorithm given previously in this section (but now priority queue is not needed) takes time $O(m(n + r))$. It should again be obvious, that the expected performance can be much better whenever k is relatively small as compared to m. Then the filtration would take expected time $O(kn)$. This would dominate the total running time for small r if the checking is implemented carefully.

5   Longest Common Time

Let us denote the line segments of P as $\pi_i = [p_i, p_i']$ for $1 \le i \le m$, and the line segments of T as $\tau_j = [t_j, t_j']$ for $1 \le i \le n$.

Our problem in this section is to find a translation f such that the line segments of $P + f$ intersects T as much as possible. For any horizontal line segments L and M, let $c(L, M)$ denote the length of their intersection line segment $L \cap M$. Then let

$$C(f) = \sum_{i,j} c(\pi_i + f, \tau_j).$$

Our problem is to maximize this function. The value of $C(f)$ is nonzero only if the vertical component f_y of $f = (f_x, f_y)$ brings some π_i to the same vertical position as some τ_j, that is, only if $f_y = (t_j)_y - (p_i)_y$ for some i, j.

Let H be the set of different values $(t_j)_y - (p_i)_y$ for $1 \le i \le m$, $1 \le j \le n$. Note that H here is a standard set, not a multiset; the size of H is $O(mn)$.

As $C(f)$ gets maximum when $f_y \in H$ we obtain that

$$\max_f C(f) = \max_{f_y \in H} \max_{f_x} C((f_x, f_y)). \tag{1}$$

We will now explicitly construct the function $C((f_x, f_y)) = C_{f_y}(f_x)$ for all fixed $f_y \in H$. To this end, assume that $f_y = (t_j)_y - (p_i)_y$ and consider the value of $c_{ij}(f_x) = c(\pi_i + (f_x, f_y), \tau_j)$. This is the contribution of the intersection of $\pi_i + (f_x, f_y)$ and τ_j to the value of $C_{f_y}(f_x)$. The following elementary analysis characterizes $c_{ij}(f_x)$. When f_x is small enough, $c_{ij}(f_x)$ equals 0. When f_x grows, at some point the end point of $\pi_i + (f_x, f_y)$ meets the starting point of τ_i and then $c_{ij}(f_x)$ starts to grow linearly with slope 1 until the starting points and the end points of $\pi_i + (f_x, f_y)$ and τ_j meet, whichever comes first. After that, $c_{ij}(f_x)$ has a constant value (minimum of the lengths of the two line segments) until the starting points or the end points (i.e., the remaining pair of the two alternatives) meet, from which point on, $c_{ij}(f_x)$ decreases linearly with slope -1 until it becomes zero at the point where the starting point of π_i hits the end point of τ_j. An easy exercise shows that the only turning points of $c_{ij}(f_x)$ are the four points described above and their values are

$$
\begin{aligned}
f_x &= (t_j)_x - (p'_i)_x & &\text{slope 1 starts} \\
f_x &= \min\left((t_j)_x - (p_i)_x, (t'_j)_x - (p'_i)_x\right) & &\text{slope 0 starts} \\
f_x &= \max\left((t_j)_x - (p_i)_x, (t'_j)_x - (p'_i)_x\right) & &\text{slope } -1 \text{ starts} \\
f_x &= (t'_j)_x - (p_i)_x & &\text{slope 0 starts.}
\end{aligned}
$$

Hence the slope changes by $+1$ at the first and the last turning point, while it changes by -1 at the second and the third turning point.

Now, $C_{f_y}(f_x) = \sum_{i,j} c_{ij}(f_x)$, hence C_{f_y} is a sum of piecewise linear continuous functions and therefore it gets its maximum value at some turning point of the c_{ij}'s. Let $g_1 \le g_2 \le \cdots \le g_K$ be the turning points in increasing order, each point listed according to its multiplicity; note that different functions c_{ij} may have the same turning point. So, for each i, j, the four values

$$
\begin{aligned}
(t_j)_x - (p'_i)_x & \qquad &\text{(type 1)} \\
(t_j)_x - (p_i)_x & \qquad &\text{(type 2)} \\
(t'_j)_x - (p'_i)_x & \qquad &\text{(type 3)} \\
(t'_j)_x - (p_i)_x & \qquad &\text{(type 4)}
\end{aligned}
$$

are in the lists of the g:s, and each knows its "type" shown above.

To evaluate $C_{f_y}(f_x)$ at its all turning points we scan the turning points g_k and keep track of the changes of the slope of the function C_{f_y}. Then it is easy to evaluate $C_{f_y}(f_x)$ at the next turning point from its value at the previous one. Let v be the previous value and let s represent the slope. The evaluation is given below.

(1) $v \leftarrow 0; s \leftarrow 0$
(2)   **for** $k \leftarrow 1, \ldots, K$ **do**
(3)   **if** $g_k \ne g_{k-1}$ **then** $v \leftarrow v + s(g_k - g_{k-1})$
(4)   **if** g_k is of type 1 or type 4 **then** $s \leftarrow s + 1$
(5)   **else** $s \leftarrow s - 1$.

This should be repeated for all different $f_y \in H$. We next describe a method that generates the turning points in increasing order simultaneously for all different f_y. The method will traverse T using four pointers (the four "types") per an element of P. A priority queue is again used for sorting the translations given by the $4m$ pointers; the x-coordinates of the translations then give the turning points in ascending order. At each turning point we update the counters v_h and s_h where h is given by the y-coordinate of the translation.

We need two traversal orders of T. The first is the one we have used so far, the lexicographic order of the starting points t_j. This order is given as $\overline{T}$. The second order is the lexicographic order of the end points t'_j. Let $\overline{T}'$ be the end points in the sorted order.

Let q_i^1, q_i^2, q_i^3, and q_i^4 be the pointers of the four types, associated with element π_i of P. Pointers q_i^1 and q_i^2 traverse $\overline{T}$, and pointers q_i^3 and q_i^4 traverse $\overline{T}'$. The translation associated with the current value of each pointer is given as

$$tr(q_i^1) = q_i^1 - p'_i$$
$$tr(q_i^2) = q_i^2 - p_i$$
$$tr(q_i^3) = q_i^3 - p'_i$$
$$tr(q_i^4) = q_i^4 - p_i.$$

So, when the pointers refer to t_j or t'_j, the x-coordinate of these translations give the turning points, of types 1, 2, 3, and 4, associated with the intersection of π_i and π_j. The y-coordinate gives the vertical translation $(t_j)_y - (p_i)_y$ that is needed to classify the turning points correctly.

During the traversal, all $4m$ translations tr given by the $4m$ pointers are kept in a priority queue. By repeatedly extracting the minimum from the queue (and updating the pointer that gives this minimum tr) we get the translations in ascending lexicographic order, and hence the x-coordinate of the translations gives all turning points in ascending order.

Let $f = (f_x, f_y)$ be the next translation obtained in this way. Then we retrieve the slope counter s_{f_y} and the value counter v_{f_y}. Assuming that we have also stored the last turning point z_{f_y} at which v_{f_y} was updated, we can now perform the following updates. If $f_x \neq z_{f_y}$, then let $v_{f_y} \leftarrow v_{f_y} + s_{f_y}(f_x - z_{f_y})$ and $z_{f_y} \leftarrow f_x$. Moreover, if f is of types 1 or 4, then let $s_{f_y} \leftarrow s_{f_y} + 1$ otherwise $s_{f_y} \leftarrow s_{f_y} - 1$.

In this way we obtain the values v_h for each function C_h and each $h \in H$, at each turning point. By (1), the maximum value of C must be among them.

The described procedure needs $O(n \log n)$ time for sorting T into $\overline{T}$ and $\overline{T}'$, $O(mn \log m)$ time for generating the $O(mn)$ turning points in increasing order. At each turning point we have to retrieve the corresponding slope and value counters using the vertical translation $h \in H$ as the search key. Hence we need in general time $O(\log |H|) = O(\log(mn))$. This gives a total time requirement $O(mn \log(mn))$ and space requirement $O(mn)$.

When P and T represent music, the size of H is limited independently of m and n. Certainly $|H|$ is less than 300 and often much smaller. We can use bucketing to manage the slope and value counters. The resource bounds become $O(n \log n + mn \log m)$ for time and, more importantly, $O(m + n)$ for space.

6 Conclusion

We presented efficient algorithms for three pattern matching problems. Their motivation comes from music but as computational problems they have a clear geometric nature. Also our algorithms are geometric, based on simple sweepline techniques. The algorithms adapt themselves easily to different variations of the problems such as to weighted matching or to patterns that consist of rectangles instead of line segments.

Experimentation with the algorithms on real music data would be interesting, to see whether these techniques can compete in accuracy, flexibility and speed with the dynamic programming based methods.

References

1. H. Alt and L. Guibas. Discrete geometric shapes: Matching, interpolation, and approximation. In J.-R. Sack and J. Urrutia, editors, *Handbook of Computational Geometry*, pages 121 – 153. Elsevier Science Publishers B.V. North-Holland, Amsterdam, 1999.
2. H. Alt, K. Mehlhorn, H. Wagener, and E. Welzl. Congruence, similarity and symmetries of geometric objects. *Discrete Comput. Geom.*, 3:237–256, 1988.
3. M.D. Atkinson. An optimal algorithm for geometric congruence. *J. Algorithms*, 8:159– 172, 1997.
4. J. L. Bentley and T. A. Ottmann. Algorithms for reporting and counting geometric intersections. *IEEE Transactions on Computers*, C-28:643–647, September 1979.
5. L.P. Chew and K. Kedem. Improvements on geometric pattern matching problems. In *Proceedings of the Scandinavian Workshop Algomthm Theory (SWAT)*, pages 318–325, 1992.
6. M. Clausen, R. Engelbrecht, D. Meyer, and J. Schmitz. Proms: A web-based tool for searching in polyphonic music. In *Proceedings of the International Symposium on Music Information Retrieval (ISMIR'2000)*, 2000.
7. R. Cole and R. Hariharan. Verifying candidate matches in sparse and wildcard matching. In *Proceedings of the 34th ACM Symposium on Theory of Computing*, pages 592–601. ACM Press, 2002.
8. M.J. Dovey. A technique for "regular expression" style searching in polyphonic music. In *the 2nd Annual International Symposium on Music Information Retrieval (ISMIR'2001)*, pages 179–185, 2001.
9. A. Efrat and A. Itai. Improvements on bottleneck matching and related problems using geometry. In *Proceedings of the twelfth annual symposium on Computational geometry*, pages 301–310. ACM Press, 1996.
10. A. Ghias, J. Logan, D. Chamberlin, and B.C. Smith. Query by humming - musical information retrieval in an audio database. In *ACM Multimedia 95 Proceedings*, pages 231– 236, 1995. Electronic Proceedings: http://www.cs.cornell.edu/Info/Faculty/bsmith/query-by-humming.
11. J. Holub, C.S. Iliopoulos, and L. Mouchard. Distributed string matching using finite automata. *Journal of Automata, Languages and Combinatorics*, 6(2):191–204, 2001.
12. D. Huttenlocher and S. Ullman. Recognizing solid objects by alignment with an image. *Intern. J. Computer Vision*, 5:195–212, 1990.
13. K. Lemström. *String Matching Techniques for Music Retrieval*. PhD thesis, University of Helsinki, Department of Computer Science, 2000. Report A-2000-4.
14. K. Lemström and S. Perttu. SEMEX - an efficient music retrieval prototype. In *Proceedings of the International Symposium on Music Information Retrieval (ISMIR'2000)*, 2000.
15. K. Lemström and J. Tarhio. Detecting monophonic patterns within polyphonic sources. In *Content-Based Multimedia Information Access Conference Proceedings (RIAO'2000)*, pages 1261–1279, 2000.
16. K. Lemström and E. Ukkonen. Including interval encoding into edit distance based music comparison and retrieval. In *Proceedings of the AISB'2000 Symposium on Creative & Cultural Aspects and Applications of AI & Cognitive Science*, pages 53–60, 2000.
17. V. Mäkinen, G. Navarro, and E. Ukkonen. Algorithms for transposition invariant string matching. Technical Report TR/DCC-2002-5, Department of Computer Science, University of Chile, 2002.

18. R.J. McNab, L.A. Smith, D. Bainbridge, and I.H. Witten. The New Zealand digital library MELody inDEX. *D-Lib Magazine*, 1997. http://www.nzdl.org/musiclib.
19. D. Meredith, G.A. Wiggins, and K. Lemström. Pattern induction and matching in polyphonic music and other multi-dimensional data. In *the 5th World Multi-Conference on Systemics, Cybernetics and Informatics (SCI'2001)*, volume X, pages 61–66, 2001.
20. M. Mongeau and D. Sankoff. Comparison of musical sequences. *Computers and the Humanities*, 24:161–175, 1990.
21. G.A. Wiggins, K. Lemström, and D. Meredith. SIA(M) — a family of efficient algorithms for translation invariant pattern matching in multidimensional datasets. Manuscript (submitted), September 2002.

Teaching an Old Course New Tricks:
A Portable Courseware Language Based on XML

Lutz Wegner

Universität Kassel, Fachbereich Mathematik/Informatik,
D-34109 Kassel, Germany
`wegner@db.informatik.uni-kassel.de`

Abstract. Producing eLearning material remains a costly and challenging task. Reflecting on a course written 20 years ago and still in use today, the author tries to identify the essentials for successful computer-supported teaching material. One apparent need is to make the contents more portable. In particular, the Web is seen as the universal distribution and presentation channel. To allow for interaction, animation, vector graphics and display independence, a Portable CourseWare Language (PCWL) is proposed. PCWL is based on XML, SVG, and Java. An editor for PCWL is provided. The paper leads from past to future with screenshots from Teletext to IE5 with SVG plug-in.

1 The Courseware Lifecycle

Ever so often a paper appears in the popular press which laments about eLearning and its apparent lack of acceptance (see e.g. the press reviews in [1]). Having attended video-based instructions in Karlsruhe around 1972 (that is 30 years ago!) and as both an author for computer-assisted learning material and as an operator of a CAL-lab for almost twenty years, these frustrations come as no surprise.

It seems, promoters of eLearning continuously fall into one (or all) of the four following traps.

The production cost trap. A reasonably complete course requires around 1500 pages (frames) which corresponds to about 20 - 30 h of documentary film. Like the film it requires an author for the script, a director, a cutter (graphics programmer), etc. Underestimating the effort, producers and authors soon realize that the product will not run up to their expectations. However, at that point nobody dares to openly admit that the project is heading for disaster as all involved are totally underpaid and overworked. At the end, frustrations on all sides prevail and the result is either a jerky "slide show for guinea- pigs" or a dry copy of a book.

The "come-as-you-please" trap. From the very beginning eLearning (originally also called "programmed instructions") was designed to replace the instructor/lecturer, even though it was often not openly admitted. Indeed it can be argued that public funding goes into eLearning only because the donators hope that - in the long run - it will save on the cost for education.

In reality, asynchronous learning (anytime-anywhere) never worked as learning is a social process which needs immediate feed-back from fellow students and guidance

R. Klein et al. (Eds.): Comp. Sci. in Perspective (Ottmann Festschrift), LNCS 2598, pp. 343-355, 2003.

from a dedicated teacher. This can be overcome by running the courseware in a synchronous mode, e.g. in a learning theatre. Then, however, the effort for lecturing in this theatre (usually assisted by one or two assistants), the cost of maintaining the lab, and the trouble with keeping course material and actual teaching well balanced, renders this form of blended eLearning rather cost-intensive. As a matter of fact, the old fashioned professor in front of a large audience in a lecture hall - provided he/she is up to the challenge - is hard to beat both in terms of cost-effectiveness, knowledge transfer, and social guidance.

The technology trap. Courseware technology ages fast. One might consider it a side-effect of "Moore's Law" which claims that the hardware doubles the performance at constant prices about every 18 months. With increased display resolutions, fancy graphics, audio effects and elaborate Web-designs, the look&feel of courseware quickly becomes outdated. In contrast, courseware contents and didactic components do not necessarily age fast.

Much against popular belief, computer science core material remains current for a long time. To prove the point, consider relational theory for DBMS lectures, process synchronization and interprocess communication methods for operating systems courses, algorithms and data structures material, graph theory for networking lectures, mathematical foundations of computer graphics, etc. - all well over 30 years old and still current. Similarly, a minimalistic approach for presenting the material avoids fast aging. Such an approach relies on simple animations where useful to demonstrate dynamic changes. It also restricts itself to solid core topics and linguistic excellence - including wit and sparingly used humor. Not surprisingly, these are ingredients of any good lecturing material - textbook or electronic courseware- and are thus independent of technology.

The marketing trap. This trap is related to the "effort trap" from above. As the cost for producing course material sky-rockets, authors and producers hope to amortize the cost by "selling" it to other educational units. However, as it turns out, lecturers are very reluctant to actually use another person's course material, even simple slides or exercise sheets. One reason is what might be called "the assumed mistake and the non-understood remark".

This is a well-known phenomenon with all borrowed intellectual property. As soon as course material becomes slightly more demanding, people disagree on the proper way to present it, cast doubt about correct answers to exercises, would put emphasis on different aspects or see things in a different light, cannot follow a side remark, etc. Thus, as lecturers continue to distrust courseware from other sources, the market for eLearning material essentially reduces to the author himself.

In summery, one might claim that courseware development and employment of these technologies suffers from tunnel vision, a claim made by Brown and Duguid [2] for information technology in general. Not being able to use our peripheral vision, as we rush ahead, we ignore context, background, history, common knowledge and social resources as integral part of learning. Trusting on technology only, we loose sight of "communities, organizations, and institutions that frame human activities [2, p. 5]".

2 "Introduction to UNIX" as a Case Study

To work in the field of eLearning one has to be a lunatic. Phases of euphoria are followed by depression. Figure 1 below shows some of these ups and downs and related technologies against an approximate time-line

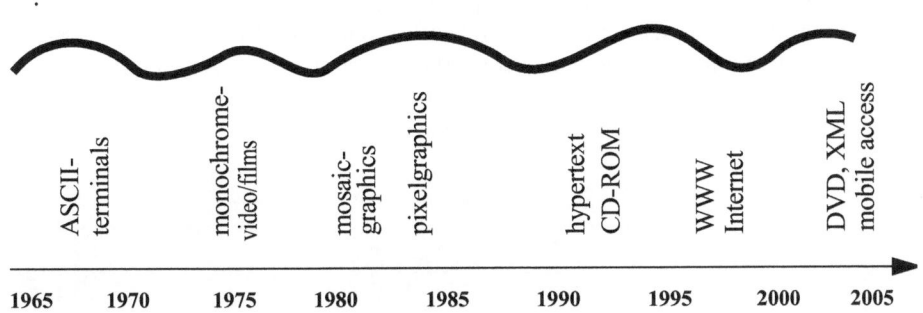

Figure 1. Trends in eLearning now and then

Each technology gave rise to expectations that this technological push would finally bring eLearning the ultimate break-through. Soon afterwards, as labs became deserted and unsold shrink-wrapped course material collected dust on shelves, investments (in intellectual effort and in financial terms) were written off.

Some experts in the eLearning field, like Ottmann [13], argue that the production of classic courseware in the form of animated books is too expensive and inflexible and should be replaced by lecture recording. Others, like Krämer and Wegner [14] and the authors from the CSCW-community [16] suggest supplementing the courseware with synchronous groupware tools to create a social context. However, this makes eLearning even more elaborate and technology dependent.

In some cases, however, a lucky choice of topics presented in a well-written form can be used in a learning theatre or even off-line. The author's own course "Introduction to UNIX", turned out to be a lucky pick and an effort was made to salvage the contents of this course and to port it to new technologies. In the following we report on these efforts and the lessons learned from this painful process. By demonstrating our current development based on XML [9] and Scalable Vector Graphics (SVG) [6, 8], we hope to have finally achieved the port to end all ports[1].

This course originated with the COSTOC project in the mid-Eighties. It was initiated by Hermann Maurer, best known for his Hyper-G and HyperWave developments [3]. COSTOC collected some thirty remarkable courses from respected researchers and teachers world-wide. Each course (or at least the majority of completed courses) was composed of about ten lectures (chapters) each with 90 - 120 frames. The technological base was BTX (Bildschirmtext, Teletext) including animation sequences which were downloaded to Microcomputers (Mupid teletext terminals equipped with a Z80

---

1. In talking about the new Constitution and its appearance that promises permanency, Benjamin Franklin remarks that "...in the world nothing can be said to be certain except death and taxes."

microprocessor) and ran a course viewer. Note that this is a very early, if not the earliest, form of mobile code later known as applets or Tclets.

Courses were written on these microcomputers as well and each package of pages (Btx-frames) had to fit into a space of at most 32 KB. This led to a minimalistic design with precise wordings and simple, yet powerful animations. A rigorous editing discipline guaranteed a totally consistent appearance as far as use of colors for headings, text, emphasis, questions, summaries, table of contents, etc. was concerned. About 100 questions, mostly multiple-choice, were part of the course and later on turned out to be the major success factor for its survival. Figure 2 shows a screenshot - explaining the link (ln) command - from the original course.

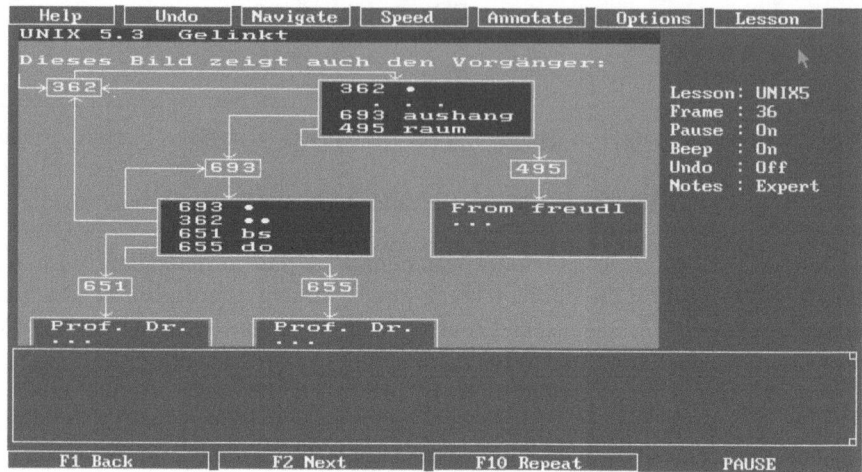

Figure 2. Screenshot from the original COSTOC course

In total, the effort that went into this course came up to 8 man years including editing and German and English text supplements [5]. The course ran in a special Btx-Lab at Kassel University. The other courses were offered there as well, but none of them was ever used for lecturing or was accessed by many students.

The lab was soon abandoned, in part also because of the poor quality of the displays attached to the Mupid microcomputers. Fortunately enough, however, Herman Maurer had his software ported to IBM-compatible PCs with EGA-graphics. In Kassel, these modified COSTOC-courses were made available in a PC-lab, consisting of 12 IBM PS/2 models connected via token ring to a courseware server [4]. Moreover, a technician managed to set up the PCs both as ASCII-terminals to our UNIX-server and as courseware players under Windows 3.1. Although not usable with overlapping windows, students could nevertheless switch between true UNIX-exercises and course material (which included simulated UNIX sessions) by means of a hot-key combination.

As mentioned before, the course was used in synchronous mode only. Offers to make use of the courses, which were actually provided free of charge, were never taken up as to our knowledge.

As far as content is concerned, the course remained surprisingly up-to-date. Naturally, it did not mention graphic interfaces or particular tools, like modern mail-interfaces, but concentrated on file structures, processes, and the Bourne Shell. Over the years, the UNIX systems which we employed in the back and which were made accessible to the students for hands-on experiences changed from AIX to Linux to Sun Solaris and back to Linux. The course worked well with all of them.

Topics originally not included were later added to the written documentation. They included mounting of external devices, PGP encryption, hard links versus soft links, foreground and background control in the bash, etc. Changes to the electronic material were not possible - there was no editor available to us! Like true legacy software, the course included a bug which would crash the PC (a coding of a character from Btx) which we were never able to eliminate but had to ask students instead to skip that frame.

3 Finally We Part - From the Hardware!

By the end of the Nineties the PC-lab was totally outdated although the IBM PS/2 machines with their 16" monitors refused to die. However, the courseware would not run properly on Windows 95 and successors. Some of the code made low-level access to the graphics board, a common style in the late Eighties, which made it not portable to operating systems which shield their hardware.

In talking about possible ports to a new environment, in particular one where the course could run in one window in parallel with another window open to a UNIX machine, it was clear that the courseware should be designed for the universal interface of today: the Internet browser. This was easier said than done.

As it turns out, this courseware from the early Eighties was more sophisticated than what HTML and the WWW could handle. Can you have decent animations on the Web? With some plug-in we were able to use animated GIFs which consumed enormous space and were not portable or scalable. Figure 3 shows a screenshot of the course running within IE5 with most pictures and animations taken from the original material and converted to GIFs. Readers may access this material at our site http://www.db.informatik.uni-kassel.de/Lehre/unix/.

How about the nice questions and limited tries for answers in the exercises? How about the fill-in questions with tolerance for spellings? Writing those CGI-scripts to handle all of that turned out to be hard, in particular since HTTP is a stateless protocol. Modern tools, like Mediator 6 Pro [15], which we bought upon recommendation from eLearning experts, turned out to be a disappointment.

We therefore looked for alternatives and decided to go for XML and SVG (Scalable Vector Graphics) as international standards. With additional funding from the BMFT project "Notebook University" we decided to design our own generic courseware language into which any course material could be embedded.

4 PCWL: A Portable CourseWare Language

PCWL is XML-based. XML is a meta-language, thus PCWL can be anything the designers want it to be. However, if we are to provide a runtime environment, it must

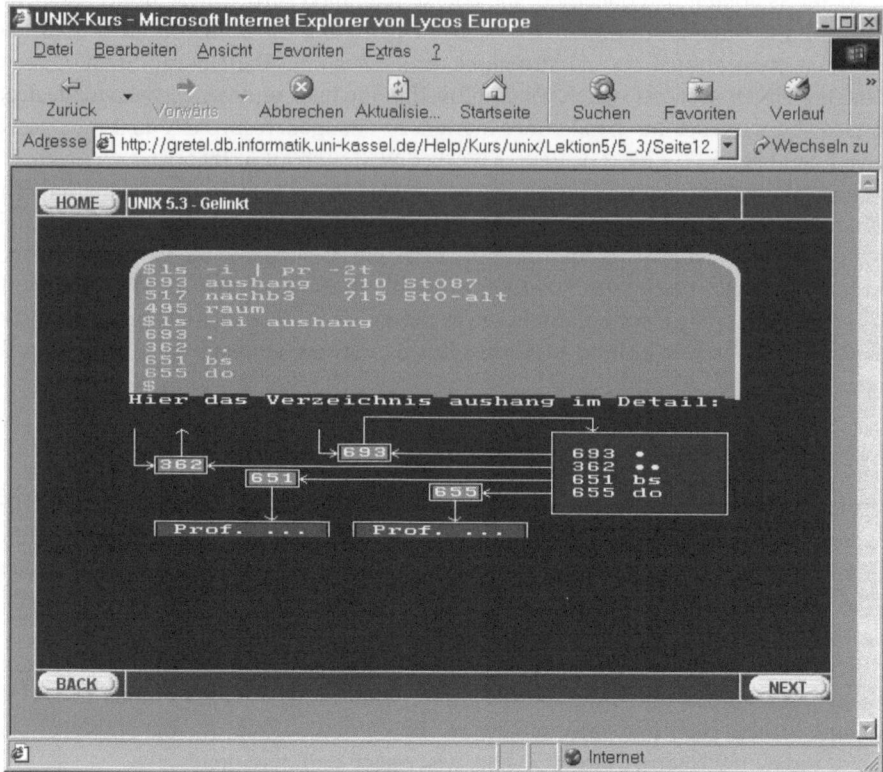

Figure 3. COSTOC ported to animated GIF for use in the Web

be compatible with the usual Internet Explorer/Netscape environment found on a student's PC, possibly including the more common plug-ins. Thus it must follow certain semantic design rules to provide consistent look&feel in that environment.

What constitutes then a course? We felt that the original COSTOC-design guides from the Eighties were still adequate. Much like a book, a course is an ordered collection of "slides" or pages, organized into lectures or chapters, possibly subdivided into sub-chapters or sections.

Like in slide shows with Presentation Manager, a page should be filled or modified in steps, controlled by the viewer, say via a Pause-Button. We thus include a modal button which is either MORE or NEXT, where more keeps filling an existing page or starts or continues an animation and NEXT jumps to a new frame (page).

Apart from "ordinary" lesson frames there should be question frames including answers and explanation texts. All of it should be properly interconnected by links and navigational aids, augmented by a table of contents and a help system.

Mapping these structural design criteria into an XML document structure with elements and useful attributes was largely a question of taste. The project team decided to have `courseIndex` as root of the course document with attributes `creationData` and `topic`, as shown below.

```xml
<?xml version="1.0" encoding="UTF-8" standalone="no" ?>
<courseIndex creationData="Universität Kassel-01.07.2002"
   topic="Einführung in UNIX">
<author mail="wegner@db.informatik.uni-kassel.de"
   name="Lutz Wegner"
   www="http://www.db.informatik.uni-kassel.de" />
<co-author mail="morad@db.informatik.uni-kassel.de"
   name="Morad Ahmed"
   www="http://www.db.informatik.uni-kassel.de" />
<co-author ... />
  <chapter directory="chapter01">Lektion 0</chapter>
  <chapter directory="chapter02">Lektion 1</chapter>
  ...
  <chapter directory="chapter11">Lektion 10</chapter>
</courseIndex>
```

Elements within CourseIndex are author, co-author and chapter with the obvious meanings. As can also be seen, chapters contain a directory-attribute which links them to the file storage system. The same mapping happens with frame-elements which map "pages" to files with a file-attribute.

Clearly, a good design should separate logical from physical organization. However, a mapping which is largely isomorphic (chapter = directory, frame = file) seems very natural and helps enormously with course organization. Also, more complex organizations make it harder for the PHP-server to navigate across the course structure.

We skip examples for the chapter organization and conclude the XML-mapping with an example frame which includes a PAUSE-step and two SVG-elements, one of them animated.

```xml
<?xml version="1.0" encoding="UTF-8" standalone="no" ?>
<frame topic="UNIX 0.5 - Beispiele">
  <lesson>
  <step id="1">
   <animation height="180" loc="M" src="image405-1.svg"
       width="420" />
   <text> Prof. Fix gibt seinen Benutzernamen
   ( <strong>login-Namen</strong> ) ein.
   </text>
  </step>
  <step id="2">
  _<text> Die Eingabe schließt er mit der Taste '
   <strong>return</strong> ' ab.
   </text>
   <image height="180" loc="M" src="image405-2.svg"
   width="420" />
  </step>
```

```
</lesson>
</frame>
```

Figure 4 below shows the actual screenshot for this frame where the letters "f", "i", and "x" appear one-by-one with a one second interval on the simulated screen. The second part of the text appears after clicking onto the MORE-button (or pressing space).

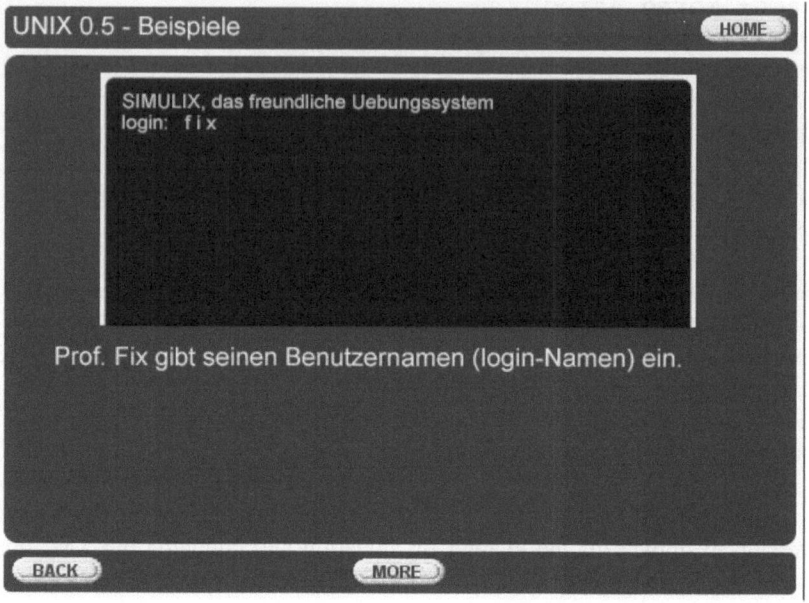

Figure 4. Screenshot from a scene animated with SVG

The actual SVG-code for the simple animated appearance of the three "typed-in" letters is shown below. The complexity of the code seems elaborate for such a small task. However, SVG is actually easy to read and with the introduction of powerful editors hand-coding of animations will eventually disappear.

```
<?xml version="1.0" encoding="iso-8859-1"?>
<!DOCTYPE svg PUBLIC "-//W3C//DTD SVG 20001102//EN"
   "http://www.w3.org/TR/2000/CR-SVG-20001102/DTD/svg-
   20001102.dtd">

<svg width="420" height="180" xml:space="preserve"
   xmlns="http://www.w3.org/2000/svg">
<g id="BGround">

  <rect
  x="0"
```

```
 y="0"
 height="240"
 width="420"
 rx= "10"
 fill= "#000000"
 stroke= "#FFFFCC"
 stroke-width="7" />

</g>

<g  id="Main">
  <text
  x="15"
  y="25"
  style="fill:LightGreen; font-size:14"
  >SIMULIX, das freundliche Uebungssystem</text>
<text
  x="15"
  y="40"
  style="fill:LightGreen; font-size:14"
  >login:    f i x</text>

  <rect id="RectElement" x="50" y="30" width="50"
     height="14" fill="rgb(0,0,0)"   >
  <animate attributeName="x" attributeType="XML"
     begin="0s" dur="2s" fill="freeze"  from="50"
     to="100" />
  <animate attributeName="width" attributeType="XML"
     begin="0s" dur="2s" fill="freeze"  from="50" to="0" />
  </rect>

  <text
  x="15"
  y="55"
  style="fill:LightGreen; font-size:14"
  ></text>
</g>

</svg>
```

The syntactic correctness of each XML document which constitutes part of a course unit can be validated against a schema written in XML Schema [10] designed for these units, e.g. a lesson-frame or a question-frame. In practice, this is not used very much as the units are produced by a course editor which - at least in theory - should produce valid units only. Secondly, units, which are XML documents after all, need a stylesheet [11, 12] to become a HTML-page with embedded SVG code. The result is then viewed and becomes a displayed page as shown in Figure 4. If a course unit is invalid, then this immediately becomes apparent with either a browser error message and/or a blank page.

Note that the use of XML with a stylesheet separates content from appearance. In the future additional stylesheets from third parties could appear, in particular for a mobile community, e.g. geared for players with lower resolutions or low bandwidth connections. Similarly, a stylesheet for output in the Formatting Objects (FO) style, now part of XSL [12], is conceivable from which printed material in pdf-format can be produced.

The general set-up is shown in Figure 5 below. As can be seen, the page which is shipped to the client is assembled first and decorated with HTML display markup according to the XSL translation (stylesheet). The browser at the client side then displays the page invoking Adobe's SVG viewer and catching interaction events through Javascript. As it turns out, this delivery process is slow independent of the transmission speed because of the scripting involved.

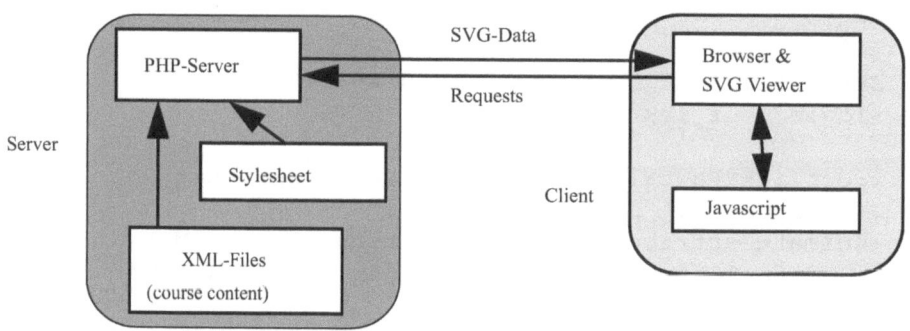

Figure 5. Courseware delivery

As a result, the display inside the browser currently switches to a white background in-between frames which some users find annoying. We hope to fix this when more sophisticated browsers appear in the future which offer options for more intelligent swapping of display buffers.

5 A Courseware Editor

Producing XML courseware documents by hand is rather awkward. Once the general design for PCWL was fixed, it was therefore decided to produce an editor for the authoring of courses.

The editor, written in Java, guides the user in defining chapters and sections, in inserting frames into these sections and linking them in the order desired. It offers menus and a text pad to assemble text and images for a frame. Unfortunately, it is not a WYSIWYG editor which "works inside" the actual display. Technically speaking, there is a separate preview step which requires first inserting newly edited parts into the current DOM-tree[7] (which is then turned into a XML/SVG document). The resulting tree is then viewed after translating it with an appropriate XSL stylesheet.

Figure 6 below shows the editor's interface, here in the process of defining the answers and comments for a multiple choice question with a single correct answer (as opposed to MC questions where more than one alternative is correct and where the student has to click on all correct answers before he or she may progress to the next question). The example offers 4 alternatives, with C being correct, and at most 3 tries with incorrect answers before the correct answer is revealed. For German readers: the actual question asks what topic is <u>not</u> covered in the UNIX course.

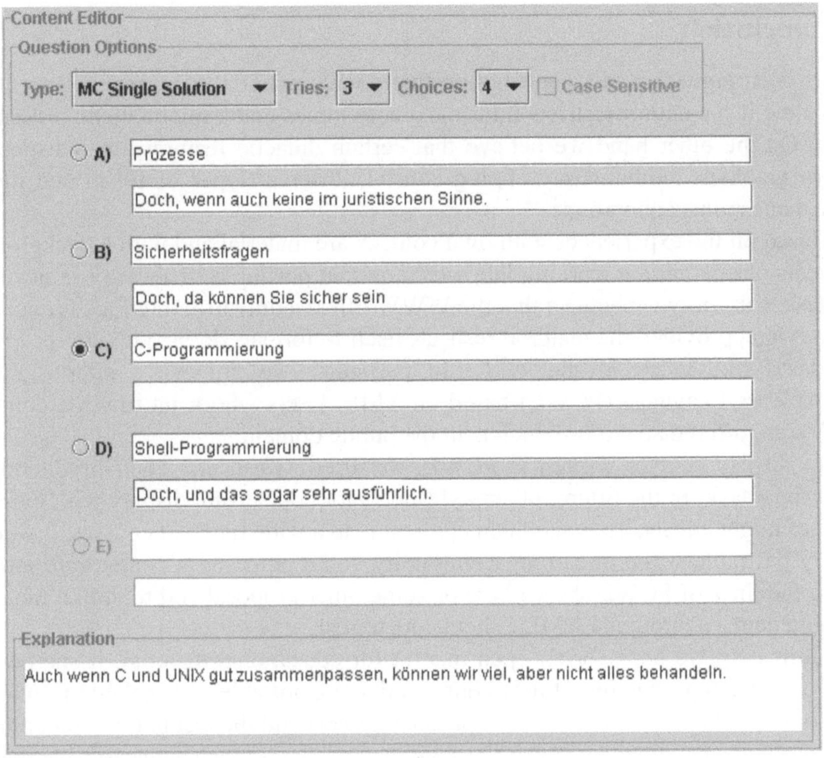

Figure 6. Editor interface to generate a multiple choice question

It should also be mentioned that the answers are shipped with the questions, i.e. there is no re-connecting to the server to complete viewing and answering the question.

This also raises the point of pre-compiling the entire course (or at least a lecture) such that the XML sources are translated under suitable stylesheets into a linked sequence of HTML/SVG-pages which is either shipped to the browser or could be stored on a CD-ROM and distributed over off-line channels. Alternatively, a user could load a PHP-server onto his own PC and let server and client run locally, even on a notebook, without being connected to the Web and without have to fear that the server could be down. Although these options are there, we feel that "always on" with a flat rate and high reliability will be the common mode for Internet users in the next few years.

Note also that the editor assumes that SVG animations have been produced by a separate editor, e.g. Adobe Illustrator or any of the many native editors listed in [17, 18]. Thus the PCWL-editors only inserts links to these files into the appropriate location of the XML document. In the long run we would expect powerful XML/SVG editors to include features for courseware production much like today's desktop publishing systems support book writing.

6 Conclusion

It has been argued in this contribution that course material can remain current for a long time if the author restricts himself to core topics and a minimalistic presentation style. On the other hand we believe that certain didactic features are indispensable: vector graphics, animations, stepping control, different types of fill-in and multiple choice questions, annotations.

Based on the experiences with own courseware material and with a background of 15 years of operating a learning lab, we claim that porting courseware is a never ending task. Currently we believe that the WWW offers a universal interface for accessing courseware, provided the material restricts itself to formats defined by non-proprietary standards like those of the W3C. In particular we propose a simple Portable CourseWare Language (PCWL) based on XML. Users who want to write courses in PCWL are offered an editor which is in the public domain.

To display courses written in PCWL, we offer stylesheets. Their production is a non-trivial task. In the future, other stylesheets for courseware display with individual courses might appear. To make them applicable to a wide range of courses, it would be highly desirable if the eLearning community could agree on a courseware language along the lines of PCWL shown here or some other proposal and to define the syntax of valid courses through a XML Schema document.

Ultimately we hope for an open market for educational material. It should allow easy and cheap production of new course units in a portable and scalable format. This way one should be able to avoid the production cost and the technology trap, although - as has been shown with software production - technology changes don't lead to significant cost reductions and the Web itself is no guard against technological absolescence. Finally, courses written in an open format like XML should foster modular interconnection and free distribution of teachware units in the Web. On the other hand it touches on digital rights issues. Thus, the marketing trap and the question of how to create incentives to write high-quality courseware remain challenges for the future.

Acknowledgement

Morad Ahmad advised the project members on the use of XML and SVG. Andreas Nestmann designed and wrote the course editor, Nabil Benamar designed PCWL and wrote the runtime environment. Stefan Fröhlich and Christian Schmidt had earlier on ported the UNIX-course to HTML. Their contributions are gratefully acknowledged.

The presented results form part of the "Notebook University" project within the zip-Initiative (Zukunftsinvestitionsprogramm) of the German Federal Government (BMFT).

References

1. http://www.learningcenter.unisg.ch/Learningcenter/
2. John Seeley Brown and Paul Duguid. The Social Life of Information, Harvard Business School Press, Boston, Mass., 2000
3. H. Maurer. Hypermedia Systems as Internet Tools, in: Computer Science Today - Recent Trends and Developments, Jan van Leeuwen (Ed.), Springer LNCS 1000, Springer-Verlag, Berlin-Heidelberg (1995) 608-624.
4. Lutz Wegner: Zehn Jahre computerunterstützter Unterricht am FB Mathematik/Informatik der GhK, Preprint 9/96 (Dezember 1996) http://www.db.informatik.uni-kassel.de/~wegner/TR9-96.ps.gz
5. Lutz Wegner: Introduction to UNIX, Kursdokumentation HyperCOSTOC, Computer Science Vol. 16, Hofbauer Verlag, Chur, 1989, ISBN 3 906382 80 X, ca. 200 S.
6. Marcel Salanthé, SVG Scalable Vector Graphics, Markt + Technik Verlag, 2001
7. Document Object Model (DOM) Level 2 Core Specification, Version 1.0, W3C Recommendation 13. Nov. 2000
8. Scalable Vector Graphics (SVG) 1.0 Specification, W3C Recommendation 4. Sept. 2001
9. Extensible Markup Language (XML) 1.0 (Second Edition), W3C Recommendation 6. Oct. 2000
10. XML Schema Part 1: Structures, W3C Recommendation 2. May 2001
11. XSL Transformations (XSLT) Version 1.0, W3C Recommendation 16. Nov. 1999
12. Extensible Stylesheet Language (XSL) Version 1.0, W3C Recommendation 15 Oct. 2001
13. Amitava Datta and Thomas Ottmann. Towards a Virtual University, J.UCS, Vol. 7, No. 10, 2001
14. Bernd Krämer and Lutz Wegner: From Custom Text Books to Interactive Distance Teaching. Proc. ED-MEDIA/ED-TELECOM 98, World Conference on Educational Multimedia and Hypermedia & World Conference on Educational Telecommunications, Freiburg, Germany, June 20-25, 1998 (Th. Ottmann and Ivan Tomek eds.), AACE, pp. 800-805
15. www.matchware.net
16. Reza Hazemi, Stephen Hailes and Steve Wilbur (Eds): The Digital University - Reinventing the Academy (CSCW Series), Springer-Verlag London (1998)
17. Andreas Neumann, Peter Sykora, and Andreas Winter: Das Web auf neuen Pfaden - SVG und SMIL: Graphik und Animation fürs Web (in German), c't 20 (2002) 218 - 225
18. http://www.w3.org/Graphics/SVG/SVG-Implementations.htm8

Author Index

Lecture Notes in Computer Science

For information about Vols. 1–2495

please contact your bookseller or Springer-Verlag

Vol. 2535: N. Suri (Ed.), Mobile Agents. Proceedings, 2002. X, 203 pages. 2002.

Vol. 2536: M. Parashar (Ed.), Grid Computing – GRID 2002. Proceedings, 2002. XI, 318 pages. 2002.

Vol. 2537: D.G. Feitelson, L. Rudolph, U. Schwiegelshohn (Eds.), Job Scheduling Strategies for Parallel Processing. Proceedings, 2002. VII, 237 pages. 2002.

Vol. 2538: B. König-Ries, K. Makki, S.A.M. Makki, N. Pissinou, P. Scheuermann (Eds.), Developing an Infrastructure for Mobile and Wireless Systems. Proceedings 2001. X, 183 pages. 2002.

Vol. 2539: K. Börner, C. Chen (Eds.), Visual Interfaces to Digital Libraries. X, 233 pages. 2002.

Vol. 2540: W.I. Grosky, F. Plášil (Eds.), SOFSEM 2002: Theory and Practice of Informatics. Proceedings, 2002. X, 289 pages. 2002.

Vol. 2541: T. Barkowsky, Mental Representation and Processing of Geographic Knowledge. X, 174 pages. 2002. (Subseries LNAI).

Vol. 2544: S. Bhalla (Ed.), Databases in Networked Information Systems. Proceedings 2002. X, 285 pages. 2002.

Vol. 2545: P. Forbrig, Q, Limbourg, B. Urban, J. Vanderdonckt (Eds.), Interactive Systems. Proceedings 2002. X, 269 pages. 2002.

Vol. 2546: J. Sterbenz, O. Takada, C. Tschudin, B. Plattner (Eds.), Active Networks. Proceedings, 2002. XIV, 267 pages. 2002.

Vol. 2547: R. Fleischer, B. Moret, E. Meineche Schmidt (Eds.), Experimental Algorithmics. XVII, 279 pages. 2002.

Vol. 2548: J. Hernández, Ana Moreira (Eds.), Object-Oriented Technology. Proceedings, 2002. VIII, 223 pages. 2002.

Vol. 2549: J. Cortadella, A. Yakovlev, G. Rozenberg (Eds.), Concurrency and Hardware Design. XI, 345 pages. 2002.

Vol. 2550: A. Jean-Marie (Ed.), Advances in Computing Science – ASIAN 2002. Proceedings, 2002. X, 233 pages. 2002.

Vol. 2551: A. Menezes, P. Sarkar (Eds.), Progress in Cryptology – INDOCRYPT 2002. Proceedings, 2002. XI, 437 pages. 2002.

Vol. 2552: S. Sahni, V.K. Prasanna, U. Shukla (Eds.), High Performance Computing – HiPC 2002. Proceedings, 2002. XXI, 735 pages. 2002.

Vol. 2553: B. Andersson, M. Bergholtz, P. Johannesson (Eds.), Natural Language Processing and Information Systems. Proceedings, 2002. X, 241 pages. 2002.

Vol. 2554: M. Beetz, Plan-Based Control of Robotic Agents. XI, 191 pages. 2002. (Subseries LNAI).

Vol. 2555: E.-P. Lim, S. Foo, C. Khoo, H. Chen, E. Fox, S. Urs, T. Costantino (Eds.), Digital Libraries: People, Knowledge, and Technology. Proceedings, 2002. XVII, 535 pages. 2002.

Vol. 2556: M. Agrawal, A. Seth (Eds.), FST TCS 2002: Foundations of Software Technology and Theoretical Computer Science. Proceedings, 2002. XI, 361 pages. 2002.

Vol. 2557: B. McKay, J. Slaney (Eds.), AI 2002: Advances in Artificial Intelligence. Proceedings, 2002. XV, 730 pages. 2002. (Subseries LNAI).

Vol. 2558: P. Perner, Data Mining on Multimedia Data. X, 131 pages. 2002.

Vol. 2559: M. Oivo, S. Komi-Sirviö (Eds.), Product Focused Software Process Improvement. Proceedings, 2002. XV, 646 pages. 2002.

Vol. 2560: S. Goronzy, Robust Adaptation to Non-Native Accents in Automatic Speech Recognition. Proceedings, 2002. XI, 144 pages. 2002. (Subseries LNAI).

Vol. 2561: H.C.M. de Swart (Ed.), Relational Methods in Computer Science. Proceedings, 2001. X, 315 pages. 2002.

Vol. 2562: V. Dahl, P. Wadler (Eds.), Practical Aspects of Declarative Languages. Proceedings, 2003. X, 315 pages. 2002.

Vol. 2566: T.Æ. Mogensen, D.A. Schmidt, I.H. Sudborough (Eds.), The Essence of Computation. XIV, 473 pages. 2002.

Vol. 2567: Y.G. Desmedt (Ed.), Public Key Cryptography – PKC 2003. Proceedings, 2003. XI, 365 pages. 2002.

Vol. 2568: M. Hagiya, A. Ohuchi (Eds.), DNA Computing. Proceedings, 2002. XI, 338 pages. 2003.

Vol. 2569: D. Gollmann, G. Karjoth, M. Waidner (Eds.), Computer Security – ESORICS 2002. Proceedings, 2002. XIII, 648 pages. 2002. (Subseries LNAI).

Vol. 2570: M. Jünger, G. Reinelt, G. Rinaldi (Eds.), Combinatorial Optimization – Eureka, You Shrink!. Proceedings, 2001. X, 209 pages. 2003.

Vol. 2571: S.K. Das, S. Bhattacharya (Eds.), Distributed Computing. Proceedings, 2002. XIV, 354 pages. 2002.

Vol. 2572: D. Calvanese, M. Lenzerini, R. Motwani (Eds.), Database Theory – ICDT 2003. Proceedings, 2003. XI, 455 pages. 2002.

Vol. 2574: M.-S. Chen, P.K. Chrysanthis, M. Sloman, A. Zaslavsky (Eds.), Mobile Data Management. Proceedings, 2003. XII, 414 pages. 2003.

Vol. 2575: L.D. Zuck, P.C. Attie, A. Cortesi, S. Mukhopadhyay (Eds.), Verification, Model Checking, and Abstract Interpretation. Proceedings, 2003. XI, 325 pages. 2003.

Vol. 2576: S. Cimato, C. Galdi, G. Persiano (Eds.), Security in Communication Networks. Proceedings, 2002. IX, 365 pages. 2003.

Vol. 2578: F.A.P. Petitcolas (Ed.), Information Hiding. Proceedings, 2002. IX, 427 pages. 2003.

Vol. 2580: H. Erdogmus, T. Weng (Eds.), COTS-Based Software Systems. Proceedings, 2003. XVIII, 261 pages. 2003.

Vol. 2583: S. Matwin, C. Sammut (Eds.), Inductive Logic Programming. Proceedings, 2002. X, 351 pages. 2003. (Subseries LNAI).

Vol. 2588: A. Gelbukh (Ed.), Computational Linguistics and Intelligent Text Processing. Proceedings, 2003. XV, 648 pages. 2003.

Vol. 2594: A. Asperti, B. Buchberger, J.H. Davenport (Eds.), Mathematical Knowledge Management. Proceedings, 2003. X, 225 pages. 2003.

Vol. 2598: R. Klein, H.-W. Six, L. Wegner (Eds.), Computer Science in Perspective. X, 357 pages. 2003.

Vol. 2600: S. Mendelson, A.J. Smola, Advanced Lectures on Machine Learning. Proceedings, 2002. IX, 259 pages. 2003. (Subseries LNAI).